VOLUME FIFTEEN

ADVANCES IN CHEMICAL POLLUTION, ENVIRONMENTAL MANAGEMENT AND PROTECTION

Climate-Smart Water Resource Management: A Paradigm Shift from Ancient to Modern Practices with Integrated Technologies

VOLUME FIFTEEN

ADVANCES IN CHEMICAL POLLUTION, ENVIRONMENTAL MANAGEMENT AND PROTECTION

Climate-Smart Water Resource Management: A Paradigm Shift from Ancient to Modern Practices with Integrated Technologies

Edited by

RAJIB MAITY
Department of Civil Engineering, Indian Institute of Technology Kharagpur, Kharagpur, West Bengal, India

AMAN SRIVASTAVA
Department of Civil Engineering, Indian Institute of Technology Kharagpur, Kharagpur, West Bengal, India

ELSEVIER

ACADEMIC PRESS
An imprint of Elsevier

Academic Press is an imprint of Elsevier
50 Hampshire Street, 5th Floor, Cambridge, MA 02139, United States
125 London Wall, London, EC2Y 5AS, United Kingdom

First edition 2026

ISBN: 978-0-443-34601-9
ISSN: 2468-9289

For information on all Academic Press publications
visit our website at https://www.elsevier.com/books-and-journals

Publisher: Zoe Kruze
Acquisitions Editor: Jason Mitchell
Editorial Project Manager: Devwart Chauhan
Production Project Manager: Bharatwaj Varatharajan
Cover Designer: Muthukumaran Thangaraj

Typeset by MPS Limited, India

Contents

Section D Synthesis and future directions

Contributors

Jubikali
Department of Agricultural Engineering and Technology, School of Engineering and Technology, Nagaland University, Kohima, Nagaland, India

Abdulhalim Musa Abubakar
Department of Chemical Engineering, Faculty of Engineering, Modibbo Adama University, Yola, Adamawa State, Nigeria

Mohamed Khalid AlOmar
Department of Civil Engineering, College of Engineering, University of Al-Maarif, Ramadi City, Iraq

G. Sri Bala
Department of Civil Engineering; Centre for Sustainable Inland Aquaculture, SRKR Engineering College, Bhimavaram, India

Tridipa Biswas
Govind Ballabh Pant 'National Institute of Himalayan Environment' (NIHE), North-East Regional Centre (NERC), Itanagar, Arunachal Pradesh, India

Chandranath Chatterjee
Agricultural and Food Engineering Department, Indian Institute of Technology Kharagpur, Kharagpur, India

Noureddine Daif
Laboratory of Optimizing Agricultural Production in Subhumid Zones (LOPAZS), Agronomy Department, Faculty of Science, 20 August 1955 University of Skikda: Universite du 20 Août 1955 de Skikda, Route El Hadaik, BP 26, Skikda, Algeria

Abhijeet Das
Department of Civil Engineering, C.V. Raman Global University (CGU), Bhubaneswar, Odisha, India

Dhrubajyoti Das
Forest Ecology and Climate Change Division, Rain Forest Research Institute, Jorhat, Assam, India

Manimoy Das
Forest Ecology and Climate Change Division, Rain Forest Research Institute, Jorhat, Assam, India

Salah Difi
Vegetal Chemistry-Water-Energy Laboratory (LCV2E), Department of Hydraulic, Civil Engineering and Architecture Faculty, University of Hassiba Benbouali, Chlef, Algeria

Nibedita Guru
Forest Ecology and Climate Change Division, Rain Forest Research Institute, Jorhat, Assam, India

Mohammed Majeed Hameed
Upper Euphrates Centre for Sustainable Development Research, University of Anbar, Ramadi City, Iraq

Salim Heddam
Agronomy Department, Faculty of Science, University 20 Août 1955 Skikda, Route El Hadaik, BP 26, Skikda, Algeria

Siraj Hussain
Department of Geological Engineering, Balochistan University of Technology, Engineering and Management Sciences (BUITEMS), Quetta, Pakistan

Neena Isaac
River Rejuvenation Division, Central Water & Power Research Station, Pune, India

Shubham Jain
Department of Civil Engineering, Indian Institute of Technology Bombay, Mumbai, India

Saroj Raj Kafle
Department of Chemical Engineering, Chungbuk National University, Cheongju, South Korea

Yateesh Ketholia
Geosciences & Geohazards Department, Indian Institute of Remote Sensing, Dehradun, Uttarakhand, India

Amina Khatun
Department of Natural Resource Management, College of Horticulture & Farming System Research, Assam Agricultural University, Nalbari, India

Liza G. Kiba
Department of Agricultural Engineering and Technology, School of Agricultural Sciences, Nagaland University, Medziphema, Nagaland, India

Ashish Kumar
Agricultural and Food Engineering Department, Indian Institute of Technology Kharagpur, Kharagpur, India

Dudekula Nikhil Kumar
River Rejuvenation Division, Central Water & Power Research Station, Pune, India

Rajib Maity
Department of Civil Engineering, Indian Institute of Technology Kharagpur, Kharagpur, West Bengal, India

Adil Masood
Department of Natural and Applied Sciences, TERI School of Advanced Studies, New Delhi, India

Arjita Maurya
RMSI Pvt. Ltd., Dehradun, Uttarakhand, India

Shilpi Maurya
River Rejuvenation Division, Central Water & Power Research Station, Pune, India

Khaled Merabet
Hydraulics Division, Agronomy Department, Faculty of Science, University 20 Août 1955, Route El Hadaik, BP 26, Skikda, Algeria

Ashrumochan Mohanty
School of Water Resources, Indian Institute of Technology Kharagpur, Kharagpur, India

Godknows I.J. Mugabe
Zimbawe National Geospatial Space Agency, University of Zimbabwe, Harare, Zimbabwe

Never Mujere
Department of Geography, Geospatial Sciences and Earth Observation, University of Zimbabwe, Harare, Zimbabwe

Ahmad Abubakar Mustapha
School of Computer Science and Engineering, Vellore Institute of Technology, Andhra Pradesh (VIT-AP) University, Andhra Pradesh, India

T. Vamsi Nagaraju
Department of Civil Engineering; Centre for Sustainable Inland Aquaculture, SRKR Engineering College, Bhimavaram, India

Grace Nengzouzam
Department of Agricultural Engineering and Technology, School of Engineering and Technology, Nagaland University, Kohima, Nagaland, India

Eze Cordelia Nneka
Department of Chemical Engineering, Faculty of Engineering, Chukwuemeka Odumegwu Ojukwu University (COOU), Uli, Anambra State, Nigeria

Y. Nyamnyei K
Department of Agricultural Engineering and Technology, School of Engineering and Technology, Nagaland University, Kohima, Nagaland, India

Akshay Pachore
River Rejuvenation Division, Central Water & Power Research Station, Pune, India

Vipin Chander Pant
Council of India, New Delhi, India

Mark Prabhakar Vuppati
River Rejuvenation Division, Central Water & Power Research Station, Pune, India

Krishna Rawat
School of Water Resources, Indian Institute of Technology Kharagpur, Kharagpur, India

Bhabagrahi Sahoo
School of Water Resources, Indian Institute of Technology Kharagpur, Kharagpur, India

Achala Shakya
Department of Computer Science and Engineering, School of Engineering and Sciences, SRM University-AP, Guntur, Andhra Pradesh, India

Aman Srivastava
Department of Civil Engineering, Indian Institute of Technology Kharagpur, Kharagpur, West Bengal, India

Gaurav Tripathi
Department of Civil Engineering, Indian Institute of Technology Bombay, Powai, Mumbai, Maharashtra, India

Ritambhara K. Upadhyay
Department of Geology, Central University of Himachal Pradesh, Dharamshala, Himachal Pradesh, India

Abubakar Yusuf Waziri
Department of Chemical Engineering, Faculty of Engineering, Modibbo Adama University, Yola, Adamawa State, Nigeria

Sheihnon R. Yeanganghu
Department of Agricultural Engineering and Technology, School of Engineering and Technology, Nagaland University, Kohima, Nagaland, India

Preface

Water, the very elixir of life, is under unprecedented stress. Across the globe, rising temperatures, shifting precipitation regimes, and intensifying extreme events are rewriting the hydrological rules that civilizations have relied upon for millennia. At the same time, population growth, urbanization, and industrial expansion continue to amplify the demand for this finite resource. In this crucible of change, the question is no longer whether we must adapt, but how, and with what tools, wisdom, and collaborative spirit, we can forge a water-secure future. The genesis of this edited book volume lies in a conviction: that the path to climate-resilient water management does not require a wholesale rejection of the past in favour of the new, nor a romantic return to tradition alone. Rather, it calls for a deliberate, respectful, and creative paradigm shift: one that bridges the deep ingenuity of ancient practices with the dazzling capabilities of modern technologies. It is this synthesis that we term *climate-smart water resource management.*

When we first conceived this book, we envisioned a collection that would not merely catalogue methods but would tell a story. A story of how indigenous communities in the Indian Himalayas engineered paddy-cum-fish cultivation and bamboo irrigation networks that have sustained both people and ecosystems for centuries. A tale of how satellite constellations now peer through clouds to map floods in near real-time, and how machine learning algorithms learn the language of rivers to forecast their moods. A story of how Large Language Models, trained on the accumulated knowledge of our species, might soon become conversational partners in the control rooms of water utilities. Most importantly, a story of convergence, where these diverse threads are woven into a coherent fabric of practice, policy, and hope. The response from the scholarly community exceeded our expectations. Researchers from across disciplines, viz., hydrology, remote sensing, artificial intelligence, environmental engineering, anthropology, and policy studies, answered our call. Their contributions, distilled into the sixteen chapters of this volume, collectively demonstrate that the *integration* of knowledge domains is not an abstract ideal but an operational imperative.

Section A lays the foundation by grounding the reader in both the philosophical and empirical roots of climate-smart water management. Chapter 1 introduces the technological triumvirate of AI-, ML, RS-, GIS, and LLMs, framing them as facilitators of adaptive governance. It is followed by an immersive study of the Apatani tribe's indigenous water systems, a living laboratory of low-carbon, nature-based infrastructure. The section then transitions to quantitative assessments: a Water Quality Index analysis of the Mula-Mutha River and an entropy-based investigation of precipitation complexity in the Brahmani River Basin. Together, these chapters remind us that innovation is not always about invention; sometimes it is about *rediscovery* and *rigour*.

Section B plunges into the frontier of computational methods. Here, artificial neural networks are trained to forecast seasonal rainfall over arid landscapes; bidirectional LSTM and GRU models compete to estimate dissolved oxygen with ever-greater fidelity; stacked and bidirectional LSTM architectures are tested for their ability to predict daily streamflow into a major reservoir. The section culminates in a forward-looking examination of Large Language Models, not as futuristic speculation, but as tools already demonstrating the capacity to synthesise heterogeneous data, support decision-makers, and democratise access to complex information. These chapters do not merely report accuracy metrics; they illustrate a profound shift in how we model, interpret, and communicate hydrological processes.

Section C showcases the spatial and spectral eyes that now observe our planet. From the prioritisation of surface water quality using multi-criteria decision analysis and the mapping of flood inundation with Sentinel-1 SAR, to the morphometric dissection of watersheds in North-East India and the analysis of long-term interactions between land use and climate, this section is a testament to the power of geospatial technologies. It includes critical reviews of LiDAR advancements for precision water management and two decades of remote sensing and machine learning applications in Indian aquaculture. Each chapter demonstrates that the resolution of our questions has improved in lockstep with the resolution of our sensors.

Section D is a single, synthesising chapter—a deliberate choice. After traversing the diverse landscapes of indigenous knowledge, artificial intelligence, and earth observation, Chapter 16 returns to the whole. It recapitulates the cross-cutting themes, distils implementable strategies for policy and governance, and issues a call to action. It reminds us that technology, no matter how sophisticated, remains a tool; its value is realised only when it serves human and ecological well-being equitably.

Producing a book of this scope is a collective endeavour. We extend our deepest gratitude to each author whose rigorous scholarship and patience made this volume possible. Your willingness to share data, methods, and inferences across disciplinary boundaries is the living embodiment of the paradigm shift we advocate. We are equally indebted to the reviewers, whose anonymous yet meticulous critiques sharpened every chapter. We also gratefully acknowledge the Indian Institute of Technology Kharagpur (IIT Kharagpur) for fostering an intellectually vibrant environment and for its sustained institutional support, which provided the foundation for conceptualizing and editing this work. To end on a high note, the editorial team at Elsevier provided us with unwavering support and professionalism, guiding us from proposal to print with grace and efficiency.

This preface is being written at a moment when the World Meteorological Organization reports that the global hydrological cycle is spinning out of the patterns that we once considered stable. It is easy to feel daunted. Yet, as the chapters that follow amply demonstrate, the human capacity for ingenuity is immense. We possess traditional knowledge refined over centuries and computational tools advancing at an exponential speed. What we require now is the will to integrate them and the humility to learn from both the past and the future. Therefore, we invite you, the reader, to engage with this book not as a passive recipient of information, but as an active participant in the journey it describes. Whether you are a researcher seeking new methodologies, a policymaker designing adaptation strategies, a practitioner in the field, or a student discovering the richness of water science, we trust that you will find within these pages both practical guidance and intellectual inspiration. The water challenges of the twenty-first century are formidable, but so are the alliances we can forge between ancient and modern, between data and intuition, between disciplines and cultures. This book is one such alliance. May it serve as a catalyst for many more.

Editors

Rajib Maity and Aman Srivastava

SECTION A

Foundations and indigenous perspectives: Integrating traditional knowledge with modern water assessment

CHAPTER ONE

Introduction to climate-smart water resource management with AI-ML, RS-GIS, and LLMs

Rajib Maity* and **Aman Srivastava**
Department of Civil Engineering, Indian Institute of Technology Kharagpur, Kharagpur, West Bengal, India
*Corresponding author. e-mail address: rajib@civil.iitkgp.ac.in

Contents

Abstract

This introductory chapter examines global water resource challenges and how advanced technologies can support sustainable management strategies in a climate-stressed world. It begins by addressing the urgent issue of water scarcity, driven by factors like population growth, urbanization, and industrialization, and further intensified by climate change. Rising global temperatures, unpredictable rainfall patterns, and increased frequency of extreme events such as droughts and floods all

Advances in Chemical Pollution, Environmental Management and Protection, Volume 15
ISSN 2468-9289, https://doi.org/10.1016/bs.apmp.2025.11.014

underscore the need for adaptive, resilient water management systems. In response to these challenges, the chapter explores the revolutionary role of Artificial Intelligence and Machine Learning (AI-ML), Remote Sensing and Geographical Information Systems (RS-GIS), and Large Language Models (LLMs). These technologies improve water quality monitoring, resource allocation, predictive modeling, and stakeholder communication, providing powerful tools for real-time decision-making and long-term planning. The chapter also highlights the importance of interdisciplinary collaboration across climate science, engineering, data science, and policy, urging stakeholders to adopt technology-driven, climate-smart solutions. Finally, the chapter sets the stage for the detailed analyses to follow, providing a foundation for understanding how traditional water management practices can be effectively integrated with modern technologies. This book ultimately aims to bridge gaps between theory and practical applications, guiding researchers, and policymakers in implementing sustainable water resource management strategies in an era marked by unprecedented environmental challenges.

Keywords: Sustainable water governance, Predictive analytics in water management, Climate adaptation strategies, Data-driven decision making, Water security in semi-arid regions, Environmental monitoring systems, Stakeholder engagement tools.

1. Introduction

Global water scarcity and stress have become critical challenges, exacerbated by increasing population, rapid industrialization, and accelerated urbanization.[1] About 2.2 billion people lack access to safe drinking water, and nearly 4 billion people experience severe water scarcity for at least one month each year.[2,3] As the world population is projected to reach 9.7 billion by 2050, the demand for freshwater is expected to rise by 55 %, placing unprecedented stress on already limited resources.[4,5] Urbanization and industrial expansion further intensify this crisis by increasing water pollution and shifting demand toward concentrated urban areas, where infrastructure may be inadequate to meet needs.[6] Agriculture, which consumes approximately 70 % of global freshwater resources, is pressured to increase food production, further straining water availability.[7] These combined factors create complex challenges for water resource management, demanding a shift from conventional approaches to more integrated, sustainable, and technology-driven solutions.

Climate change significantly disrupts global hydrological cycles, altering water availability and exacerbating extreme weather events such as droughts and floods.[1,8,9] Rising global temperatures have intensified evaporation rates and altered precipitation patterns, leading to uneven water distribution.[10]

For instance, studies suggest that global drought conditions have increased by 29 % since 2000, primarily affecting regions already vulnerable to water scarcity.[11,12] Concurrently, extreme rainfall events have become more frequent, causing flooding and contamination of water supplies. These climate-induced shifts impact surface and groundwater resources, leading to unpredictable seasonal flows and decreased water security.[2,13,14] Moreover, the frequency of extreme events disrupts agricultural systems, stresses urban infrastructure, and complicates water management practices.[15] Such impacts underscore the urgency for adaptive water resource management approaches that integrate climate resilience into long-term strategies.

Advanced technologies, including Artificial Intelligence and Machine Learning (AI-ML), Remote Sensing and Geographic Information Systems (RS-GIS), and Large Language Models (LLMs), are increasingly vital in transforming water resource management.[16] AI-ML algorithms enable predictive modeling for critical water-related issues, from forecasting droughts and floods to optimizing water distribution networks.[17–20] RS-GIS technologies are pivotal in mapping water availability, tracking land-use changes, and monitoring real-time climate impacts on water systems. Satellites and sensors, for example, provide spatially explicit data that improve our understanding of surface and groundwater resources, allowing for proactive water management at local and global scales.[21–23] Meanwhile, LLMs like GPT-3 (Generative Pre-trained Transformer 3), LaMDA (Language Model for Dialogue Applications), Jurassic-1 Jumbo, BLOOM, etc., improve decision support by synthesizing vast amounts of research and policy data, translating scientific understanding into actionable strategies for diverse stakeholders.[24,25] Integrating these advanced technologies with traditional water management systems is imperative for a more adaptive, precise, and resilient approach, which is crucial for managing the multi-dimensional water challenges in a changing climate. This hybrid approach is needed to facilitate sustainable resource allocation, improve water use efficiency, and support climate-smart adaptation strategies for long-term water security.

This book thus aims to provide a comprehensive framework for climate-smart water resource management by bridging traditional practices with cutting-edge technologies. The primary objective is to present scientifically grounded strategies for sustainable water management, focusing on AI-ML, RS-GIS, and LLMs as revolutionary tools for improving water security. This resource is intended for a wide audience—including researchers, policymakers, and students—as both an educational guide and a strategic tool. Through rigorous analysis, the book emphasizes practical applications and innovations that can help address contemporary water management challenges. Ultimately, it underscores the importance of integrating technology with conventional methods to foster resilient, adaptive, and efficient water resource strategies that are relevant in an era of climatic and environmental uncertainty.

2. Historical evolution of water resource management

2.1 Ancient and traditional water management practices

Ancient civilizations developed diverse water management systems inclined to regional climates and resources, many of which remain relevant today (see Table 1). In Mesopotamia, constructing canals and levees helped manage the Tigris and Euphrates rivers, supporting one of the earliest known agricultural societies.[26] Ancient Egypt utilized the Nile's seasonal floods for basin irrigation, constructing dikes and canals that enabled crop cultivation even in dry seasons.[27] In South Asia, the Indus Valley civilization devised sophisticated drainage systems, wells, and reservoirs to capture and distribute water efficiently.[28] In arid regions of the Middle East and Persia, qanat systems—underground channels that transported water from aquifers to surface settlements—sustained communities across kilometers, minimizing evaporation and preserving water quality.[29] Meanwhile, the Inca civilization developed terraced fields and aqueducts in the Andean highlands to maximize limited rainfall, prevent soil erosion, and effectively distribute water.[30] Traditional water conservation practices, like India's step-wells, baolis, and tank cascade systems, store monsoon rainwater for year-round use.[31–33] Japan's age-old yusui systems distributed mountain spring water to lowland agriculture,[34] while African societies used sand dams to trap seasonal rains in riverbeds, providing water through droughts.[35] These systems emphasize sustainable practices, such as harnessing natural hydrological cycles and utilizing gravity for water distribution, which provide visions for modern water conservation. Contemporary water management can learn from these systems' low-energy, community-oriented design, as their efficiency and adaptability exemplify sustainable approaches that are increasingly relevant in the face of climate variability and water scarcity.

2.2 The shift to modern water management

The advent of industrialization marked a shift towards engineering-intensive approaches in water management, emphasizing large-scale infrastructure for

Table 1 Evolution of water management practices across regions from ancient techniques to modern engineering and technology-driven innovations.

Region/practice	Ancient and traditional methods	Modern engineering-based approaches	Integrated, technology-driven management
India	Stepwells, rainwater harvesting, and traditional tank cascade systems.	Large dams (e.g., Bhakra Nangal) and extensive canal systems.	RS-GIS is used for water level monitoring in reservoirs; AI is used to optimize irrigation scheduling.
Middle East and North Africa	Qanats, falaj systems for groundwater channeling, cisterns.	Desalination plants and large-scale water transfer projects (e.g., Nile Basin).	RS for real-time mapping of aquifers; AI-ML for desalination efficiency and energy optimization.
China	Dujiangyan Irrigation System, terraced fields.	Major dam projects (e.g., Three Gorges Dam) and flood control systems.	RS-GIS for flood prediction; AI in dynamic dam operation; LLMs for policy scenario analysis.
Europe (Mediterranean)	Roman aqueducts and rainwater cisterns in Crete and other areas.	Reservoir networks and cross-border water agreements (e.g., Danube River).	AI-driven demand forecasting; RS-GIS for cross-border water resource monitoring.
Latin America	Terraces and reservoirs in the Andes, Maya reservoirs.	Hydropower dams (e.g., Itaipú), irrigation networks.	AI for hydropower optimization; RS-GIS for drought-prone areas; LLMs for community engagement.

(continued)

Table 1 Evolution of water management practices across regions from ancient techniques to modern engineering and technology-driven innovations. *(cont'd)*

Region/practice	Ancient and traditional methods	Modern engineering-based approaches	Integrated, technology-driven management
Sub-Saharan Africa	Earthen dams, rainwater harvesting, small-scale irrigation.	Large dam projects (e.g., Aswan Dam), water transfer schemes.	RS-GIS for watershed health; AI-ML for small-scale irrigation improvement; LLMs for knowledge sharing among communities.
North America	Indigenous rainwater collection and water-sharing agreements.	Massive dam networks (e.g., Hoover Dam) and aqueducts for urban supply.	RS-GIS for water quality monitoring; AI for reservoir management; LLMs for regulatory guidance.
Southeast Asia	Rice terrace irrigation systems (e.g., Banaue), bamboo aqueducts.	Dams and extensive irrigation systems for agriculture.	RS-GIS is used for soil moisture monitoring, and AI is used for optimized water resource distribution.

resource control and economic growth (see Table 1). During the 20th century, dams, reservoirs, and canals were constructed worldwide, with iconic examples like the Hoover Dam in the United States[36] and the Aswan High Dam in Egypt,[37] aimed at flood control, irrigation, and hydropower generation. Between 1950 and 2000, global dam construction surged, with approximately 50,000 large dams built globally, impounding vast volumes of freshwater.[38] While these structures provided economic benefits, their environmental and social costs are now widely recognized. Dams alter natural river flows, impacting aquatic ecosystems, reducing sediment transport, and disrupting habitats.[39] Moreover, large-scale irrigation projects have often led to soil salinization, declining groundwater levels, and water logging, affecting agricultural sustainability.[40] The environmental consequences of modern water management approaches have spurred critiques, emphasizing the need for holistic management that accounts for ecosystem dynamics and long-term sustainability. Modern methods, while resource-intensive, lack the adaptability seen in traditional systems, which functioned harmoniously with natural water cycles. The limitations of conventional infrastructure reveal the importance of integrating sustainable practices with advanced technology to effectively address today's complex water challenges.

2.3 Transitioning to integrated, technology-driven management

Climate change pressures have intensified the need to combine historical water management knowledge with advanced technologies to create resilient, adaptable systems (see Fig. 1).[1] AI-ML now enables real-time analysis of vast datasets, supporting precise water resource modeling and demand forecasting.[41–43] Additionally, RS-GIS technologies provide high-resolution, time-sensitive data, which is crucial for tracking hydrological changes, mapping watershed health, and assessing soil moisture levels.[44–47] These spatial tools can identify drought-prone areas or regions at risk of floods, guiding pre-emptive water management decisions.[48–50] LLMs represent a new frontier in managing water resources by streamlining communication between scientists, policymakers, and the public. These models aid in generating reports, answering technical inquiries, and simulating policy impacts, which supports more transparent, data-driven decision-making. Integrating LLMs into water management also has significant implications for policy development, providing visions based on vast scientific literature and previous case studies.[24,25] Ultimately, bridging traditional knowledge with AI-ML, RS-GIS, and LLMs facilitates a climate-smart approach to water

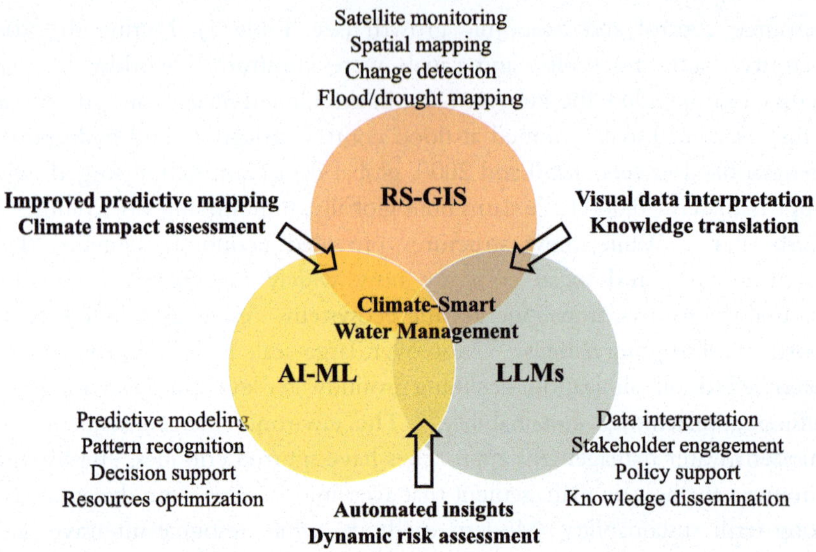

Fig. 1 Basin Venn diagram illustrating the integration of AI-ML, RS-GIS, and LLMs in climate-smart water resource management.

management, addressing both current and future water challenges with the adaptability required in a changing climate. To this end, Table 1 summarizes the evolution of water management practices globally, showcasing the transition from ancient and traditional techniques to modern engineering and, more recently, to technology-driven approaches.

3. Technological integration in water resource management

3.1 Artificial intelligence and machine learning (AI-ML)

AI and ML, including advanced subsets like Deep Learning (DL), are increasingly transforming water resource management. AI involves simulating human intelligence processes, empowering systems to analyze vast datasets, identify patterns, and make data-driven decisions autonomously. In water management, ML—a branch of AI—employs statistical algorithms that can learn from data and make predictions or classifications, which are essential for applications like resource allocation and quality monitoring. Advanced DL algorithms using neural network architectures are especially effective for complex tasks such as interpreting satellite imagery for water body detection and classifying water pollutants in real-time.[51–53]

Several AI-ML-DL models are proving invaluable in hydrological applications. For instance, supervised learning models like Support Vector Machines (SVMs) and Random Forests (RFs) are commonly used for flood (allied disasters) forecasting by analyzing parameters such as rainfall intensity, river discharge, and soil moisture.[54–58] Reinforcement Learning (RL), which learns by maximizing cumulative rewards, and DL is employed in optimizing water resource management to minimize the impacts of droughts and floods.[59,60] Through clustering algorithms like K-means, unsupervised learning groups similar data points, helping identify vulnerable regions prone to extreme climatic events like cyclones and heatwaves.[61] Meanwhile, DL models such as Convolutional Neural Networks (CNNs) excel in remote sensing applications, analyzing high-resolution satellite imagery to track changes in water bodies, detect land degradation, and monitor the impacts of extreme weather events.[62,63] Case studies from global regions exemplify these applications. In Australia, ML algorithms are used to predict and mitigate urban water shortages by integrating hydrological and climate data.[64] The U.S. uses AI for flood prediction and early warning systems, providing a significant advantage in flood-prone areas.[65] Moreover, in China, DL models are applied to optimize water quality monitoring across extensive river networks, providing real-time feedback to environmental agencies.[66] These applications underscore AI-ML's potential to boost water resource management efficiency, precision, and adaptability, supporting sustainable, climate-responsive practices worldwide.

3.2 Remote sensing and GIS (RS-GIS) technologies

RS and GIS have become foundational technologies in modern water resource management, allowing accurate monitoring, mapping, and analysis of water systems on local, regional, and global scales. Remote sensing utilizes satellite- and drone-based sensors to gather real-time and historical data on various water parameters such as water extent, quality, and seasonal dynamics. This data is processed and analyzed using GIS platforms, which integrate spatial information to create detailed, layered maps that support informed decision-making. Satellite-based monitoring allows for consistent and comprehensive coverage of water resources, essential for tracking seasonal fluctuations, drought impacts, and water quality shifts.[67,68] Advanced multispectral and hyperspectral sensors, such as those on NASA's Landsat and ESA's Sentinel missions, capture high-resolution imagery that aids in assessing chlorophyll levels, suspended sediments, and turbidity.[69,70] These indicators are crucial for determining water quality and detecting pollution sources. GIS complements RS by providing spatial analysis tools that allow for the integration of RS data with other datasets—such as hydrological and climatic records—facilitating a multi-dimensional view of water resources.

Case studies highlight the effectiveness of RS-GIS integration in water management worldwide. In India, RS-GIS tools are used in the Ganga Basin to monitor water quality and identify pollution hotspots, allowing timely intervention.[71] In the drought-prone regions of Sub-Saharan Africa, satellite data supports early warning systems by detecting soil moisture levels and groundwater depletion, informing drought preparedness measures.[72] Similarly, GIS-based data integration models in the Maheshwaram watershed, India, help manage groundwater resources by analyzing recharge rates, aquifer depletion, and land use.[73] When coupled with AI-ML algorithms, RS-GIS technologies further improve predictive modeling, providing governments and stakeholders powerful tools to assess water availability, optimize allocation, and develop sustainable water management policies aligned to shifting climatic conditions.[74,75] These tools are pivotal for devising adaptive water resource strategies in a rapidly changing global environment.

3.3 Large language models (LLMs) in water management

LLMs, such as OpenAI's GPT series and Google's AI (Gemini), represent an emerging frontier in water resource management by yielding advanced Natural Language Processing (NLP) capabilities.[24,25] These models are designed to process and analyze large volumes of text data, which can come from diverse sources, including policy documents, scientific literature, environmental reports, and public opinion data. The application of LLMs in water management remains in its early stages, yet their potential for improving decision-making, communication, and stakeholder engagement is vast. LLMs can generate intuitions by processing complex water management datasets, identifying trends, and synthesizing relevant knowledge across disciplines. For instance, these models can assist in summarizing regulatory frameworks for water quality standards, providing quick access to legal requirements across regions, which is invaluable for policymakers. Additionally, LLMs are increasingly adept at answering technical questions posed by water management professionals, who may require fast, accurate responses on a range of topics, from hydrological modeling to policy analysis. The primary challenge remains adapting LLMs to process highly technical, domain-specific language and account for regional dialects, water management jargon, and complex environmental contexts. As this field grows, integrating LLMs with real-time data and multilingual capabilities

Table 2 Comparative applications of AI-ML-DL, RS-GIS, and LLMs in water resource management.

Purpose	AI-ML-DL	RS-GIS	LLMs
Water Quality Monitoring	Machine learning models predict contaminant levels and detect anomalies in real-time	Satellite imagery and GIS data track pollution sources and map water quality trends	Summarizes research on water pollutants and compares water standards across regions
Resource Allocation and Optimization	Deep learning optimizes water distribution networks, reducing waste and energy costs	GIS identifies areas with high demand and potential scarcity, supporting allocation planning	Processes vast documentation on water rights and suggests optimal allocation frameworks
Flood and Drought Forecasting	AI algorithms analyze historical weather patterns to improve forecasting accuracy	Satellite data provides hydrological and topographical mapping to assess flood-risk areas	Integrates past emergency responses, offering guidelines for managing water during crises
Irrigation Management	Predictive ML models help farmers manage irrigation schedules, reducing overuse	RS imagery tracks soil moisture and vegetation health for precise irrigation control	Provides real-time recommendations for crop-specific water requirements, aggregating agronomic data
Climate Impact Assessment on Water Resources	AI models simulate the impacts of various climate scenarios on water resources	GIS-based climate models analyze land and water interactions under changing climate conditions	Synthesizes international climate policies to recommend localized adaptation strategies
Public Awareness and Engagement	Social media sentiment analysis detects community concerns regarding water issues	GIS maps highlight local water resources and issues, enhancing public awareness	Engages the public by generating easy-to-understand summaries of technical reports on water management

(continued)

Table 2 Comparative applications of AI-ML-DL, RS-GIS, and LLMs in water resource management. (*cont'd*)

Purpose	AI-ML-DL	RS-GIS	LLMs
Policy and Regulatory Analysis	ML reviews historical policy impacts, supporting predictive policy analysis	GIS evaluates the spatial impact of regulatory frameworks on water resources and distribution	Analyzes and translates policy language for decision-makers, providing accessible summaries
Disaster Response and Recovery	AI classifies damage levels post-disaster, guiding efficient resource allocation	Real-time satellite imaging aids in the immediate assessment of flood and drought impacts	Integrates lessons learned from past events, offering structured action plans for emergency responses
Stakeholder Communication and Knowledge Sharing	NLP-based chatbots answer common questions on water resource management	Visual RS-GIS maps facilitate discussions among stakeholders on resource management issues	Generates multilingual summaries, bridging communication gaps between global and local stakeholders

will unlock even greater potential for managing water resources in a climate-smart, inclusive, and responsive manner. The use of LLMs promises to transform not only data access but also the effectiveness of stakeholder communication in addressing water challenges worldwide.

Table 2 highlights the varied applications of AI-ML-DL, RS-GIS, and LLMs in water resource management, underscoring their complementary strengths across diverse domains. AI-ML-DL models excel in predictive analytics, improving water quality monitoring, resource allocation, and climate impact simulations. RS-GIS provides real-time spatial information that is crucial for flood mapping, drought assessment, and resource distribution. Meanwhile, LLMs support decision-making by synthesizing complex data and facilitating stakeholder communication. Together, these technologies address water management challenges holistically, blending predictive precision, spatial analysis, and accessible language, thus fostering data-driven, responsive, and inclusive strategies for sustainable water use.

4. Scope of the book

4.1 Aspect 1: Climate change and advancing technologies: Evolution of water management infrastructure

Section A of this book delves into the intricate relationship between climate change and the evolving water management landscape, comprehensively examining how advancing technologies have revolutionized this critical sector. The section begins by exploring the profound impacts of climate change on water resources, emphasizing the urgency for innovative solutions that address growing uncertainties. It then traces the historical evolution of water management practices, drawing lessons from ancient methodologies while showcasing how these understandings inform modern approaches. The narrative transitions to a forward-looking perspective, introducing integrated technologies such as AI-ML, RS-GIS, and LLMs. Together, this section provides a cohesive foundation for understanding the need for advanced, interdisciplinary strategies in climate-smart water management.

4.2 Aspect 2: Smart water management through AI-ML applications

Section B explores how AI-ML revolutionizes water resource management, providing cutting-edge tools for achieving climate resilience and sustainability. The section begins by establishing foundational concepts of AI-ML, laying the groundwork for their potential in addressing complex water-related challenges. It delves into the role of ML in improving water quality assessment, empowering precision and efficiency in identifying and mitigating contamination. Building on this, the section highlights predictive modeling techniques for climate-resilient water resource planning, showcasing their ability to forecast future scenarios and guide proactive interventions. Furthermore, AI-driven strategies for optimizing water distribution networks are discussed, emphasizing improved efficiency and reduced wastage. Finally, the section examines AI-ML innovations promoting sustainable water use and conservation, presenting practical applications that balance current demands with future needs.

4.3 Aspect 3: Remote sensing and GIS techniques for water resource monitoring

Section C highlights the potential of RS-GIS technologies in monitoring and managing water resources amidst changing climatic conditions. It begins by exploring satellite-based monitoring techniques; geospatial technologies are examined for their applications in watershed planning, providing innovative approaches to sustainable management at regional and local scales. The section also delves into integrating RS and GIS for advanced hydrological modeling, enabling detailed analysis and prediction of water flow dynamics. Additionally, the use of RS to monitor the impacts of climate change on water resources is discussed, providing an understanding of the evolving challenges and potential adaptive strategies. Finally, advancements in LiDAR technology are presented, highlighting its precision capabilities for effective water resource assessment and management.

4.4 Aspect 4: Harnessing the power of large language models (LLMs) in water management

Section D explores the transformative role of LLMs in water management, highlighting their potential to revolutionize decision-making and communication. It begins by introducing LLMs, emphasizing their ability to process vast datasets and provide implementable actions for efficient resource allocation. NLP applications are showcased for their utility in assessing water quality, thereby providing scalable real-time monitoring and reporting solutions. The section also examines the optimization of water resource planning through LLM-driven analytical capabilities, ensuring precision in managing competing demands. Furthermore, the potential of LLMs to improve communication and stakeholder engagement is discussed, particularly in bridging technical complexities and nurturing collaborative governance. Lastly, emerging trends in integrating LLMs into climate-smart water governance are outlined, emphasizing their adaptability to future challenges. This section provides a forward-looking perspective on LLMs as pivotal tools for sustainable and equitable water management in the face of climate change.

5. Concluding remarks for the introduction chapter

As water scarcity intensifies, interdisciplinary collaboration is more critical than ever. Addressing climate impacts on water resources requires joint efforts from climate scientists, engineers, data scientists, and policymakers. Climate science provides foundational visions into hydrological cycles, while engineering expertise is crucial in designing adaptive infrastructure. Data science, particularly through AI and ML, allows for predictive analytics and enriched resource allocation, while policy provides the regulatory frameworks to implement effective solutions. Together, these disciplines can produce a holistic approach to water resource management that is both responsive and sustainable, creating a system resilient to environmental and anthropogenic pressures. The urgency of sustainable water management calls for climate-smart, technology-driven solutions that can anticipate, monitor, and mitigate water-related challenges. Stakeholders—government agencies, local communities, and industry leaders—must prioritize strategies integrating advanced technologies with traditional water conservation practices. AI-ML, RS-GIS, and LLMs yield powerful tools to improve decision-making, resource allocation, and crisis management, ensuring that water resources are safeguarded against climate variability and extreme events. By adopting these climate-resilient approaches, stakeholders can better manage water resources and ensure equitable distribution. This introductory chapter has laid the groundwork for understanding the interplay of technology and climate action in water resource management. Upcoming chapters will dive deeper into specific areas, starting with a closer look at how climate change impacts water systems. From there, the book will explore how AI-ML, RS-GIS, and LLMs contribute uniquely to water management challenges, equipping readers with theoretical and practical understandings. This progression aims to provide an all-rounded view of how these advanced tools, when combined with interdisciplinary expertise, can transform water resource management in an era of climate uncertainty.

Author's contribution

Both authors contributed to the study's conception and design. They collected the data, reports, documents, and research papers and conducted the analysis. Both authors drafted and approved the final manuscript.

Conflict of interest

The authors declare that the research was conducted without any commercial or financial relationships that could be construed as a potential conflict of interest.

Declaration of generative AI and AI-assisted technologies in the writing process

While preparing this work, the authors used the 'OpenAI - Cloud Computing Platform Version 4′ to improve language. After using this service, the authors reviewed and edited the content as needed and took full responsibility for the publication's content.

References

1. IPCC. Summary for policymakers. IPCC: Geneva, Switzerland, 2023; pp. 1–34. https://doi.org/10.59327/IPCC/AR6–9789291691647.001
2. Mishra RK. Fresh water availability and its global challenge. *Br J Multidiscip Adv Stud* 2023;**4**(3):1–78. https://doi.org/10.37745/bjmas.2022.0207.
3. UN. *United Nations children's fund, & World Health Organization on progress on household drinking water, sanitation and hygiene 2000-2022: special focus on gender;* World Health Organization, 2024.
4. Jones ER, Bierkens MF, van Vliet MT. Current and future global water scarcity intensifies when accounting for surface water quality. *Nat Clim Change* 2024:1–7. https://doi.org/10.1038/s41558-024-02007-0.
5. Liu J, Li D, Chen H, Wang H, Wada Y, Kummu M, et al. Timing the first emergence and disappearance of global water scarcity. *Nat Commun* 2024;**15**(1):7129. https://doi.org/10.1038/s41467-024-51302-z.
6. Dilanchiev A, Nuta F, Khan I, Khan H. Urbanization, renewable energy production, and carbon dioxide emission in BSEC member states: implications for climate change mitigation and energy markets. *Environ Sci Pollut Res* 2023;**30**(25):67338–67350. https://doi.org/10.1007/s11356-023-27221-9.
7. Ingrao C, Strippoli R, Lagioia G, Huisingh D. Water scarcity in agriculture: an overview of causes, impacts and approaches for reducing the risks. *Heliyon* 2023;**9**(8):e18507. https://doi.org/10.1016/j.heliyon.2023.e18507.
8. Maity R, Kashid SS. Importance analysis of local and global climate inputs for basin-scale streamflow prediction. *Water Resour Res* 2011;**47**(11) https://doi.org/10.1029/2010WR009742.
9. Srivastava A, Maity R, Desai VR. Assessing global-scale synergy between adaptation, mitigation, and sustainable development for projected climate change. In *Ecological footprints of climate change;* Chatterjee, U., Akanwa, A. O., Kumar, S., Singh, S. K., Dutta Roy, A., Eds.; Springer Climate: Cham, 2022, https://doi.org/10.1007/978-3-031-15501-7_2.
10. Chanda K, Maity R. Uncovering global climate fields causing local precipitation extremes. *Hydrological Sci J* 2016;**61**(7):1227–1237. https://doi.org/10.1080/02626667.2015.1006232.

11. Konapala G, Mishra AK, Wada Y, Mann ME. Climate change will affect global water availability through compounding changes in seasonal precipitation and evaporation. *Nat Commun* 2020;**11**(1):3044. https://doi.org/10.1038/s41467-020-16757-w.
12. Ndehedehe CE, Ferreira VG, Adeyeri OE, Correa FM, Usman M, Oussou FE, et al. Global assessment of drought characteristics in the anthropocene. *Resources, Environ Sustain* 2023;**12**:100105. https://doi.org/10.1016/j.resenv.2022.100105.
13. Srivastava A, Maity R, Desai VR. Investigating spatio-temporal trends and anomalies in long-term meteorological variables to determine if maharashtra is an emerging warming state in India. Springer: Singapore, 2024https://doi.org/10.1007/978-981-99-9610-0_25
14. Broyles LM, Pakhtigian EL, Rosinger AY, Mejia A. Climate and hydrological seasonal effects on household water insecurity: a systematic review. *Wiley Interdiscip Reviews: Water* 2022;**9**(3):e1593. https://doi.org/10.1002/wat2.1593.
15. Raymond C, Horton RM, Zscheischler J, Martius O, AghaKouchak A, Balch J, et al. Understanding and managing connected extreme events. *Nat Clim Change* 2020;**10**(7):611–621. https://doi.org/10.1038/s41558-020-0790-4.
16. Maity R, Srivastava A, Sarkar S, Khan MI. Revolutionizing the future of hydrological science: impact of machine learning and deep learning amidst emerging explainable AI and transfer learning. *Appl Comput Geosci* 2024. https://doi.org/10.1016/j.acags.2024.100206. 100206 (in press).
17. Kumar S, Srivastava A, Maity R. Modeling climate change impacts on vector-borne disease using machine learning models: case study of visceral leishmaniasis (Kala-azar) from Indian state of Bihar. *Expert Syst Appl* 2024;**237**:121490. https://doi.org/10.1016/j.eswa.2023.121490.
18. Maity R, Nagesh Kumar D. Basin-scale stream-flow forecasting using the information of large-scale atmospheric circulation phenomena. *Hydrological Processes: An Int J* 2008;**22**(5):643–650. https://doi.org/10.1002/hyp.6630.
19. Srivastava A, Maity R. Assessing the potential of AI–ML in urban climate change adaptation and sustainable development. *Sustainability* 2023;**15**(23):16461. https://doi.org/10.3390/su152316461.
20. Sun AY, Scanlon BR. How can big data and machine learning benefit environment and water management: a survey of methods, applications, and future directions. *Environ Res Lett* 2019;**14**(7):073001. https://doi.org/10.1088/1748-9326/ab1b7d.
21. Abdelhaleem FS, Basiouny M, Ashour E, Mahmoud A. Application of remote sensing and geographic information systems in irrigation water management under water scarcity conditions in Fayoum, Egypt. *J Environ Manag* 2021;**299**:113683. https://doi.org/10.1016/j.jenvman.2021.113683.
22. Du X, Tariq A, Islam F, Aziz S, Waseem LA, Ahmad MN, et al. Integrated study of GIS and remote sensing to identify potential sites for rainwater harvesting structures. *Phys Chem Earth, Parts A/B/C* 2024;**134**:103574. https://doi.org/10.1016/j.pce.2024.103574.
23. Zagade ND, Umrikar BN. Drought severity modeling of upper Bhima river basin, western India, using GIS–AHP tools for effective mitigation and resource management. *Nat Hazards* 2021;**105**:1165–1188. https://doi.org/10.1007/s11069-020-04350-9.
24. Egbemhenghe AU, Ojeyemi T, Iwuozor KO, Emenike EC, Ogunsanya TI, Anidiobi SU, et al. Revolutionizing water treatment, conservation, and management: harnessing the power of AI-driven ChatGPT solutions. *Environ Chall* 2023;**13**:100782. https://doi.org/10.1016/j.envc.2023.100782.
25. Zhu JJ, Jiang J, Yang M, Ren ZJ. ChatGPT and environmental research. *Environ Sci & Technol* 2023;**57**(46):17667–17670. https://doi.org/10.1021/acs.est.3c01818.

26. Alivernini S. "Let's move the earth and build a canal!" The management of water infrastructures in a Sumerian city at the end of the third millennium. *Water History* 2020;**12**:93–104. https://doi.org/10.1007/s12685-020-00242-8.
27. Luo P, Sun Y, Wang S, Wang S, Lyu J, Zhou M, et al. Historical assessment and future sustainability challenges of Egyptian water resources management. *J Clean Prod* 2020;**263**:121154. https://doi.org/10.1016/j.jclepro.2020.121154.
28. Gilmartin D. *Blood and water: the Indus river basin in modern history;* University of California Press, 2020.
29. Pazwash H. *Qanats and historic structures in Persia: potential modern applications;* CRC Press, 2024.
30. Seligmann LJ, Fine-Dare KS. *The andean world;* Routledge: Abingdon, 2019https://doi.org/10.4324/9781315621715
31. Eslamian, S., Eslamian, F. A., Eds. *Handbook of water harvesting and conservation.* John Wiley & Sons: Incorporated, 2021, https://doi.org/10.1002/9781119776017.
32. Srivastava A, Chinnasamy P. Understanding declining storage capacity of tank cascade system of madurai: potential for better water management for rural, Peri-Urban, and Urban catchments. In *Rural technology development and delivery. design science and innovation;* Chigullapalli, S., Susha Lekshmi, S. U., Deshpande, A. P., Eds.; Springer: Singapore, 2023, https://doi.org/10.1007/978-981-19-2312-8_14
33. Srivastava A, Chinnasamy P. Tank cascade system in Southern India as a traditional surface water infrastructure: a review. In *Rural technology development and delivery. design science and innovation;* Chigullapalli, S., Susha Lekshmi, S. U., Deshpande, A. P., Eds.; Springer: Singapore, 2023https://doi.org/10.1007/978-981-19-2312-8_15
34. Luo P, He B, Takara K, Xiong YE, Nover D, Duan W, et al. Historical assessment of Chinese and Japanese flood management policies and implications for managing future floods. *Environ Sci & Policy* 2015;**48**:265–277. https://doi.org/10.1016/j.envsci.2014.12.015.
35. Ritchie H, Eisma JA, Parker A. Sand dams as a potential solution to rural water security in drylands: existing research and future opportunities. *Front Water* 2021;**3**:651954. https://doi.org/10.3389/frwa.2021.651954.
36. Di Baldassarre G, Mazzoleni M, Rusca M. The legacy of large dams in the United States. *Ambio* 2021;**50**(10):1798–1808. https://doi.org/10.1007/s13280-021-01533-x.
37. Eldardiry H, Hossain F. A blueprint for adapting high Aswan dam operation in Egypt to challenges of filling and operation of the grand Ethiopian Renaissance dam. *J Hydrol* 2021;**598**:125708. https://doi.org/10.1016/j.jhydrol.2020.125708.
38. Zhang AT, Gu VX. Global Dam tracker: a database of more than 35,000 dams with location, catchment, and attribute information. *Sci Data* 2023;**10**(1):111. https://doi.org/10.1038/s41597-023-02008-2.
39. Larsen A, Larsen JR, Lane SN. Dam builders and their works: beaver influences on the structure and function of river corridor hydrology, geomorphology, biogeochemistry and ecosystems. *Earth-Science Rev* 2021;**218**:103623. https://doi.org/10.1016/j.earscirev.2021.103623.
40. Singh A. Soil salinization management for sustainable development: a review. *J Environ Manag* 2021;**277**:111383. https://doi.org/10.1016/j.jenvman.2020.111383.
41. Elbeltagi A, Al-Saeedi AH, Raza A, Abd-Elaty I, El-Rawy M. Forecasting long-series daily reference evapotranspiration based on best subset regression and machine learning in Egypt. *Water* 2023;**15**(6):1149. https://doi.org/10.3390/w15061149.
42. Pande CB, Al-Ansari N, Kushwaha NL, et al. Forecasting of SPI and meteorological drought based on the artificial neural network and M5P model tree. *Land* 2022;**11**(11):2040. https://doi.org/10.3390/land11112040.
43. Senapati U, Srivastava A, Maity R. High-resolution agricultural drought hazard mapping using the potential of geospatial data and machine learning approaches. *Environ Monit Assess* 2025;**197**(11):1–25. https://doi.org/10.1007/s10661-025-14538-w.

44. Liladhar Rane N, Purushottam Choudhary S, et al. Delineation of environmentally sustainable urban settlement using GIS-based MIF and AHP techniques. *Geocarto Int* 2024;**39**(1):2335249. https://doi.org/10.1080/10106049.2024.2335249.
45. Pal M, Maity R, Suman M, Das SK, Patel P, Srivastava HS. Satellite-based probabilistic assessment of soil moisture using C-band quad-polarized RISAT1 data. *IEEE Trans Geosci Remote Sens* 2016;**55**(3):1351–1362. https://doi.org/10.1109/TGRS.2016.2623378.
46. Pande CB, Moharir KN, Radwan N, Mohd Sidek L, Alshehri F, et al. Characterizing land use/land cover change dynamics by an enhanced random forest machine learning model: a Google Earth Engine implementation. *Environ Sci Europe* 2024;**36**(1):84. https://doi.org/10.1186/s12302-024-00901-0.
47. Rane NL, Achari A, Choudhary SP, Mallick SK, Pande CB, et al. A decision framework for potential dam site selection using GIS, MIF and TOPSIS in Ulhas river basin, India. *J Clean Prod* 2023;**423**:138890. https://doi.org/10.1016/j.jclepro.2023.138890.
48. Srivastava A, Maity R. Unveiling an environmental drought Index and its applicability in the perspective of drought recognition amidst climate change. *J Hydrol* 2023;**627**:130462. https://doi.org/10.1016/j.jhydrol.2023.130462.
49. Srivastava A, Pandey A, Maity R. Comparative multi-index analysis of existing drought typology and environmental droughts in a climate-stressed region. *Sci Rep* 2025;**15**(1):34349. https://doi.org/10.1038/s41598-025-16859-9.
50. Srivastava A, Maity R, Desai VR. Synthesising causal loop between environmental and compound droughts: A systems-driven framework for climate-resilient governance in monsoon-sensitive agroecosystems. *Next Sustain* 2025;**6**:100159. https://doi.org/10.1016/j.nxsust.2025.100159.
51. Khan MI, Maity R. Hybrid deep learning approach for multi-step-ahead daily rainfall prediction using GCM simulations. *IEEE Access* 2020;**8**:52774–52784. https://doi.org/10.1109/ACCESS.2020.2980977.
52. Soori M, Arezoo B, Dastres R. Artificial intelligence, machine learning and deep learning in advanced robotics, a review. *Cognit Robot* 2023;**3**:54–70. https://doi.org/10.1016/j.cogr.2023.04.001.
53. Yang L, Driscol J, Sarigai S, Wu Q, Lippitt CD, Morgan M. Towards synoptic water monitoring systems: a review of AI methods for automating water body detection and water quality monitoring using remote sensing. *Sensors* 2022;**22**(6):2416. https://doi.org/10.3390/s22062416.
54. Kumar V, Kedam N, Sharma KV, Mehta DJ, Caloiero T. Advanced machine learning techniques to improve hydrological prediction: a comparative analysis of streamflow prediction models. *Water* 2023;**15**(14):2572. https://doi.org/10.3390/w15142572.
55. Maity R, Kashid SS, Bhatnagar A. Hydrometerological modeling approaches using support vector regression (SVR) and genetic programming (GP). *ISH J Hydraulic Eng* 2009;**15**(sup1):244–257. https://doi.org/10.1080/09715010.2009.10514978.
56. Maity R, Bhagwat PP, Bhatnagar A. Potential of support vector regression for prediction of monthly streamflow using endogenous property. *Hydrological Processes: An Int J* 2010;**24**(7):917–923. https://doi.org/10.1002/hyp.7535.
57. Pal M, Maity R, Ratnam JV, Nonaka M, Behera SK. Long-lead prediction of ENSO modoki index using machine learning algorithms. *Sci Rep* 2020;**10**(1):365. https://doi.org/10.1038/s41598-019-57183-3.
58. Pham LT, Luo L, Finley A. Evaluation of random forests for short-term daily streamflow forecasting in rainfall-and snowmelt-driven watersheds. *Hydrol Earth Syst Sci* 2021;**25**(6):2997–3015. https://doi.org/10.5194/hess-25-2997-2021.
59. Hajgató G, Paál G, Gyires-Tóth B. Deep reinforcement learning for real-time optimization of pumps in water distribution systems. *J Water Resour Plan Manag* 2020;**146**(11):04020079. https://doi.org/10.1061/(ASCE)WR.1943-5452.0001287.

60. Khan MI, Maity R. Development of a long-range hydrological drought prediction framework using deep learning. *Water Resour Manag* 2024;**38**(4):1497–1509. https://doi.org/10.1007/s11269-024-03735-w.
61. Văduva AG, Munteanu M, Oprea SV, Bâra A, Niculae AM. Understanding climate change and air quality over the last decade. Evidence from news and weather data processing. *IEEE Access* 2023. https://doi.org/10.1109/ACCESS.2023.3345466.
62. Elbeltagi A, Ehsan M, Sharma G, Yu J, Khadke L, et al. Advanced stacked integration method for forecasting long-term drought severity: CNN with machine learning models. *J Hydrology: Regional Stud* 2024;**53**:101759. https://doi.org/10.1016/j.ejrh.2024.101759.
63. Liang J, Deng Y, Zeng D. A deep neural network combined CNN and GCN for remote sensing scene classification. *IEEE J Sel Top Appl Earth Observations Remote Sens* 2020;**13**:4325–4338. https://doi.org/10.1109/JSTARS.2020.3011333.
64. Zubaidi SL, Kumar P, Al-Bugharbee H, Ahmed AN, Ridha HM, Mo KH, et al. Developing a hybrid model for accurate short-term water demand prediction under extreme weather conditions: a case study in Melbourne, Australia. *Appl Water Sci* 2023;**13**(9):184. https://doi.org/10.1007/s13201-023-01995-2.
65. Byaruhanga N, Kibirige D, Gokool S, Mkhonta G. Evolution of flood prediction and forecasting models for flood early warning systems: a scoping review. *Water* 2024;**16**(13):1763. https://doi.org/10.3390/w16131763.
66. Huang J, Zhang Y, Bing H, Peng J, Dong F, Gao J, et al. Characterizing the river water quality in China: recent progress and on-going challenges. *Water Res* 2021;**201**:117309. https://doi.org/10.1016/j.watres.2021.117309.
67. Inoue Y. Satellite-and drone-based remote sensing of crops and soils for smart farming–a review. *Soil Sci Plant Nutr* 2020;**66**(6):798–810. https://doi.org/10.1080/00380768.2020.1738899.
68. Kucharczyk M, Hugenholtz CH. Remote sensing of natural hazard-related disasters with small drones: global trends, biases, and research opportunities. *Remote Sens Environ* 2021;**264**:112577. https://doi.org/10.1016/j.rse.2021.112577.
69. Kwon S, Shin J, Seo IW, Noh H, Jung SH, You H. Measurement of suspended sediment concentration in open channel flows based on hyperspectral imagery from UAVs. *Adv Water Resour* 2022;**159**:104076. https://doi.org/10.1016/j.advwatres.2021.104076.
70. Maciel FP, Haakonsson S, Ponce de León L, Bonilla S, Pedocchi F. Challenges for chlorophyll-a remote sensing in a highly variable turbidity estuary, an implementation with sentinel-2. *Geocarto Int* 2023;**38**(1):2160017. https://doi.org/10.1080/10106049.2022.2160017.
71. Dandge KP, Patil SS. Spatial distribution of ground water quality index using remote sensing and GIS techniques. *Appl Water Sci* 2022;**12**(1):7. https://doi.org/10.1007/s13201-021-01546-7.
72. Bhaga TD, Dube T, Shekede MD, Shoko C. Impacts of climate variability and drought on surface water resources in Sub-Saharan Africa using remote sensing: a review. *Remote Sens* 2020;**12**(24):4184. https://doi.org/10.3390/rs12244184.
73. Fauzia SL, Rahman A, Ahmed S. Distributed groundwater recharge potentials assessment based on GIS model and its dynamics in the crystalline rocks of South India. *Sci Rep* 2021;**11**(1):11772. https://doi.org/10.1038/s41598-021-90898-w.
74. El-Rawy M, Wahba M, Fathi H, Alshehri F, Abdalla F, El Attar RM. Assessment of groundwater quality in arid regions utilizing principal component analysis, GIS, and machine learning techniques. *Mar Pollut Bull* 2024;**205**:116645. https://doi.org/10.1016/j.marpolbul.2024.116645.
75. Arabameri A, Pal SC, Rezaie F, Chakrabortty R, Chowdhuri I, Blaschke T, et al. Comparison of multi-criteria and artificial intelligence models for land-subsidence susceptibility zonation. *J Environ Manag* 2021;**284**:112067. https://doi.org/10.1016/j.jenvman.2021.112067.

CHAPTER TWO

Indigenous water resource management systems of the Apatani tribe: Bridging traditional practices with modern solutions in the Indian Himalayan Region

Tridipa Biswas*

Govind Ballabh Pant 'National Institute of Himalayan Environment' (NIHE), North-East Regional Centre (NERC), Itanagar, Arunachal Pradesh, India

*Corresponding author. e-mail address: tridipa.b@gbpihed.nic.in; tridipabiswas@gmail.com

Contents

Abstract

Water resource management is vital for maintaining ecological balance, particularly in fragile, resource-dependent, and climate-sensitive regions. Indigenous systems have garnered increasing attention for their potential to support sustainable and adaptive water usage. This chapter explores the traditional water management practices of the Apatani tribe in Ziro Valley, Arunachal Pradesh, India, and their relevance for

Advances in Chemical Pollution, Environmental Management and Protection, Volume 15
ISSN 2468-9289, https://doi.org/10.1016/bs.apmp.2025.11.015

addressing contemporary environmental challenges. Key indigenous practices documented in this study include the paddy-cum-fish cultivation system (Aji-ngyii), bamboo irrigation networks (Borang), gravity-fed water channels (Bogo), water harvesting structures (Dlong), and community-based governance (Bulyang). Field surveys conducted in eight representative villages demonstrated that these systems exhibit notable climate resilience through ecosystem-based adaptation strategies that optimize water use while maintaining ecological balance. The study found that Apatani practices effectively integrate agro-hydrological systems, utilize low-carbon, nature-based infrastructure, incorporate ecosystem-based field designs, and foster adaptive community governance, contributing to water security, biodiversity conservation, and climate adaptation. Despite the mounting pressures of modernization, population growth, and climate change, the Apatani system offers valuable insights into sustainable water governance in the Himalayas. This study bridges traditional ecological knowledge with modern solutions, offering a replicable framework for climate-resilient water management in other ecologically vulnerable regions.

Keywords: Apatani tribe, Ecosystem-based adaptation, Climate resilience, Sustainable water governance, Indian Himalayan region.

1. Introduction

Water is a critical resource, and its management is essential for human survival, especially in regions heavily dependent on agriculture. Over the last decade, efforts to protect and manage global freshwater systems have shifted from conventional approaches to indigenous water resource management systems that recognize natural processes.[1,2] Among them, the Apatani tribe of Arunachal Pradesh, India, takes up the case study of their ingenious paddy fish cultivation system as an example of sustaining water governance in the traditional mode.[3] This combined treatment shows not only the success of Traditional Ecological Knowledge (TEK) but also delivers lessons learned for the stewardship of community-based natural resources.[4]

The Apatani's water management practices developed through generations in the Ziro Valley represent a harmonious blend of agricultural innovation and cultural preservation.[5] Integrating ingenious bamboo-based water diversion structures and optimizing water-carrying canals for maximum irrigation efficiency supports both crop and aquaculture.[6] This integrated approach is an ecosystem-based adaptation (EbA) process that underpins sound concerns for climate resilience and biodiversity conservation.[7]

Indigenous knowledge systems for natural resource management have become increasingly important as global environmental challenges have become more intense.[2,8] Traditional ecological knowledge proved to complement modern conservation efforts regarding water resource management practices among the Apatani, and it did not disrupt cultural integrity.[5,9] The sustainability lessons provided by this system are valuable for sustainable water resource management in regions experiencing similar environmental pressures and agricultural challenges.[10]

This study focuses on the evolution and effectiveness of Apatani's water management practices as a contribution to sustainable agriculture and ecosystem services. Additionally, it is necessary to understand how indigenous knowledge systems can inform modern approaches to water resource management and climate adaptation strategies by analysing how they manage water traditionally and how their social frameworks operate.

2. Research rationale and objectives

The study of the Apatani tribe's indigenous water resource management systems holds paramount significance as it offers valuable insights into sustainable water management practices that have evolved over centuries in the challenging terrain of the Indian Himalayan region. This study bridges the critical gap between traditional ecological knowledge and contemporary water management solutions, which is particularly relevant in an era of climate change and water scarcity.[11,12] This study aims to document the Apatani's indigenous water management techniques, particularly their distinctive wet rice cultivation and fish culture system, to understand how these practices can be preserved and adapted to meet contemporary challenges. By examining the integration of traditional ecological knowledge with modern conservation approaches, this study seeks to develop a framework that can inform sustainable water resource policies in mountainous regions while protecting valuable cultural heritage. Furthermore, this study's documentation of indigenous knowledge systems serves as a model for other mountainous regions facing similar water management challenges, while also supporting the growing recognition of traditional ecological knowledge in global environmental conservation efforts. The findings will contribute to both the theoretical understanding of indigenous resource management and practical solutions for water security in the Indian Himalayan Region.

The specific objectives of this research are:

I. Document the traditional water management practices of the Apatani tribe
II. Evaluate the effectiveness of these systems in modern contexts
III. Assess the sustainability and adaptability of these practices in the face of climate change

3. Materials and methods

3.1 Study area

The Ziro Valley situated in the Lower Subansiri district of Arunachal Pradesh, India, spans coordinate between 27° 25′ 33.44″ – 27° 39′ 27.52″ N latitude and 93° 45′ 40.24″ – 94° 0′ 11.29″ E longitude (Fig. 1). The valley which is divided by the meandering Kille River, is characterized by its distinctive topography of high hills and a unique agricultural landscape. The region experiences a subtropical monsoon climate with pronounced seasonal variations, with temperatures ranging from 6.3 °C to 28.1 °C during the summer months and dropping to between 1.0 °C and 18.4 °C in winter.

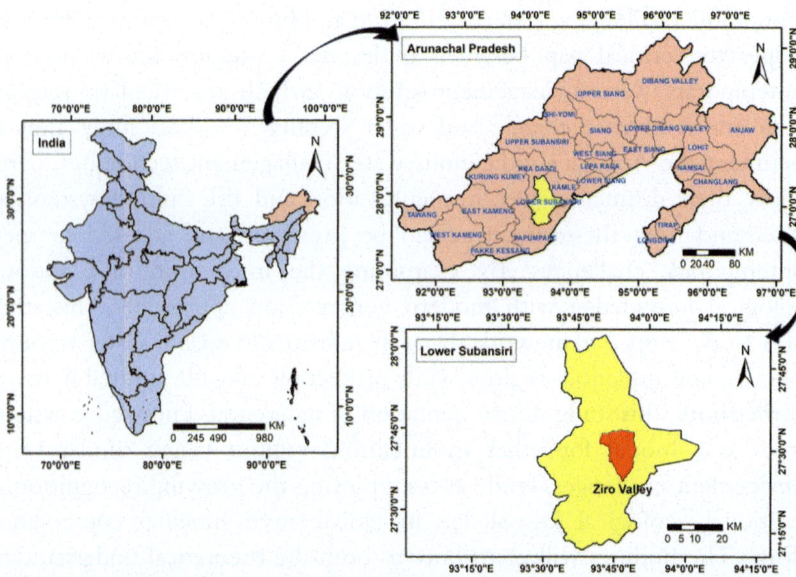

Fig. 1 The study area map. Map lines delineate study areas and do not necessarily depict accepted national boundaries.

The annual precipitation averages 1500 mm, with the majority of rainfall occurring during the monsoon season (May-August), whereas the winter months (November-February) remain relatively dry. The Ziro Valley is renowned for the sustainable agricultural practices of the Apatani tribe, particularly the rice-fish farming system and a unique bamboo-pine home garden agroforestry method, both adapted to the valley's geographical and climatic features. A highly evolved local water management system supports these practices and optimizes land use. As part of the Eastern Himalaya biodiversity hotspot, Ziro Valley is rich in flora and fauna, showcasing the Apatani people's profound ecological knowledge. This landscape, noted for its cultural heritage and resilience, is being considered for UNESCO's World Heritage List, emphasizing its ecological and cultural significance.

3.2 The Apatani tribe: Background and cultural context

The Apatani Tribe resides in the Ziro Valley of Arunachal Pradesh. The tribe's agricultural practices are centered on the wetland-based farming system known as 'Bari', primarily involving rice cultivation. This farming method has been practiced for centuries and showcases a deep understanding of the local ecosystems.

Their social structure is closely tied to agricultural and water management systems, with the community collectively managing water resources. In Apatani society, water is considered a shared resource, and its use and distribution are governed by traditional norms and customs. These systems highlight the tribe's deep ecological knowledge, which has evolved in response to their environment.

3.2.1 Historical background and evolution of Apatani water resource management

The origins of the Apatani water resource management system can be traced back several centuries, when the tribe first settled in the Ziro Valley of Arunachal Pradesh. According to oral traditions passed down through generations, the Apatani people migrated from Tibet through various regions before establishing themselves in the current location around the 13th-14th century CE. Upon settling in the Ziro Valley, they encountered a unique topographical challenge – a bowl-shaped valley surrounded by hills, which required innovative water management solutions for survival. Initially, they developed a basic system of water channeling using locally available bamboo and wood, which gradually evolved into the sophisticated wet rice cultivation system we see today.

The evolution of their water management system can be divided into distinct historical phases. In the early settlement phase (13th-15th centuries), the focus was primarily on basic irrigation channels and hillside terracing. During the consolidation phase (16th-18th centuries), they developed the intricate 'Bogo' system, a network of primary and secondary channels that efficiently distributed water throughout the valley. The community governance system, known as the 'Bulyang', emerged during this period to manage water resources equitably. The 19th century marked a significant advancement in their techniques, with the introduction of the paddy-cum-fish culture system, which maximized land and water use efficiency. This period also saw the establishment of sophisticated water-sharing agreements between villages and the codification of customary laws regarding water rights.

The 20th century has brought new challenges and adaptations. The post-independence period (after 1947) saw the first interaction between traditional practices and modern agricultural techniques in India. Despite external influences, the Apatani community has maintained their core water management principles while selectively incorporating beneficial modern elements. The late 20th century (1970s-1990s) witnessed increased recognition of their sustainable practices, with researchers and international organizations documenting their methods. The system demonstrated remarkable resilience to various climate fluctuations, proving its adaptability. In recent decades (2000s onwards), the community has faced new challenges due to climate change, population growth, and modernization pressures. However, they have successfully maintained their traditional system while adapting to contemporary needs, such as incorporating modern materials like PVC pipes alongside traditional bamboo channels, and establishing formal documentation of their traditional knowledge systems.

A crucial aspect of this historical evolution is the community's ability to maintain social cohesion around water management. Traditional institutions have evolved while retaining their core principles of equity, sustainability, and community participation. The system has shown remarkable adaptability by incorporating new technologies and practices without compromising its fundamental ecological principles. This historical resilience and adaptability have made the Apatani water management system a model for sustainable resource management, studied and admired by researchers and practitioners worldwide. This evolution continues today as the community faces new challenges from climate change and modernization, demonstrating how traditional ecological knowledge can remain relevant and vital in the contemporary world.

Today, the Apatani water management system stands as a testament to centuries of indigenous innovation and adaptation. It represents a living heritage that continues to evolve while maintaining its core principles of sustainability, community management, and ecological balance. The historical journey of this system provides valuable insights into how traditional knowledge systems can adapt to changing circumstances while maintaining their essential characteristics and effectiveness.

3.3 Methodology

The methodology adopted in this study is shown in Fig. 2.

3.3.1 Field survey and site selection

The comprehensive field survey was strategically conducted from September to October 2024, encompassing eight representative villages in the Ziro Valley of Lower Subansiri district, Arunachal Pradesh: Kalung,

Fig. 2 Methodology of research study.

Hari, Hija, Hong, Bamin Michi, Reru, Lempia, and Mudang Tage. Village selection was purposively determined through three critical criteria: (a) geographical accessibility, (b) documented presence of traditional water resource management practices, and (c) representative coverage of Apatani tribal water management techniques within the Ziro Valley ecosystem.

3.3.2 Data collection techniques

A multi-modal qualitative research approach was employed, primarily utilizing semi-structured interviews with diverse community stakeholders, including village elders, local leadership, and specialized water resource management practitioners. The interview protocol was meticulously designed to explore three core thematic domains: historical contextual understanding of water management practices, contemporary water sourcing, storage, and distribution techniques, and the intricate interplay between community collectives and individual household water management strategies.

3.3.3 On-field observations

Systematic physical documentation and verification of water management infrastructure were conducted through comprehensive site visits, focusing on capturing detailed insights into traditional technological innovations. Key physical elements, such as intricate bamboo water channeling systems, community water storage lakes, ingeniously designed terraced agricultural fields, and sophisticated drainage mechanisms were thoroughly examined, photographed, and analytically documented.

3.3.4 Documentation and data analysis with validation

Field observations and interview transcripts were meticulously recorded in specialized field notebooks, and the data were systematically categorized into structured thematic frameworks, including technique typologies, community practice patterns, and seasonal variation analyses. Rigorous cross-verification methodologies were implemented, involving multiple community member consultations to ensure data accuracy and to minimize potential research biases. The synthesized research approach integrated interviews, direct observational evidence, and photographic documentation to construct a holistic and nuanced understanding of Apatani traditional water resource management practices, ultimately bridging indigenous knowledge systems with potential contemporary water management strategies.

4. Results & discussion

4.1 Traditional water resource management practices of the Apatani tribe

4.1.1 Paddy-cum-fish cultivation system (Aji-ngyii)

The Apatani tribe practices an innovative integrated farming system combining wet rice cultivation with fish farming, locally known as 'Aji-ngyii' shown in Fig. 3. This practice involves cultivating two varieties of rice (Mipya and Emoh) alongside fish (Ngyii) in the same field. This system maximizes water use efficiency by cultivating rice in waterlogged fields while simultaneously rearing fish, which provides an additional source of nutrition and income. The paddy fields are designed with strong bunds (Agher) and deeper water channels (Khele) around the perimeter to retain water and prevent fish from escaping during floods. The channels (Siikho/Parkho) within the fields facilitate water management, providing habitats for fish while also serving irrigation needs. These channels, typically 30–45 cm deep, run through the terraced fields and serve multiple purposes: they provide habitat for fish, facilitate water circulation, and create microenvironments that enhance both rice and fish production. The fish feed on insects and

Fig. 3 Paddy-cum-fish cultivation system (Aji-ngyii).

weeds, naturally controlling pests, while their movements aerate the soil and their excreta fertilize the paddy fields. This system maximizes water use efficiency while creating a symbiotic relationship between rice cultivation and fish rearing. The system demonstrates remarkable water efficiency, as the same water simultaneously supports rice cultivation and fish farming, while the fish waste serves as a natural fertilizer for the rice crops. The water depth is carefully managed throughout the growing season, with farmers using traditional markers to monitor and adjust levels. This method has evolved over centuries and is characterized by the use of organic inputs, such as poultry droppings and decomposed plant materials, which enhance soil fertility and support aquatic life. The Apatanis have developed this system to ensure year-round productivity while maintaining an ecological balance. Fish help in natural pest control and field fertilization, while the regulated water flow ensures optimal conditions for both rice and fish.

4.1.2 Traditional irrigation channels (Bogo system)

The 'Bogo' system represents the tribe's indigenous irrigation network, consisting of hierarchically arranged water channels shown in Fig. 4. The terrain of the Ziro Valley, characterized by gentle slopes, allows the Apatani people to employ gravity-driven irrigation systems. Water is diverted from higher elevations through a network of canals and channels that flow downward into the fields. This approach minimizes the need for artificial pumps or external energy sources, making the system environmentally friendly. Primary channels draw water from natural streams, feed secondary channels, and distribute water to different village sectors. Tertiary channels further divide the water among the individual fields. This hierarchical system ensures equitable water distribution and efficient resource utilization in the region. The channels are strategically designed with varying depths and widths, incorporating natural gradients for optimal water flow. This gravity-based system requires minimal external input and demonstrates sophisticated engineering knowledge. The channels are strategically designed with appropriate gradients and dimensions to ensure optimal water flow and distribution throughout the agricultural landscape. Regular maintenance rituals and collective cleaning activities ensure the continued efficiency of the system.

4.1.3 Bamboo irrigation system (Borang)

One of the most remarkable aspects of the Apatani Tribe's water resource management is its use of bamboo irrigation systems shown in Fig. 5.

Fig. 4 Traditional irrigation channels (Bogo system).

A remarkable feature is the use of bamboo pipes (Borang) for water conveyance, which prevents water loss through seepage and maintains a consistent flow. The 'Borang' system utilizes bamboo pipes of varying diameters to transport water across different elevations and distances. These bamboo pipes are carefully selected and treated to ensure durability, with joints sealed using local materials. The system includes primary distribution pipes (Danth) and secondary distribution pipes (Siiri) that deliver water directly to the crops. Bamboo pipes are installed at precise angles and support structures to maintain consistent water flow while preventing erosion. This system is particularly effective in areas where conventional channel irrigation is challenging due to terrain constraints. This gravity-based system is remarkable for its efficiency and minimal environmental impact. The system includes features like pressure-reduction chambers and flow regulators made entirely from local materials. The tribe has developed a sophisticated understanding of topography and uses gravity-based systems

Fig. 5 Bamboo irrigation system (Borang).

to ensure efficient water distribution. Bamboo, abundant in the region, is used to construct channels that divert water from natural springs and streams to agricultural fields. These channels are designed to ensure minimal water loss by preventing evaporation and seepage. The bamboo irrigation system serves as a sustainable and low-cost solution for channeling water. This system exemplifies eco-friendly engineering, as bamboo, a renewable resource, effectively diverts and regulates the water flow across the terraces. The system is labor-intensive but has been maintained for generations, showcasing the tribe's commitment to sustainable practices.

4.1.4 Water harvesting and storage (Dlong)

The Apatani employ various water harvesting techniques, including the construction of reservoirs (Dlong) at strategic locations shown in Fig. 6. These structures capture and store rainwater and spring water for use during dry periods. The placement and design of these storage systems reflect a deep understanding of local hydrology and topography of the area. The community also maintains sacred pools (Bhubin) that serve both religious and water conservation purposes.

Fig. 6 Water harvesting and storage (Dlong).

4.1.5 Spring management (Siirpi)

Springs (Siirpi) are considered sacred and are carefully managed through traditional conservation practices. The tribe has developed specific protocols for spring protection, including maintaining vegetation cover in catchment areas and regulating water extraction. Each spring has designated caretakers who ensure its sustainable use and protection. The spring management system includes traditional methods for maintaining water quality and regulating flow.

4.1.6 Traditional wetland management (Kele)

The Apatani manage natural wetlands (Kele) through customary practices that maintain ecosystem services while supporting agricultural activities shown in Fig. 7. These wetlands act as natural water purification systems and help in flood control. The community has developed specific guidelines for wetland use, including seasonal restrictions and maintenance requirements.

4.1.7 Community-based water management

Water management among the Apatani Tribe is not just a technical or agricultural endeavor but also a social one. The community follows a well-

Fig. 7 Traditional wetland management (Kele).

structured system for water allocation based on traditional rights and the needs of the residents. Water rights are communally owned, and their distribution is carefully monitored by village councils, known as 'Bulyang'. The village council (Bulyang) oversees water distribution, with specific rules governing the timing and quantity of water allocation to different fields in the village. Water rights are typically hereditary but can be transferred through established procedures. The system includes provisions for dealing with water scarcity, with priority given to fields based on their location and the traditional water-sharing agreements between different sectors of the village. The councils ensure that water is equitably distributed among households, preventing overuse and conflict.

This traditional governance system fosters a sense of collective responsibility and environmental stewardship among the community members.

4.2 The Apatani tribe's water resource management exemplifies modern sustainable practices

4.2.1 Case study 1: Sustainable hydro-agricultural practices in Apatani valley

The Apatani tribe's indigenous water resource management system in the Ziro Valley of Arunachal Pradesh represents a highly sophisticated integration of traditional knowledge with ecological principles. The system features an intricate network of irrigation channels, where water is conveyed through primary and secondary channels using bamboo or wooden pipes installed 15–25 cm above the bed, ensuring an optimal water depth (25–35 cm) in perfectly leveled terraces. Modern water quality analysis revealed favorable conditions with dissolved oxygen (8.11 ± 0.13), pH (6.2–7.1), and controlled nutrient levels (ammonium <0.01 mg/l, nitrate 3.8 ± 0.3 mg/l), supporting both rice cultivation and aquaculture. The system incorporates innovative features, like dual-outlet mechanisms (Hubur) with bamboo screens for water control and fish retention, while bio-fencing along primary canals prevents soil erosion. This traditional system achieves impressive productivity (3–5 tons rice/ha/year and 200–500 kg fish/ha/year) without chemical inputs through natural nutrient management, including the integration of Azolla and Lemna as nitrogen fixers. To modernize while preserving these effective traditional practices, recommendations include establishing local fish hatcheries, introducing improved fish strains, and implementing modern monitoring systems, demonstrating how indigenous knowledge can be successfully bridged with contemporary agricultural solutions for sustainable water resource management in mountainous regions.[3]

4.2.2 Case study 2: Innovative water allocation strategies in traditional agricultural systems

The Apatani tribe of Arunachal Pradesh exemplifies a sophisticated understanding of water management that integrates hydraulic engineering principles with social resource management to create a sustainable agricultural system. Their traditional water distribution mechanism utilizes a network of bamboo pipes and gravity-fed channels to ensure equitable access to water across their rice and fish cultivation fields. This system is meticulously designed to allocate water based on the specific needs of each farmer's plot, considering factors such as field size and crop requirements. The cooperative management of these resources is underpinned by community-agreed principles, reflecting an advanced ecological awareness and communal responsibility that aligns closely with modern sustainable practices in resource management. The Apatani's approach not only enhances agricultural productivity but also promotes environmental conservation, making it a model for contemporary water management strategies.[13]

4.2.3 *Case study 3: Climate-resilient sustainable water resource management*

The Apatani tribe's traditional water management system exemplifies an adaptive resilience strategy that effectively addresses environmental challenges through a sophisticated understanding of climate variability and ecosystem dynamics. Their innovative approach integrates predictive strategies, risk mitigation, and sustainable resource utilization, reflecting a deep-rooted scientific knowledge that parallels modern climate resilience frameworks. Central to their system is a network of terraced fields and earthen dams, which allows for precise water allocation and management, adapting to seasonal variations in rainfall and temperature. This method not only supports the cultivation of rice and fish in a symbiotic relationship but also involves community-based decision-making processes that enhance local ecological knowledge and awareness. By utilizing meteorological observations and ecological indicators, the Apatani have developed a localized climate adaptation strategy that underscores the importance of traditional practices in contemporary environmental management discussions. Their system serves as a model for sustainable agriculture, demonstrating how indigenous knowledge can inform modern practices in climate adaptation and resource management.[14]

4.3 Challenges to traditional water management

Despite the success and sustainability of traditional water management systems, the Apatani Tribe faces several challenges:

4.3.1 *Modernization and development pressures*

As Arunachal Pradesh undergoes rapid development, there is an increasing push for the Apatani to adopt modern agricultural techniques. These methods often require greater water inputs and external resources compared to traditional practices. The shift towards mechanized farming can lead to conflicts over water resources because modern systems tend to be less efficient in water usage. This transition not only threatens the sustainability of local water resources but also increases the risk of resource depletion as the demand for water escalates. The introduction of high-yielding crop varieties that require more water than traditional varieties further compounds this challenge.

4.3.2 *Population growth and resource strain*

The growing population in the Ziro Valley has led to an increased demand for water, placing a strain on the traditional systems. While these systems have historically been able to meet the needs of the community, the increasing population requires additional infrastructure and resources to maintain water availability. Increased population density within the valley puts additional pressure on resources, potentially leading to the over-exploitation of water.

4.3.3 Environmental degradation

Deforestation and changes in land use patterns pose significant threats to water resources in the region. The loss of forest cover has affected water catchment areas, leading to reduced water availability for agricultural purposes. Additionally, climate change has intensified the risk of droughts and floods, further threatening the sustainability of traditional water management systems. Unpredictable rainfall patterns are straining the traditional irrigation methods, requiring adaptive strategies to manage water scarcity.

4.3.4 Policy interventions and recognition

Despite the recognized sustainability of traditional practices among indigenous communities like the Apatani, national and state-level water policies often overlook these systems. The lack of integration of indigenous knowledge into formal policy frameworks hampers efforts to address contemporary challenges effectively. This disconnect can lead to a reliance on unsustainable practices that do not align with local ecological conditions or cultural practices.

4.4 Implications and policy recommendations

To harness the benefits of Apatani water resource management, policies should encourage the integration of traditional knowledge into contemporary water management frameworks. Key recommendations include:

4.4.1 Documentation and preservation of traditional ecological knowledge (TEK)

4.4.1.1 Systematic documentation

The first step in preserving Apatani water practices is systematic documentation of the TEK. This involves recording the methods, practices, and underlying principles that guide the Apatani's water management systems. Such documentation will serve as a valuable resource for future generations and can facilitate the adaptation of these practices in other regions facing similar environmental challenges. By creating comprehensive records, policymakers can ensure that this knowledge is not lost and can be utilized to inform contemporary water management strategies.

4.4.1.2 Adaptation to other regions

The insights gained from documenting TEK can be instrumental in adapting these practices to different ecological contexts in the future. For instance, the Apatani's wet rice cultivation and fish farming techniques, which effectively utilize local water resources while maintaining soil fertility, can inspire similar sustainable practices in other agricultural communities.

4.4.2 Community-based participatory management

4.4.2.1 Integration into local policies

Implementing community-based participatory management practices akin to those used by the Apatani can significantly improve local water governance. This approach emphasizes the involvement of local communities in decision-making processes related to water distribution and conservation. By empowering communities to manage their resources, conflicts over water use can be minimized, and conservation efforts can be more effectively tailored to local needs.[15–17] Policy frameworks should recognize traditional governance systems, such as 'Bulyang' and provide them with legal standing in water management decisions.

4.4.2.2 Collaborative water management

The cooperative efforts observed in the Apatani system highlight the importance of collective action in managing water resources. Policies should encourage collaborative frameworks that bring together various stakeholders, including government agencies, local communities, and NGOs to work towards common goals in water management. This multi-stakeholder approach can leverage both traditional knowledge and modern scientific understanding to develop comprehensive strategies for sustainable water use. Successful examples of such collaborations include springshed and watershed development programs that incorporate traditional knowledge while addressing contemporary challenges.[18,19]

4.4.3 Climate adaptation measures

4.4.3.1 Research on resilient practices

As climate change poses significant threats to traditional agricultural systems, it is essential to invest in research that explores climate adaptation measures relevant to TEK. This includes identifying resilient crop varieties that can withstand changing climatic conditions and improving canal designs to enhance water delivery efficiency during dry periods. Experimental studies combining traditional water management principles with climate-adaptive technologies have shown promising results in enhancing system resilience while maintaining cultural continuity in the region. Such research should be prioritized and funded through appropriate scientific institutions.

4.5 Lessons from indigenous water management for climate-smart solutions

The Apatani tribe's approach fundamentally works with natural processes rather than against them. Their indigenous water management practices present an empirically validated example of climate-smart water resource management. These systems offer profound insights for developing climate-smart water resource solutions that can be adapted and scaled for application in other regions facing similar environmental challenges. Their approach demonstrates how traditional ecological knowledge systems can be effectively integrated with modern approaches to address contemporary water resource challenges in a changing climate. The system's long-term sustainability under environmental stress and socio-economic change offers several scientifically grounded lessons, especially relevant to mountainous and ecologically fragile landscapes.

4.5.1 Integration of agro-hydrological systems enhances climate resilience

The Apatani practice of integrated rice-fish cultivation (Aji-ngyii) exemplifies a multifunctional agroecosystem that aligns with climate-smart agriculture (CSA) principles outlined by the FAO.[20] This system not only maximizes water use efficiency but also increases resource productivity, soil health, and ecosystem stability.

Scientific evidence shows that paddy-cum-fish cultivation systems enhance soil fertility through fish waste (natural nutrient cycling), reduce pest incidence (biocontrol), and provide dietary protein, thereby improving food and nutritional security while minimizing external inputs.[3,7]

This multi-functional water use represents a model for climate-smart approaches that maximize productivity per unit of water. This principle can be applied in other regions practicing monoculture farming, where it can be adapted into similar multi-layered production systems tailored to local ecological conditions. These types of practices are now being promoted as viable climate-resilient approaches under programs like India's National Innovations on Climate Resilient Agriculture (NICRA).

4.5.2 Low-carbon nature-based infrastructure reduces climate footprint

The Apatani's bamboo-based irrigation system (Borang) and gravity-fed channels (Bogo) exemplify nature-based solutions (NbS) that utilize local materials and topography without reliance on fossil fuels or electrical pumps. It is a classic example of green infrastructure that also demonstrates low water loss through evaporation or seepage. According to the IUCN and IPCC frameworks, such NbS are key to mitigating climate change impacts while promoting biodiversity and social equity.[21] Bamboo is a renewable and biodegradable resource with low embodied energy, and gravity systems reduce energy-related emissions, thus contributing to low-carbon development.

The 'Borang' system offers a sustainable alternative to PVC or concrete water transport systems in regions with similar ecological conditions. Recent field observations indicate that bamboo pipes treated with traditional preservatives last 3–5 years before requiring replacement, making them economically viable when compared to the high initial costs of modern materials. The natural flexibility of bamboo provides inherent earthquake resistance, which is an important consideration in mountainous regions. This approach can inspire the development of region-specific solutions using locally available indigenous materials, reduce dependence on expensive imported technologies, and foster community self-reliance.

The 'Bogo' system showcases how natural topography can be leveraged for efficient water distribution without relying on external energy sources. This principle is particularly relevant for hilly or mountainous regions, where gravitational force can be harnessed to reduce dependence on pumping systems that require fossil fuels or electricity. The hierarchical arrangement of distribution channels optimizes resource allocation while minimizing maintenance requirements, providing a low-carbon alternative to mechanized irrigation systems. The efficient design of these systems showcases indigenous engineering that is both environmentally sustainable and economically viable.

4.5.3 Ecosystem-based field design reduces impact of climate change

The structural aspects of Apatani terraced fields with integrated water channels (Siikho/Parkho) can be adapted to various topographical contexts. These designs maximize water retention and minimize soil erosion, which are critical functions in the context of increasingly erratic rainfall patterns due to climate change.

The intricate channel network and terraced design inherently function as a distributed flood control system, slowing water movement during intense rainfall events. This dispersed approach to flood management offers alternatives to centralized infrastructure in mountainous regions, where conventional flood control structures may be impractical or ecologically damaging. The specific gradient calculations and terrace dimensions developed by the Apatani provide models that can be scaled and adapted to other sloped agricultural landscapes in the Himalayas.

Similarly, Apatani's integrated rice-fish cultivation system mimics natural wetland ecosystems, enhancing biodiversity while optimizing resource use. This EbA approach demonstrates how agricultural systems can be designed to withstand climate variability by maintaining these ecosystem services.[22,23] The integration of aquaculture with agriculture creates a self-regulating system that enhances resilience to water-related stresses, including both drought and flood conditions. Additionally, the wetland paddy systems provide multiple ecosystem services, including carbon sequestration, flood mitigation, and microclimate regulation.

4.5.4 Community-based governance models foster equity and sustainability

The 'Bulyang' system (village council-based governance) reflects a form of polycentric and adaptive governance that promotes equitable water sharing, collective decision-making, and conflict resolution. Also, this system demonstrates how traditional authority structures can ensure modern water management initiatives can incorporate similar inclusive governance mechanisms that respect local leadership while ensuring the representation of all stakeholders, including marginalized groups and women.[24] Such structures align with Elinor Ostrom's principles for the effective management of common-pool resources.[25]

Decentralized governance enhances institutional flexibility and community ownership—two critical components of adaptive capacity in socio-ecological systems facing climate stress.[1] Moreover, customary water-sharing agreements within and between villages foster cooperative resource management that enhances resilience to climate change.

The community's water allocation system includes flexible management protocols that can be adjusted based on seasonal rainfall variations or drought conditions. This adaptive capacity is a critical component of climate resilience that has been increasingly recognized in modern climate adaptation frameworks. This decentralized governance can serve as a model for community-led water management in other climate-vulnerable areas, promoting collective stewardship and resilience-building.[26,27]

4.5.5 Traditional hydro-ecological knowledge enables localized adaptation

The Apatani people's deep understanding of spring hydrology, seasonal water availability, and soil-water relationships enables ecosystem-based adaptation (EbA) at the local level. Their maintenance of spring sources (Siirpi), rainwater harvesting (Dlong), and wetlands (Kele) for both ecological and agricultural functions mirrors scientifically endorsed EbA strategies that emphasize preserving natural infrastructure to buffer climate impacts.[28]

The holistic approach to managing water resources from source to agricultural fields demonstrates effective watershed management principles. This approach can inform integrated watershed development programs in other regions, emphasizing the connections between upland conservation, midland water distribution, and lowland utilization.

Also, the Apatani's practices of maintaining vegetation around water sources and treating water bodies as sacred entities have resulted in naturally protected water quality. These approaches can complement modern water treatment technologies, potentially reducing costs while enhancing community engagement in water protection. Such knowledge systems, refined over centuries, complement scientific models by providing context-specific solutions that are adaptive and sustainable.[8,9]

Additionally, the Apatani's intergenerational knowledge transmission systems ensure continuity of expertise. Modern water management initiatives can incorporate similar mechanisms for preserving and transferring knowledge by combining traditional apprenticeship approaches with formal education and documentation efforts.[29,30]

4.5.6 Scalability and transferability of practices

Several Apatani techniques—such as bamboo irrigation, community water reservoirs (Dlong), and paddy-fish culture systems—are inherently modular and replicable. These techniques have been adapted in other parts of the Eastern Himalayas and Southeast Asia, demonstrating their ecological and cultural adaptability.[2,5] For instance, terraced irrigation systems and spring rejuvenation approaches have been mainstreamed into government programs like MGNREGS and Jal Shakti Abhiyan, with growing emphasis on incorporating TEK in watershed development and springshed management.

4.5.7 Bridging traditional practices with climate policy frameworks

The Apatani water system provides a blueprint for integrating TEK with formal climate adaptation and water security policies. Studies have shown that blended frameworks—those that incorporate both traditional and scientific knowledge—offer superior resilience outcomes than either system alone.[8,14] Policymakers should consider mechanisms such as Payments for Ecosystem Services (PES), community water rights recognition, and knowledge co-production platforms to incentivize the protection and adaptation of such systems. It also, enables hybrid governance systems that value both indigenous wisdom and scientific innovation. These efforts would support India's National Action Plan on Climate Change (NAPCC) and SDG 6 (Clean Water and Sanitation) while respecting cultural heritage.

The Apatani tribe's water management system represents a living laboratory of climate-smart water solutions that combine centuries of ecological knowledge with adaptive management strategies. By extracting core principles, adapting technologies, and respecting governance structures, these indigenous practices can inform contemporary water management approaches globally, creating more resilient and sustainable water security systems in the face of increasing climate challenges.[31–33]

5. Conclusion

The Apatani tribe's indigenous water resource management system demonstrates an effective blend of traditional ecological knowledge and modern adaptations, rooted in a deep understanding of the Ziro Valley's ecosystem. This system integrates bamboo-based irrigation, terraced paddy fields, and rice-fish cultivation, which collectively enhance water use efficiency and ecosystem health. Despite modern pressures and environmental challenges, the Apatani have maintained these sustainable practices, enriching them with selective technological interventions, such as improved crop varieties and resilient infrastructure. This chapter illustrates how such indigenous practices provide valuable lessons for contemporary water management by optimizing resource use, fostering biodiversity, and contributing to climate resilience. The Apatani model exemplifies a community-driven approach that integrates cultural values with environmental stewardship, offering actionable insights into sustainable water resource management in the Indian Himalayan region.

Acknowledgements

The author expresses sincere gratitude to Dr. I.D. Bhatt, Director Incharge of G.B. Pant National Institute of Himalayan Environment (GBPNIHE), for his continual guidance and encouragement. This research was carried out under the In-House Project-1 and HI-REAP project of the GBPNIHE. The author gratefully acknowledges the financial support received from the Ministry of Environment, Forest and Climate Change (MoEFCC), Government of India, and International Centre for Integrated Mountain Development (ICIMOD), Nepal. The author extends her heartfelt appreciation to the Apatani community of Ziro Valley, Arunachal Pradesh, for generously sharing their indigenous knowledge, lived experiences, and insights on traditional water resource management systems. Their cooperation and wisdom formed the foundation of this chapter and made the documentation of these valuable practices possible.

Conflicts of interest

The author declares no conflict of interest.

Declaration of generative AI and AI-assisted technologies in the writing process

During the preparation of this research work, the author used ChatGPT 4.0 (https://chat.openai.com) partially in order to improve the English language of the manuscript. After using this tool/service, the author reviewed and edited the content as needed and takes full responsibility for the publication's content.

References

1. Parsons M, Fisher K. Indigenous peoples and transformations in freshwater governance and management. *Curr Opin Environ Sustain* 2020;**44**:124–139. https://doi.org/10.1016/j.cosust.2020.03.006.
2. Borthakur A, Singh P. *Indigenous knowledge systems in sustainable water conservation and management. Water conservation and wastewater treatment in BRICs nations;* Elsevier, 2020; pp. 321–328. http://doi.org/10.1016/B978-0-12-818339-7.00016-3.
3. Baruah D, Singh ND. Rice-FIsh cultivation of apatanis: a high altitude farming system in Arunachal Pradesh. *J Krishi Vigyan* 2018;**7**(1):187–191. https://doi.org/10.5958/2349-4433.2018.00130.7.
4. Tangjang S, Nair PKR. Rice + fish farming in homesteads: sustainable natural-resource management for subsistence in Arunachal Pradesh, India. *J Environ Sci Eng* 2015;**4**:545–557. https://doi.org/10.17265/2162-5298/2015.10.007.
5. Kala CP, Dollo M, Farooquee NA, Choudhury DC. Land-use management and wet-rice cultivation (Jebi Aji) by the apatani people in Arunachal Pradesh, India: traditional knowledge and practices. *Outlook Agriculture* 2008;**37**(2):125–129. https://doi.org/10.5367/000000008784648906.
6. Kumar A, Ramakrishnan PS. Energy flow through an Apatani village ecosystem of Arunachal Pradesh in northeast India. *Hum Ecol* 1990;**18**(3):315–336. https://doi.org/10.1007/BF00889158.
7. Das DN, Saikia SK, Das AK. Periphyton in rice–fish culture system: a case study from Arunachal Pradesh, India. *Renew Agriculture Food Syst* 2007;**22**(4):316–319. https://doi.org/10.1017/S1742170507001901.

8. Imoro ZA, Imoro AZ, Duwiejuah AB, Abukari A. Harnessing indigenous technologies for sustainable management of land, water, and food resources amidst climate change. *Front SustaFood Syst* 2021;**5**:691603. https://doi.org/10.3389/fsufs.2021.691603.
9. Rai SC. Traditional ecological knowledge and community-based natural resource management in northeast India. *J Mt Sci* 2007;**4**(3):248–258. https://doi.org/10.1007/S11629-007-0248-4.
10. Sundriyal R, C, Dollo M. Integrated agriculture and allied natural resource management in northeast mountains—transformations and assets building. *Agroecology Sustain Food Syst* 2013;**37**(6):700–726. https://doi.org/10.1080/10440046.2012.720655.
11. Jain S, Srivastava A, Vishwakarma DK, Rajput J, Rane NL, Salem A, Elbeltagi A. Protecting ancient water harvesting technologies in India: strategies for climate adaptation and sustainable development with global lessons. *Frontiers in Water* 2024;**6**:1441365. https://doi.org/10.3389/frwa.2024.1441365.
12. Srivastava A, Chinnasamy P. Water management using traditional tank cascade systems: a case study of semi-arid region of Southern India. *SN Applied Sciences* 2021;**3**(3):281. https://doi.org/10.1007/s42452-021-04232-0.
13. Dollo Mihin. Traditional irrigation system: a case of apatani tribe in Arunachal Himalaya, North East India. *Mt Forum Bull* 2009:9–11. Retrieved from https://lib.icimod.org/record/14096.
14. Lotha TN, Ritse V, Nakro V, et al. Climate change impact and traditional adaptation practices in Northeast India: a review. *Curr World Environ* 2024;**19**(2) https://doi.org/10.12944/CWE.19.2.5.
15. Kiss B, Sekulova F, Hörschelmann K, Salk CF, Takahashi W, Wamsler C. Citizen participation in the governance of nature-based solutions. *Environmental Policy and Governance* 2022;**32**(3):247–272. https://doi.org/10.1002/eet.1987.
16. Fritsch O. Public participation in water policy and governance. Edward Elgar Publishing, 2024; pp. 33–51. https://doi.org/10.4337/9781800887909.00011
17. Srivastava A, Chinnasamy P. Developing Village-Level Water Management Plans Against Extreme Climatic Events in Maharashtra (India)—A Case Study Approach. In *Water Safety, Security and Sustainability. Advanced Sciences and Technologies for Security Applications;* Vaseashta, A., Maftei, C., Eds.; Springer: Cham, 2021, https://doi.org/10.1007/978-3-030-76008-3_27
18. Nagata K, Shoji I, Arima T, Otsuka T, Kato K, Matsubayashi M, Omura M. Practicality of integrated water resources management (IWRM) in different contexts. *International Journal of Water Resources Development* 2022;**38**(5):897–919. https://doi.org/10.1080/07900627.2021.1921709.
19. Singh S, Goyal MK. A review of India's water policy and implementation toward a sustainable future. *Journal of Water and Climate Change* 2025;**16**(2):493–510. https://doi.org/10.2166/wcc.2025.560.
20. Food and Agriculture Organization (FAO) of the United Nations. Climate-smart agriculture sourcebook. Rome: FAO; 2013. ISBN 978–92-5–107720-7 (print), E-ISBN 978–92-5–107721-4 (PDF). Retrieved from https://openknowledge.fao.org/server/api/core/bitstreams/b21f2087-f398–4718-8461-b92afc82e617/content.
21. Intergovernmental Panel on Climate Change (IPCC). Climate change 2022: impacts, adaptation and vulnerability. Contribution of Working Group II to the Sixth Assessment Report of the Intergovernmental Panel on Climate Change (Pörtner H-O, Roberts DC, Tignor M, Poloczanska ES, Mintenbeck K, Alegría A, Craig M, Langsdorf S, Löschke S, Möller V, Okem A, Rama B, Editors). Retrieved from https://www.ipcc.ch/report/ar6/wg2/downloads/report/IPCC_AR6_WGII_FullReport.pdf; 2022.
22. Chaudhary S, Adhikari BR, Chaudhary P, Dorji T, Poudel R. Ecosystem-Based Adaptation (EbA) in the Hindu Kush Himalaya: Status, Progress and Challenges. In *Ecosystem-Based Disaster and Climate Resilience. Disaster and Risk Research: GADRI Book Series;* Mukherjee, M., Shaw, R., Eds.; Springer: Singapore, 2021, https://doi.org/10.1007/978-981-16-4815-1_2

23. Srivastava A, Maity R. Unveiling an Environmental Drought Index and its applicability in the perspective of drought recognition amidst climate change. *Journal of Hydrology* 2023;**627**:130462. https://doi.org/10.1016/j.jhydrol.2023.130462.
24. Srivastava A, Chinnasamy P. Watershed development interventions for rural water safety, security, and sustainability in semi-arid region of Western-India. *Environment, Development and Sustainability* 2024;**26**(7):18231–18265. https://doi.org/10.1007/s10668-023-03387-7.
25. Ostrom E. A general framework for analyzing sustainability of social-ecological systems. *Sci (N York, NY)* 2009;**325**(5939):419–422. https://doi.org/10.1126/science.1172133.
26. Adams EA, Zulu L, Ouellette-Kray Q. Community water governance for urban water security in the Global South: Status, lessons, and prospects. *Wiley Interdisciplinary Reviews: Water* 2020;**7**(5):e1466. https://doi.org/10.1002/wat2.1466.
27. Chinnasamy P, Srivastava A. Revival of traditional cascade tanks for achieving climate resilience in drylands of South India. *Frontiers in Water* 2021;**3**:639637. https://doi.org/10.3389/frwa.2021.639637.
28. Swiderska K, King-Okumu C, Monirul Islam M. Ecosystem-based adaptation: a handbook for EbA in mountain, dryland and coastal ecosystems. London: IIED; 2018. ISBN: 978–1-78431–579-5. Retrieved from https://www.iied.org/sites/default/files/pdfs/migrate/17460IIED.pdf.
29. Das A, Gujre N, Devi RJ, Mitra S. A review on traditional ecological knowledge and its role in natural resources management: North East India, a cultural paradise. *Environmental management* 2023;**72**(1):113–134. https://doi.org/10.1007/s00267-021-01554-y.
30. Srivastava A, Maity R, Desai VR. Assessing Global-Scale Synergy Between Adaptation, Mitigation, and Sustainable Development for Projected Climate Change. In *Ecological Footprints of Climate Change. Springer Climate;* Chatterjee, U., Akanwa, A. O., Kumar, S., Singh, S. K., Dutta Roy, A., Eds.; Springer: Cham, 2022, https://doi.org/10.1007/978-3-031-15501-7_2
31. Singh PK, Dey P, Jain SK, Mujumdar PP. Hydrology and water resources management in ancient India. *Hydrology and Earth System Sciences* 2020;**24**(10):4691–4707. https://doi.org/10.5194/hess-24-4691-2020.
32. Maity R, Srivastava A, Sarkar S, Khan MI. Revolutionizing the future of hydrological science: Impact of machine learning and deep learning amidst emerging explainable AI and transfer learning. *Applied Computing and Geosciences* 2024;**24**:100206. https://doi.org/10.1016/j.acags.2024.100206.
33. Kaptijn E. Learning from ancient water management: Archeology's role in modern-day climate change adaptations. *Wiley Interdisciplinary Reviews: Water* 2018;**5**(1):e1256. https://doi.org/10.1002/wat2.1256.

CHAPTER THREE

River water quality assessment and management using water quality index

Shilpi Maurya*, Akshay Pachore, Mark Prabhakar Vuppati, Dudekula Nikhil Kumar, and Neena Isaac
River Rejuvenation Division, Central Water & Power Research Station, Pune, India
*Corresponding author. e-mail address: shilpi.maurya1994@gmail.com

Contents

Abstract

Rivers are the most critical source of fresh water, but are also at risk of different kinds of pollution from both direct and indirect discharge sources. Due to pollution load, rivers no longer support an aquatic ecological balance. The restoration of the quality of the river has remained a significant question. Information on the current state of water quality and its trends is crucial for river rejuvenation. This study focused on evaluating the Water Quality Index (WQI) to analyse the quality of the Mula-Mutha River as it flows through Pune city and its conurbations, extending from Khadakwasla Dam to Daund (127 km) in the higher Bhima Basin. This research was conducted in February 2020 as a purpose-driven study (PDS) under the National Hydrology Project (NHP). The analysis used nine parameters: pH, TDS, temperature, turbidity, DO, BOD, nitrate, Phosphate, and fecal coliform. The findings indicated that the WQI values range from 22.22 to 70.00, showing a wide variability in the river water quality. The primary study locations of the Mula-Mutha Stream from Khadakwasla Dam to Daund is divided into three zones according to WQI values. Given the high pollution levels observed in Zone 2, especially locations S3 and S4, that stretch of the river needs to be prioritized, followed by S5 to S7, for river rejuvenation activities. The division of rivers into zones, based on the ratings of WQI values, helps in the river rejuvenation program.

Keywords: Mula-Mutha river, NSF-WQI, Water pollution, Water quality.

Advances in Chemical Pollution, Environmental Management and Protection, Volume 15
ISSN 2468-9289, https://doi.org/10.1016/bs.apmp.2025.11.010

1. Introduction

A river is a naturally flowing freshwater stream that flows through a channel towards a lake, ocean, sea, or another river.[1] The rivers are the foundation of many urban and rural centres along the world's banks. The industrial revolution increased the trend of population coupled with rapid urbanization during the past few decades, depleting the health of the river ecosystem, which leads to river pollution.[2] Uncontrolled and injudicious exploration of water brings changes in the physicochemical and biological quality of the water, and due to that, water becomes unsafe for drinking and other domestic usage.[3] It is said that the polluted water in rivers affects the river ecosystem and contributes to soil pollution in its basin, where river water is used in agriculture to irrigate farmlands[4,35] The rejuvenation of the river has been a significant challenge.

The water quality at present and shifts in that river's water quality are vital for rejuvenation. Water quality is a multifaceted topic that entails assessing water's physical, chemical, hydrological, and biological properties and their intricate and sensitive interconnections. Water quality can be monitored through various physical, chemical constituents, and biological parameters.[36] The standard permissible and acceptable limits have been defined by national and international agencies for each region or area. Water quality is considered unhealthy and harmful for humans and other activities if the observed value deviates from the well-defined limits. In this paper, standards from BIS,[5] CPCB,[6,7] and WHO[8] are considered when comparing the quality of water.

The WQI represents the most common indicator of the quality of the stream flow.[9] WQI is an index that represents water quality by integrating the analytical readings of various parameters such as dissolved oxygen, pH, nitrate, phosphate, turbidity, chloride, hardness, etc.[10] It simplifies and reduces the extensive data from numerous water quality parameters into a single representative value, presenting the information in a simplified and systematic format.[11] The higher WQI (NSF) score shows better quality, and the lower score shows the degraded condition of the water. The suitability of water can be easily expressed for different uses such as drinking, bathing, irrigation, industrial uses, etc. The significance of the WQI holds substantial ecological importance, as water resources are crucial to the overall environment, and this index has been recognised as one of the indicators within the Environmental Performance Index's 32-indicator framework (EPI-2020).[10,12]

Evaluating and managing water quality using the WQI metrics can provide comprehensive insights into the status of water bodies and potential risks to various uses.[13] Different international organizations invented numerous evaluation methods, prioritizing the NSF WQI based on the categories and values of indicators to meet the required standards of a given location.[14] NSFWQI is the most operated index in the United States.

WQI - Concept: The National Sanitation Foundation (1970)[14] developed a quantitative assessment tool, commonly known as the Water Quality Index (WQI), that integrates multiple physicochemical parameters into a single metric. Applying the index offers a well-defined, reasonable unit of measure that responds to variations in water quality. This approach aims to consolidate the findings of changes in parameter variations, producing a net quality value that can be easily monitored and practically analysed. This tool effectively aggregates and presents multiple parameter values as a single figure for index formulation.

This study aimed to use the application of the WQI (NSF method) to determine the quality of the River Mula-Mutha. This study was conducted as a purpose-driven study (PDS) under the National Hydrology Project (NHP).

2. Study area

The Mutha River flows through western region of Maharashtra, India. It originates in the Western Ghats and flowing in eastward trajectory, and merging with the Mula River in Pune city and streams down as the Mula-Mutha River and combines with the Bhima River at Paragon (Fig. 1). The present status of Mutha and Mula-Mutha rivers water quality is dismal in the region of Pune city. The pollution from Pune is considered the single principal cause of the deterioration of the quality of these rivers and the Ujjani reservoir. Industrialization, civilization, and many anthropogenic activities which causes worsening pollution in the river water. Hence, a situation arises in which the river looks like a canal carrying wastewater.[15] Due to Agricultural runoff, many fertilizers and pesticides are discharged into River water. Also, a large amount of sewage water discharge pollutes the river, which causes a significant growth of microorganisms. Mutha River has been blocked twice, initially at the Panshet dam, which serves as a supply of domestic water and agricultural water for Pune city. The water passes here is blocked again at Khadakwasla, a vital drinking water source for Pune city.[16]

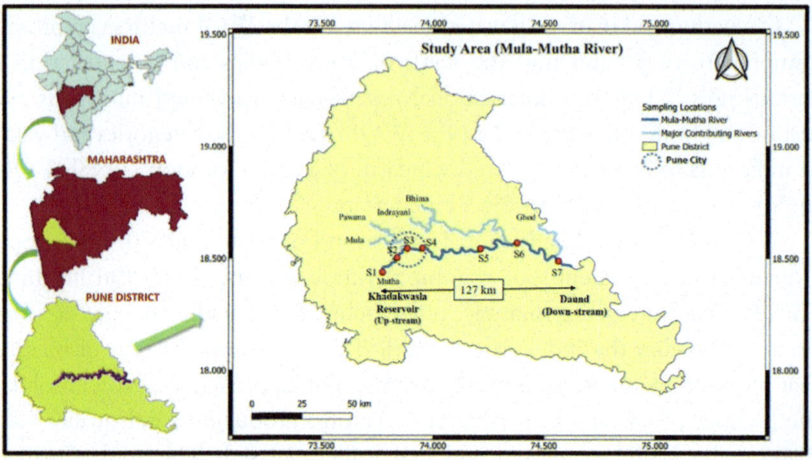

Fig. 1 Study area of Mula-Mutha river. Map lines delineate study areas and do not necessarily depict accepted national boundaries.

Table 1 Coordinates of sampling locations.

Points	Water resource	Longitude	Latitude
S1	Mutha River	E 73.77177	N 18.43758
S2	Mutha River	E 73.83707	N 18.50221
S3	Mula-Mutha River	E 73.88289	N 18.54371
S4	Mula-Mutha River	—	—
S5	Mula-Mutha River	E 74.21316	N 18.54187
S6	Bhima-Mula-Mutha River	E 74.37746	N 18.56793
S7	Ghod-Bhima-Mula-Mutha River	E 74.56446	N 18.48714

The study of the quality of the river from downstream of Khadakwasla reservoir, up to Daund, helps to identify the extent of these causative factors for deterioration in the river's quality.

In the upper Bhima Basin, a segment of 127 km of the Mula-Mutha River flows within the Pune metropolitan region and its suburbs, beginning at Khadakwasla Dam (S1) and extending up to Daund (S7), was selected for comprehensive water quality assessment and to calculate the WQI. The geospatial coordinates of the sampling locations are provided in Table 1.

3. Materials and methods

A field visit was undertaken in February 2020, to gather on-site observations, along with the collection of water samples, from seven different sites (S1 to S7) (Fig. 2). In-situ water quality data was collected for unstable parameters like Temperature, pH, Conductivity, DO, Turbidity from various depths using water quality monitor (YSI-EXO 1). Key parameters including pH and turbidity were also measured on-site using a calibrated digital meter. Temperature and TDS were measured on-site, while dissolved oxygen was preserved immediately with additives and later analyzed by Winkler's method. BOD was measured by collecting samples in two different BOD bottles and then kept under incubation in the dark at 20 °C for five days, and subsequently analyzed. Nutrients parameters including NO_3^- and PO_4^{3-} were quantified using a double-beam UV–visible spectrophotometer. A State Public Health laboratory tested Fecal coliform values using the most probable number method. All analytical procedures were performed in accordance with established standard methods as given in APHA.[17]

In this study, nine physiochemical parameters, namely pH, temperature, turbidity, TDS (total dissolved solids), DO (dissolved oxygen), BOD (biochemical oxygen demand), NO_3^-, PO_4^{3-}, Faecal coliform, were selected for generating the overall WQI for Mula-Mutha River (Table 2). The pH values of water samples were recorded using HANNA digital pH meter device, which was initially calibrated with pH 4,7 and 10 to ensure accurate readings of the samples. The temperature of the water samples is recorded in situ at different sampling points using a multiparameter probe (Sonde). The turbidity is measured by Hach, 2100 N Nephelometer, in which the suspended particles in the water samples are scattered by light at 90° and provides data expressed in nephelometric turbidity units (NTU). TDS is measured using a digital TDS meter, which was calibrated using

Fig. 2 Collection of water samples and measuring of in-situ water quality parameters.

Table 2 Brief description of water quality parameters and methods used for analysis.

Sl. No.	Parameters	Characteristics	Anthropogenic origin for pollution	Method applied
1	DO	It quantifies the oxygen concentration dissolved in water	It is affected mainly by domestic sewage & industrial effluents.	Winkler method
2	Faecal C.	It signals the potential disease-causing microorganisms that can lead to the spread of diseases through water pollution	Household sewage	MPN method
3	pH	The hydrogen ion Concentration determines the conditions of acidity, basicity or neutral state of water	It is influenced by the Industrial or household waste containing CO_2, mineral acids and hydrolysed salts	pH electrode method
4	BOD (5 day)	It measures the oxygen required for the decomposition of biodegradable organic matter present in water	Discharge of organic matter, mainly sewage	Titrimetric
5	Nitrates	Nitrogen compounds exist as ammonia (both free NH_3 and ionized NH_4^+), nitrite (NO_2^-), and nitrate (NO_3^-). These are regarded as nutrients in aquatic environments.	Agricultural wastes, Domestic effluents, industrial discharge, animal excrement	U.V-VIS
6	Phosphates	These are Vital nutrient for organisms presents in water.	Industrial discharge, fertilizers, food, dairy, and slaughtering industries	U.V-VIS

7	Temperature	It shows the amount of heat intensity	Effluents from cooling systems, Industrial discharge, Atmospheric pollution.	Thermometer AndMultiparameter Device: Sonde
8	Turbidity	It measures the degree of reduction in intensity that a beam of light experiences during passage through a pathlength	Usage of inappropriate agricultural practices, sewage discharge and effluents	Nephelometer
9	Total dissolved solids	It shows the sum of all cations and anions in water bodies	Usage of salts for road icing, drinking water treatment chemicals, agricultural run-off etc.	Electrode method

standard KCl solution and gives readings in mg/L. It estimates all the organic and inorganic components dissolved in water. The dissolved oxygen is measured using the help of the Winkler method and also from multiparameter probe (Sonde). The amount of BOD present in the sample is determined by measuring the initial and final DO using the Winkler method. The final DO is measured after a 5-day incubation period at 20 °C in the dark condition. Nitrate concentration is determined by a UV-Vis spectrophotometer (Shimadzu UV-1800), and the absorbance is measured at 220 nm and 275 nm. Phosphate was also measured in a UV-Vis spectrophotometer using the Ascorbic acid method, where there is a reaction between the orthophosphate and molybdate in acidic medium to form a blue-colored complex, read at 880 nm. Faecal coliforms analysis was conducted by the State Public Health Laboratory using the Multiple Tube Fermentation Technique (Most Probable Number method).

3.1 Methodology for development of NSF-WQI

The present methodological approach can be formulated into five steps:[18]

Indicator selection - This study deployed the Delphi technique to finalize nine indicators. Based on the conformation by water quality specialist in the US, nine variables are selected for WQI calculations (Table 3).

Table 3 Relative weightings and significance rankings for NSF WQI parameters.[19]

Parameter	Average significance score across all expert responses	Temporary weights	Final weights
DO	1.40	1.00	0.17
Faecal C.	1.50	0.90	0.16
pH	2.10	0.70	0.11
BOD (5 day)	2.30	0.60	0.11
NO_3^-	2.40	0.60	0.10
PO_4^{3-}	2.40	0.60	0.10
Temperature	2.40	0.60	0.10
Turbidity	2.90	0.50	0.08
Total solids	3.20	0.40	0.07
Total		5.90	1.00

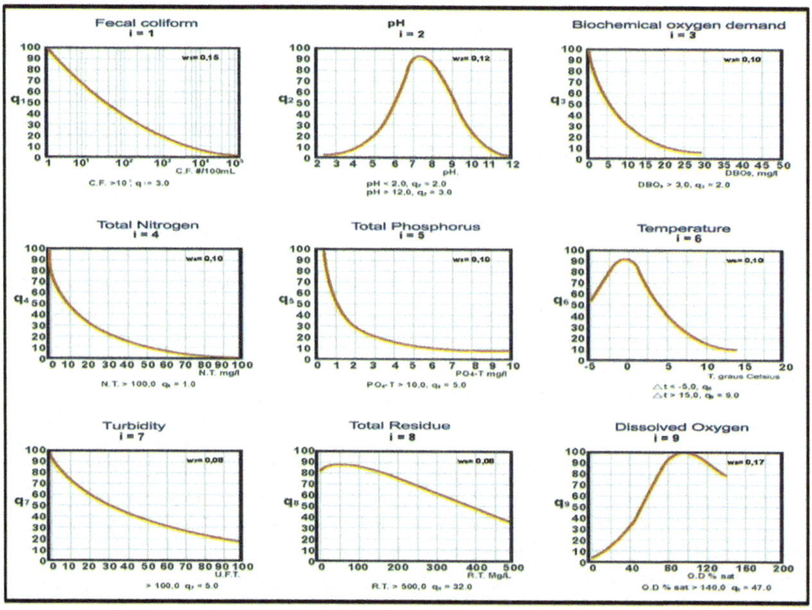

Fig. 3 Variation curves of water quality indicators for WQI calculation.[20]

Sub-indices generation - Sub-indices were developed for the calculation of the NSF water quality index, which was also recognised using the Delphi approach. The details from the experts were used to form "an average curve," that shows the representative behaviour of all sub-indices (Fig. 3).

Establishment of weights - Individual weights for the parameters have been identified on the basis of their importance in quality. The concluding weights are shown in Table 3, and the summation of all different weights equals 1.

Aggregation of index - The summation of the indices was formulated with the help of the summation method for the selected parameters.

Interpretation of final value - The concluding index values spanned from 0 (poor quality) to 100 (excellent quality). Brown and McClelland (1974) established the classification of index scores for assessing water quality in the NSF WQI: Excellent (90–100), Good (70–89), Medium (50–69), Poor (25–49), and Very Poor (0–24).

The concentration-value relationship for each indicator was represented graphically. These graphs depicted curves illustrating the variations in water quality levels based on potential assessment of each individual parameter. The collective decisions of all experts were averaged to generate a specific curve for each parameter.[19]

Table 4 Water quality index calculation for Khadakwasla sampling location.

S.no.	Variables	Measured values (units)	Sub-indexvalue, Qi	Weightage, Wi	Qi * Wi
1	D.O %	91.02 (%)	96	0.17	16.32
2	F.C.D	2300 (MPN/ml)	18	0.16	2.88
3	pH	7.64	92	0.11	10.12
4	BOD	0.549 (mg/L)	98	0.11	10.78
5	Nitrate	0.123 (mg/L)	97	0.1	9.7
6	Phosphate	0.333 (mg/L)	78	0.1	7.8
7	Temp	24.5 (°C)	16	0.1	1.6
8	Turbidity	27.0 (NTU)	55	0.08	4.4
9	TDS	37.7 (mg/L)	86	0.07	6.02
	Total			1	69.62
	WQI				70.00

The final index calculation is an additive model, expressed as follows:

$$\mathbf{WQI} = \Sigma_{i=1}^{n}\mathbf{wiSi}$$

100; 'wi' is the weighting coefficient or the weight varying between 0 and 1 such that $w1 + w2 + w3 + \cdots + wn = 1$, 'Si' represents the sub-index or quality value of the ith parameter, ranging from 0 to 100, derived from the respective quality chart based on its concentration for that specific parameter. In the graph showing the variation of parameters concerning concentration, a quality value (q) is assigned to each parameter. 'n' denotes the number of parameters involved in the WQI calculation. The following Table 4 shows the WQI calculation for sample collected from Khadakvasla dam (S1).

4. Results and discussion

The Physicochemical and biological values of nine parameters calculated for all seven sampling locations are shown in Table 5. The observed values that do not follow the standard limits are highlighted in red. These

Table 5 Water quality of Mula-Mutha river during February, 2020.

Parameters	Units	Sampling locations / code							BIS/WHO/CPCB limits (Class-C)
		S1	S2	S3	S4	S5	S6	S7	
pH		7.64	7.39	7.18	7.67	7.33	7.39	8.15	6 – 9 (CPCB)
Turbidity	**(NTU)**	27	485.3	175.5	147.8	61.4	26.6	100.7	10 (BIS)
Temp	**(°C)**	24.5	25.93	24.4	25.5	25.42	24.44	24.4	—
TDS	**(PPM)**	36.7	212.7	390.95	399.5	474.31	575.02	870.12	1500 (CPCB)
NO_3^-	**(mg/L)**	0.123	6.245	8.99	7.195	10.135	11.9	17.03	50 (CPCB)
PO_4^{3-}	**(mg/L)**	0.333	13.77	24.46	27.86	42.32	43.7	40.62	0.3 (WHO)
DO	DO %	91.02	22.47	1.82	35.1	65.36	14.16	86.3	—
	DO (mg/L)	7.52	2.27	0.15	2.9	5.4	1.17	7.13	4 mg/L or more (CPCB)
BOD	**(mg/L)**	0.549	24.3	29.01	39.34	43.13	53.32	32.93	3mg/L or less (CPCB)
F.C	**(MPN/ml)**	2.3×10^1	9.3×10^7	4.3×10^{10}	1.5×10^{12}	2.4×10^2	2.4×10^7	9.3×10^5	<5000 MPN/100ml (CPCB)

parameters are the reason for the degraded quality of the Mula-Mutha River and, hence, the lower water quality index.

After analyzing the quality of the river for these nine variables, it is found that the pH value ranges from 7.18 to 8.15. It is observed that at the S7 location, the river water is slightly alkaline. It is observed that the value of Total dissolved solids (TDS) increases gradually downstream. The high values of TDS are due to the presence of a high number of salts. The runoff from agriculture and sewage leakage contributes to a rise in TDS due to increased chloride, phosphate, and nitrate ions. Clay soil also contributes to an increase in TDS because the mineral deposits in the soil get ionized as they dissolve in water. The Urban runoff, ions from the groundwater, and ions from the erosion will also contribute to TDS. It is reliant on water temperature, Salinity, and conductivity. Temperature impacts water density, which leads to stratification. Stratified water can have a diverse TDS value at different depths.

In the study area, the observed DO range was between 0.15 mg/L to 7.13 mg/L (Table 5). As per[25] standards for Class-C (Table 5), the desirable limit of DO is 4 mg/L or more. However, in locations S2 to S4 and S6, the observed DO values were less than 4.0 mg/L (Fig. 4). This may be likely due to the result of excessive growth of algae and phytoplankton blooms, which is driven by elevated levels of phosphorus and nitrogen concentration. When these organisms die and accumulate on the river bed, microbial decomposition requires a significant amount of oxygen, which depletes the dissolved oxygen. Hence, the BOD level is increased. The

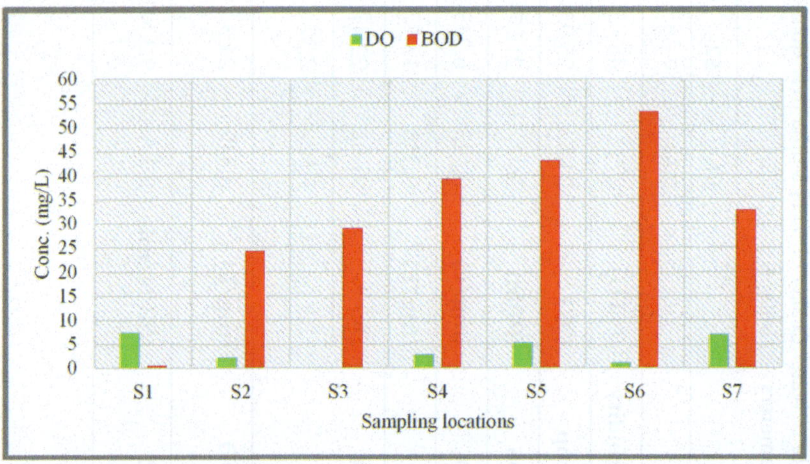

Fig. 4 Spatial variation in DO and BOD in the surface water of Mula-Mutha river.

observed BOD values in the study range from 0.549 mg/L to 53.32 mg/L. The results show that the BOD of the river water at the upstream location is within the Indian standards. However, when the river enters the city area, DO values are observed to be decreasing, and the BOD is observed to be increasing due to mixing different types of pollutants into the river. It shows elevated organic matter due to sewage disposal and eutrophic conditions.

Fecal coliform was recorded at 230 MPN/ml to 1.5×10^{12} MPN/ml. It shows a large amount of sewage contamination in river water. The concentration of nitrate was observed in the range of 0.123 mg/L to 17.030 mg/L. Nitrate levels fall within the permissible limit, i.e., 50 mg/L. The observed phosphate value ranges from 0.333 mg/L to 43.700 mg/L (Fig. 5).

Nitrate and phosphate are primary nutrients required by living organisms in the water, but they are considered pollutants if found more than the recommended limits. It creates excessive growth of aquatic plants and negatively affects river water quality by stimulating the growth of algal clumps, undesirable odor, and discoloration. It causes aeration problems; eventually, dead phytoplankton and macrophytes settle at the bottom of most of the water body. Hence, microbial growth thrives and uses DO to degrade organic waste.[20]

Finally, the Water quality index was formulated for seven sampling locations to evaluate the water quality of the selected river stretch (Table 6). The ratings have been given from very bad to excellent (0–100) water quality with the indication of colors to simplify water quality ratings (Table 7).

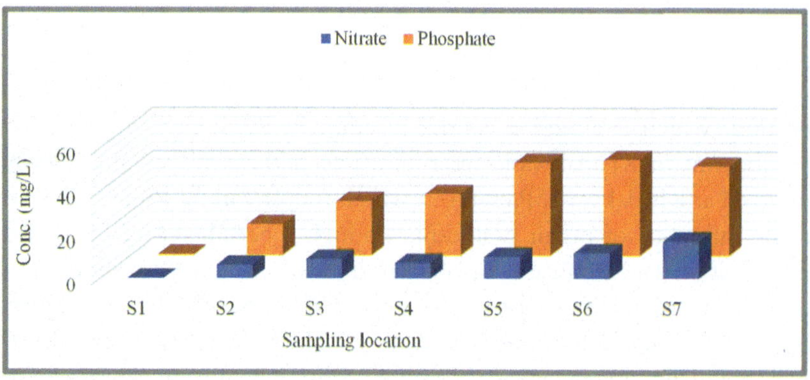

Fig. 5 Spatial variation in nutrients in the surface water of Mula-Mutha river.

Table 6 NSF - Water quality index with colour codes and designated ratings.

February-2020			
S.no.	**Locations**	**NSF-WQI**	**Water Quality Rating**
1	Khadakvasla Dam	70.00	Good quality water
2	Dattawadi	25.94	Bad quality water
3	Bundgarden	22.22	Very bad quality water
4	Mundwa	23.28	Very bad quality water
5	Khamgaon	25.97	Bad quality water
6	Pargaon	28.13	Bad quality water
7	Daund	33.48	Bad quality water

Table 7 Water quality index (WQI) with colour code, limit, rating and probable usage of water.[16]

Col our code	Index limit	Water Quality Rating	Classification of the type of water resource usage
	90-100	Excellent	Drinking, irrigational and industrial purpose
	70-89	Good	Drinking, irrigational and industrial purpose
	50-69	Moderate	Irrigational and industrial purpose
	25-49	Bad	For irrigational purpose
	0-24	Very bad	Unsuitable, proper treatment required for any kind of usage

The Water Quality Index values in the study area show considerable spatial variation and range from 22.22 to 70.00 (Fig. 6). As per the water quality rating based on WQI values, the river water quality is falling from very bad to good. The lowest WQI value (22.22) is observed at the Bund Garden location, which is almost a central location in Pune city and is 17.5 km downstream of Khadakwasla reservoir. The highest WQI value (70.0) is observed at Khadakwasla Reservoir, which is upstream of Pune and devoid of the influence of urban pollution. Though a peripheral location of the central city area, Dattawadi, has shown slightly better WQI values (25.94), locations in the central part of the city have shown the lowest observed WQI values, and the rating was in very bad range. After city limits, the river gradually improved WQI values from Khamgaon (25.97) to the downstream location Daund (33.48).

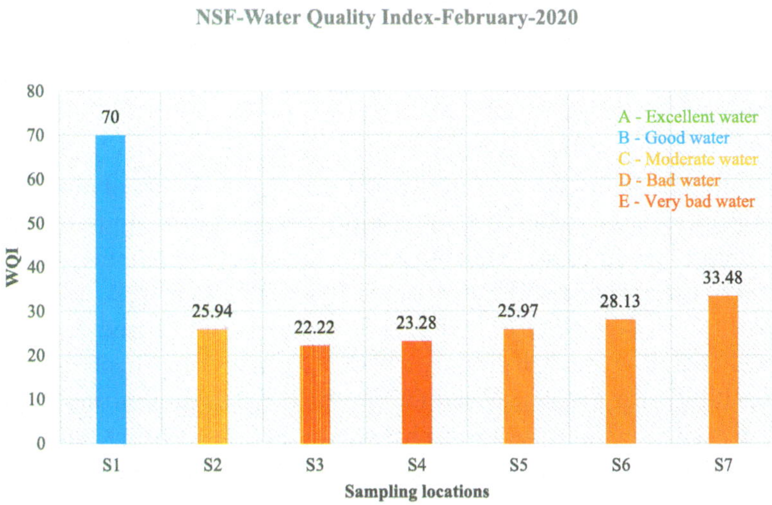

Fig. 6 Spatial variation in WQI in the study area.

The reason for the improvement in the WQI values in the river after the city limits is attributed to the combined effect of adding freshwater through tributaries and the lack of urban sewage and aeration as the river flows downstream (100 km).

River water quality can be influenced by many natural and anthropogenic disturbances, such as low flow conditions under climate change scenarios,[22,23] reservoir operation in the upstream catchment,[24] and insufficient capacity of the wastewater treatment plants (CPCB, 2005). Among these factors, changing climatic conditions can play a crucial role in disturbing the river water quality by modulating the rainfall and overall water availability in the catchment.[21,22] A recent study done by[21] on major Indian catchments has reported that, under the RCP 8.5 scenario for the future time frame from 2071 to 2100, the saturation capacity of the DO may go down by 2.3 % due to an increase in river water temperature by 70 C during the summer time. Many studies have explored the impact of climate change under different scenarios during baseline and future periods and reported the plausible degradation of various parameters such as DO, water temperature, and BOD.[21,22,26,27] Though such extensive investigation is lacking in the context of Mula – Mutha River for analyzing the impact of climate change on its water quality parameters, this can be addressed in the future scope of the present work by taking different scenarios of SSP (shared socio-economic pathways) from CMIP6 models.

Modeling water quality parameters of the river water is majorly challenged by scarcity and nature (non-linear, non-stationary, and vague interdependent relationships) of the data, which has been attempted by many scholars in the past using process-based models and data-driven approaches.[28,29] Recently, data-driven approaches are becoming more popular because of their computational efficiency and better prediction capabilities. Past studies have extensively used ANN (Artificial Neural Network), Fuzzy logic, and SVM-based AI models for modeling water quality data.[30] ANN based approach has been utilized by[32] for predicting the DO in the downstream of the Mathura city which is trained using discharge, temperature, BOD, and pH.[31] trained the Support Vector Regression (SVR) model to predict the BOD values.

Moreover, found to have a high correlation of 0.952, 0.909, and 0.907 during the training, validation, and testing period. Many previous studies have also explored hybrid AL models such as ANN-ARIMA, wavelet, neural network, wavelet, and SVR for predicting river water quality parameters.[33] Though modeling is not the focus of the present work on the Mula-Mutha River, it can be explored in future studies to predict crucial quality parameters such as DO and BOD, or prediction of WQI can be made to assess the overall quality of river water.

The present work can be extended further by incorporating numerous aspects, such as climate change and human disturbances, and constructing the prediction model.[34] The current study gives valuable insights into river water health regarding NSF-WQI computations, which help make inferences about river water health. The inference of the analysis is helpful for the stakeholders and policymakers for abstracting the information on the quality status at various studied locations and planning the management efforts accordingly.

5. Conclusions

On the basis of WQI values, the study area of the Mula-Mutha River from Khadakwasla Dam to Daund (127 KM) is divided into three zones. The sampling locations S1 (Zone 1) are less polluted, S2 to S4 (Zone 2) are highly polluted, and S4 to S7 (Zone 3) is moderately polluted. Sampling site S1 (Zone 1) represents upstream of Pune city, S2 to S4 (Zone 2) represents Pune city area, while S4 to S7 (Zone 3) represents downstream of Pune city. The central section of the river (Zone 2), spanning 15 km, was found to be severely polluted. Because of the high pollution levels observed in Zone 2, especially locations S3 and S4, that stretch of the river needs to be prioritized, followed by S5 to S7 for river rejuvenation activities.

The division of the river into zones based on the WQI values helps in the river rejuvenation program. The data gathered from field survey and the observation of the spatial variations along the Mula-Mutha River were utilized to develop a water quality model for simulating future scenarios for the river. Moreover, the outcomes of this study will support the stakeholders and government agencies in the management and betterment of the river and its ecosystem.

Acknowledgments

The authors express their gratitude and acknowledge the National Hydrology Project (NHP) for providing financial support for this study and the Director of the Central Water and Power Research Station (CW&PRS) for their guidance and support. Special thanks are given to the staff and research groups of the River Rejuvenation Division, CW&PRS, for their valuable support in conducting this work.

References

1. Kondolf GM. Hungry water: effects of dams and gravel mining on river channels. *Env Manage* 1997;**21**(4):533–551.
2. Juahir H, Zain SM, Yusoff MK, Hanidza TT, Armi AM, Toriman ME, et al. Spatial water quality assessment of Langat River Basin (Malaysia) using environmetric techniques. *Env Monit Assess* 2011;**173**(1-4):625–641.
3. Ramesh R, Purvaja R, Lakshmi A. Country report on pollution: India; 2011.
4. Trujillo-González JM, Mahecha-Pulido JD, Torres-Mora MA, Brevik EC, Keesstra SD, Jiménez-Ballesta R. Impact of potentially contaminated river water on agricultural irrigated soils in an equatorial climate. *Agriculture* 2017;**7**(7):52.
5. Bureau of Indian Standards. *Indian standard drinking water specifications (second revision). IS 10500:2012;* BIS: New Delhi, 2012.
6. Central Pollution Control Board (CPCB). *Guidelines for water quality criteria;* Government of India, 2019.
7. Central Pollution Control Board (CPCB). Classification of inland surface waters (CPCB standards). *Water Qual Parivesh* 1995;**1**(4):6.
8. World Health Organization (WHO). Guidelines for drinking-water quality. vol. 1; 2004.
9. Othman F, Alaaeldin ME, Seyam M, Ahmed AN, Teo FY, Fai CM, et al. Efficient river water quality index prediction considering minimal number of inputs variables. *Eng Appl Comput Fluid Mech* 2020;**14**(1):751–763. https://doi.org/10.1080/19942060.2020.1760942.

10. Lumb A, Sharma TC, Bibeault JF. A review of genesis and evolution of water quality index (WQI) and some future directions. *Water Qual Expo Health* 2011;**3**(1):11–24. https://doi.org/10.1007/s12403-011-0040-0.
11. Semiromi FB, Hassani AH, Torabian A, Karbassi AR, Lotfi FH. Water quality index development using fuzzy logic: a case study of the Karoon River of Iran. *Afr J Biotechnol* 2011;**10**(50):10125–10133.
12. Yale center for environmental law and policy. Environmental performance Index (EPI). Available from: http://www.yale.edu./en.
13. Lkr A, Singh MR, Puro N. Assessment of water quality status of Doyang River, Nagaland, India, using Water Quality Index. *Appl Water Sci* 2020;**10**:46. https://doi.org/10.1007/s13201-019-1133-3.
14. Brown RM, McClelland NI, Deininger RA, Tozer RG. A water quality index-do we dare. *Water Sew Work* 1970;**117**(10):339–343.
15. Jadhav SD, Jadhav MS. Study of water quality parameters of Mula-Mutha River at Pune, Maharashtra (India). *Int J Lakes Rivers* 2020;**13**(1):95–103.
16. Jagtap SS, Manivanan R. Water pollution status of Mula-Mutha Rivers in Pune City: Review. *Int J Trend Sci Res Dev* 2019;**4**(1):1080–1084.. Available from www.ijtsrd.com/papers/ijtsrd29813.pdf.
17. American Public Health Association (APHA). *Standard methods for the examination of water and waste water;* 20th ed.; APHA, AWWA, WEF: Washington DC, 1998.
18. Brown RM, McClelland NI, Deininger RA, O'Connor MF. *A water quality index—crashing the psychological barrier. Indicators of environmental quality;* Springer: Boston, MA, 1972, pp. 173–182.
19. Abbasi T, Abbasi SA. *Water quality indices;* Elsevier Science: Burlington, 2012.
20. Rodrigues IR, Geraldes MC, Ferreira AA, Salomão MSS, Anversa SV. Analysis of the water quality index of the Negro and Resende rivers in Duas Barras District, Rio de Janeiro (Brazil). *J Sediment Env* 2018;**3**(2):121–130.
21. Rajesh M, Rehana S. Impact of climate change on river water temperature and dissolved oxygen: Indian riverine thermal regimes. *Sci Rep* 2022;**12**(1):9222.
22. Rehana S, Mujumdar PP. River water quality response under hypothetical climate change scenarios in Tunga-Bhadra river, India. *Hydrol Process* 2011;**25**(22):3373–3386.
23. Jin L, Whitehead PG, Rodda H, Macadam I, Sarkar S. Simulating climate change and socio-economic change impacts on flows and water quality in the Mahanadi River system, India. *Sci Total Env* 2018;**637**:907–917.
24. Dhar A, Datta B. Optimal operation of reservoirs for downstream water quality control using linked simulation optimization. *Hydrol Process* 2008;**22**(6):842–853.
25. Central Pollution Control Board (CPCB). *Perform status common Effl treat plants India.* CPCB: India, 2005.
26. Wagner PD. Impacts of climate change and land use change on the water resources of the Mula and Mutha Rivers catchment upstream of Pune, India [dissertation]. Universität zu Köln; 2013.
27. Rehana S, Dhanya CT. Modeling of extreme risk in river water quality under climate change. *J Water Clim Change* 2018;**9**(3):512–524.
28. Shanahan P, Henze M, Koncsos L, Rauch W, Reichert P, Somlyody L, et al. River water quality modelling: II. Problems of the art. *Water Sci Technol* 1998;**38**(11): 245–252.
29. Chang FJ, Tsai YH, Chen PA, Coynel A, Vachaud G. Modeling water quality in an urban river using hydrological factors–Data driven approaches. *J Env Manage* 2015;**151**:87–96.
30. Tung TM, Yaseen ZM. A survey on river water quality modelling using artificial intelligence models: 2000–2020. *J Hydrol* 2020;**585**:124670.

31. Singh KP, Basant N, Gupta S. Support vector machines in water quality management. *Anal Chim Acta* 2011;**703**(2):152–162.
32. Sarkar A, Pandey P. River water quality modelling using artificial neural network technique. *Aquat Procedia* 2015;**4**:1070–1077.
33. Rajaee T, Khani S, Ravansalar M. Artificial intelligence-based single and hybrid models for prediction of water quality in rivers: a review. *Chemom Intell Lab Syst* 2020;**200**:103978.
34. Srivastava A, Maity R, Desai VR. Assessing Global-Scale Synergy Between Adaptation, Mitigation, and Sustainable Development for Projected Climate Change. In *Ecological Footprints of Climate Change . Springer Climate;* Chatterjee U, Akanwa AO, Kumar S, Singh SK, Dutta Roy A, Eds.; Springer: Cham, 2022, https://doi.org/10.1007/978-3-031-15501-7_2
35. Srivastava A, Maity R. Unveiling an Environmental Drought Index and its applicability in the perspective of drought recognition amidst climate change. *J. Hydrol.* 2023; **627**:130462. https://doi.org/10.1016/j.jhydrol.2023.130462.
36. Masood A, Srivastava A, Hameed MM, Tanweer MS, Ahmad K, Razali SFM. Machine learning for water quality soft sensing in wastewater treatment: State of the art and future directions. In *Innovative and hybrid technologies for wastewater treatment and recycling;* Alam P, Khan AH, Américo-Pinheiro JHP, Ahmad K, Eds.; 1st ed. CRC Press, 2024.

CHAPTER FOUR

Assessing spatiotemporal variability of precipitation complexity in Brahmani River Basin using entropy indices

Krishna Rawat, Ashrumochan Mohanty, and Bhabagrahi Sahoo*
School of Water Resources, Indian Institute of Technology Kharagpur, Kharagpur, India
*Corresponding author. e-mail address: bsahoo2003@yahoo.com; bsahoo@swr.iitkgp.ac.in

Contents

Abstract

The complexity in precipitation measurement is often overlooked in hydroclimatic time series research. To measure the complexity and spatiotemporal variability of precipitation, entropy indices are commonly used due to their advantages in capturing the intricate and nonlinear interactions of climate variables. However, different entropy indices can yield varying results regarding precipitation complexity, complicating the accurate assessment of actual complexity. To evaluate the complexity in precipitation time series, this study aims to identify and select the optimal entropy index from a set of six entropy indices based on the reliability and stability of their entropy values. The six entropy indices considered herein are: Information Entropy, Fuzzy Entropy, Approximate Entropy, Sample Entropy (SE), Wavelet Entropy, and

Advances in Chemical Pollution, Environmental Management and Protection, Volume 15
ISSN 2468-9289, https://doi.org/10.1016/bs.apmp.2025.11.007

Permutation Entropy. Further, the K-means clustering based machine learning technique is used for grouping similar precipitation patterns and identifying distinct regimes of complexity in the Brahmani River Basin, eastern India. This approach has been applied to the observed monthly precipitation gauges over the period 1954–2019 at 54 grids. The findings indicate that the SE index is the most reliable and stable, making it the best choice for assessing the complexity of the monthly precipitation in the selected basin. The K-means clustering revealed distinct spatial and temporal patterns of precipitation complexity. The results revealed that the increasing complexity of monthly precipitation in the Brahmani River Basin is strongly linked to the local industrial and agricultural activities. These findings provide valuable insights for regional precipitation predictability measurement, effective drought and flood assessment, and improved water resources management in the basin. The integration of machine learning techniques enhances the robustness of the analysis and provides a more nuanced understanding of precipitation complexity in the region.

Keywords: Climate variability, Entropy, Machine learning, Precipitation, Spatial pattern, Temporal pattern.

1. Introduction

In hydrometeorological studies, precipitation variability plays a critical role, particularly when assessing the impacts of climate change on water resources. As global temperatures continue to rise, significant alterations occur in the hydrologic cycle, leading to shifts in precipitation patterns, magnitude, timing, intensity, and distribution.[1,2] Understanding these changes is essential for effective water resource management and mitigating risks associated with water-related disasters, such as floods and droughts.[3–6] Precipitation is arguably the most influential variable in climate studies due to its direct effects on ecosystems, agriculture, and human settlements.

However, precipitation processes are inherently complex, driven by nonlinear interactions among multiple climatic factors, including temperature, humidity, wind velocity, and atmospheric pressure.[7] Traditional statistical methods such as mean, variance, and standard deviation have been widely used to analyze precipitation variability. Despite their widespread use, these measures often fail to capture the intricate dynamics that characterize precipitation. For instance, variance only reflects concentration around the mean, disregarding the distribution's shape or higher-order moments,[8] underscoring the need for more sophisticated tools to analyze precipitation variability.

Recent advances in entropy theory have offered new ways to assess the complexity and variability of precipitation. Entropy indices quantify the degree of disorder or randomness in a system and are well-suited for analyzing complex systems with nonlinear interactions.[9,10] These indices offer more comprehensive insights than traditional statistics by considering multiple parameters and avoiding assumptions about data distributions. For example, Information Entropy (IE) has been widely applied in meteorological studies to assess the complexity of climate variables.[11,12] Other entropy measures such as Fuzzy Entropy (FE), Approximate Entropy (AE), Sample Entropy (SE), Wavelet Entropy (WE), Shannon Entropy (SHE) and Permutation Entropy (PE) are developed to capture different aspects of complexity in time series data.[13–16,32] While entropy measures provide robust insights, the application of different entropy indices may lead to varying results due to their methodological differences. These differences can complicate the assessment of precipitation complexity.[17,18] Thus, a careful comparison of entropy indices is needed to identify the most reliable index for specific applications. This challenge is particularly pronounced in regions experiencing rapid urbanization, where human activities, such as land-use changes, deforestation, and dam construction exacerbate precipitation variability.[19,20] Studies have shown that urbanization alters local precipitation patterns by affecting surface characteristics and atmospheric dynamics, making it more difficult to assess the variability using traditional methods.[21]

In addition to entropy measures, machine learning techniques like K-means clustering have emerged as valuable tools for understanding precipitation complexity. Machine learning algorithms, with their ability to process large datasets and identify patterns without prior assumptions, are particularly useful for hydrological applications. K-means clustering, in particular, has been widely applied to group similar data points and identify distinct precipitation regimes.[22] For instance, K-means has been used to regionalize the entire Europe for the projected climate change impacts[22] and to improve rainfall prediction accuracy.[23] The strength of K-means clustering lies in its simplicity and efficiency, making it ideal for large datasets typical in hydrological research. When combined with entropy indices, K-means can enhance our understanding of precipitation variability by identifying spatial patterns that traditional statistical methods might overlook.

Despite the growing use of entropy indices in precipitation studies, there remains a lack of consensus on which entropy measure provides the most reliable and stable assessment of precipitation complexity. Different entropy indices often yield inconsistent results, particularly in regions where anthropogenic influences like urbanization, deforestation, and dam construction significantly alter precipitation dynamics. Additionally, while K-means clustering has proven effective in identifying precipitation regimes, its integration with entropy measures in analyzing spatiotemporal precipitation variability is still underexplored. Moreover, the precipitation variability in a coastal catchment subject to tropical cyclones has not been studied using the entropy concept. These gaps highlight the need for a comprehensive comparison of multiple entropy indices, coupled with machine learning techniques, to assess precipitation complexity in regions affected by both natural and human-induced variability.

To address these research gaps, this study focuses on the coastal Brahmani River Basin, aiming to evaluate the efficacy of the entropy indices, such as Information Entropy, Fuzzy Entropy, Approximate Entropy, Sample Entropy, Wavelet Entropy, and Permutation Entropy providing the most reliable and stable measure of precipitation complexity. Furthermore, the study investigates how integrating entropy indices with K-means clustering can enhance the understanding of spatiotemporal precipitation variability in the basin. These six entropy indices are applied to the historical monthly precipitation data collected from the India Meteorological Department (IMD) at 54 grid points across the selected basin over the period of 1954–2019. Herein the K-means clustering is also employed to group regions with similar precipitation patterns and to identify distinct precipitation regimes. The study also explores the probable factors influencing precipitation complexity, including natural variability and human-induced activities such as deforestation, dam construction, and urbanization.

By systematically addressing these objectives, this research aims to provide valuable insights into precipitation dynamics in a region increasingly impacted by climatic stress and anthropogenic pressures. The integration of entropy-based methods with machine learning techniques offers a novel and robust framework for enhancing water resource management and improving preparedness for water-related disasters in the context of changing climate.

2. Study area and data

The Brahmani River Basin (BRB) is considered as the study area for this research, located in eastern India with an area of about 39,033 km^2

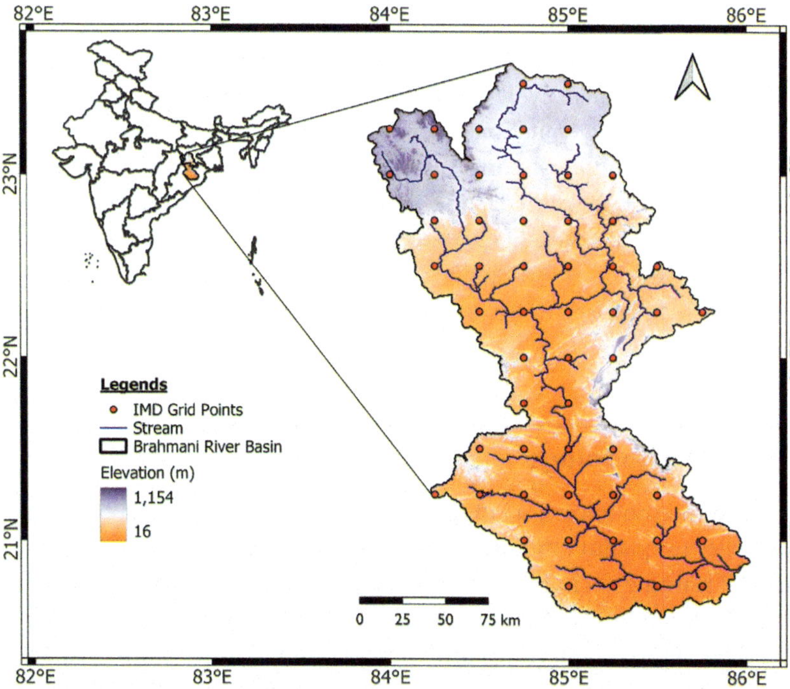

Fig. 1 Index map of Brahmani River basin showing spatial distribution of 54 IMD grid points. Map lines delineate study areas and do not necessarily depict accepted national boundaries.

(Fig. 1). It is situated between the longitudes of 83°52′–87°30′E and latitudes of 20°28′–23°35′N. The BRB experiences a tropical climate characterized by three distinct seasons: summer, monsoon, and winter. Approximately 80 % of the annual rainfall occurs during the southwest monsoon from June to September, resulting in cloudy conditions throughout this period.[24] Based on the collected precipitation data, the basin receives an average annual rainfall of 1305 mm/year, varying from 969 mm to 1574 mm.[25] Additionally, the region experiences a number of tropical cyclones and significant temperature fluctuations, with annual temperatures varying from 4 to 47 °C.[33] These climatic factors make the BRB an essential area for examining precipitation dynamics and their variability.

For this study, monthly gridded precipitation data were obtained from the IMD over 66 years (January 1954 to December 2019). This dataset provides a comprehensive view of precipitation patterns in the region, enabling a detailed analysis of spatiotemporal variability. Each station's data includes monthly records, which allows for the assessment of seasonal trends and anomalies. To enhance the robustness of our analysis, we removed the seasonal cycle from the observed data by calculating the average monthly precipitation over the 66 years. This adjustment helps mitigate seasonal influences on the assessment of precipitation complexity. Additionally, we generated series for monthly maximum and average precipitation values to facilitate a thorough investigation of variability. This extensive dataset is crucial for applying entropy indices to evaluate the complexity of precipitation dynamics in the Brahmani River Basin effectively.

3. Methodology

The methodology, as illustrated in Fig. 2, combines entropy-based approaches to quantify the spatiotemporal complexity of precipitation with machine learning techniques, particularly K-means clustering, to identify the distinct patterns and classify regions based on precipitation characteristics. The approach not only provides a comprehensive understanding of the underlying complexity of precipitation dynamics but also enhances the ability to interpret its spatial variability.

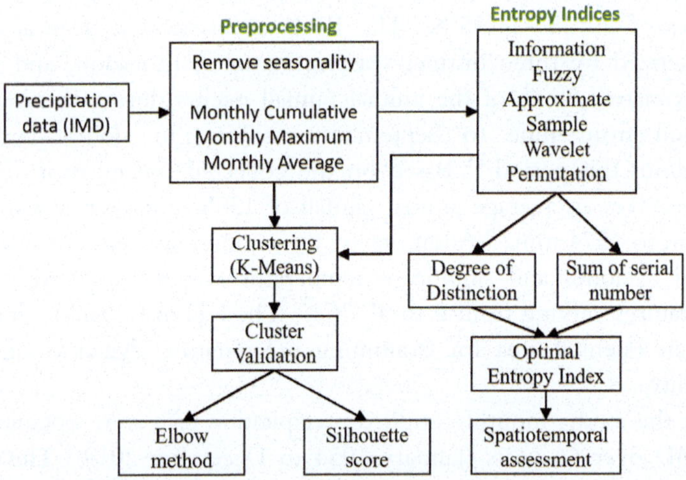

Fig. 2 Methodological framework adopted in the study.

3.1 Entropy indices for precipitation complexity measurement

Entropy is widely recognized for its ability to measure the disorder or unpredictability within a system. In the context of precipitation, entropy indices are used to assess the complexity of rainfall processes by quantifying the variability and irregularity in the precipitation time series. The six entropy indices applied in this study are summarized in Table 1. These indices provide different perspectives on the complexity of precipitation, capturing various aspects of variability and unpredictability in both the spatial and temporal domains.

3.1.1 Information entropy (IE)

Based on Shannon's information theory, Information Entropy measures the uncertainty in the distribution of precipitation values.[12] It provides a basic measure of variability and is particularly useful for detecting overall patterns in the data. IE was calculated by dividing the precipitation data into discrete intervals and computing the probability distribution of values within these intervals. The resulting entropy value indicates the overall complexity of the precipitation system, with higher values corresponding to greater variability and unpredictability.

3.1.2 Fuzzy entropy (FE)

Fuzzy Entropy extends the concept of traditional entropy by incorporating fuzziness, which allows for a more flexible measurement of uncertainty.[14] In the context of precipitation, FE captures the gradual transitions between different rainfall regimes, making it ideal for analyzing continuous data. FE was calculated by defining a fuzzy membership function for the precipitation values, which accounts for the degree of membership of each value to a particular category. The fuzziness allows for a more nuanced interpretation of complexity, particularly in regions with smooth transitions between wet and dry periods.

3.1.3 Approximate entropy (AE) and sample entropy (SE)

Both AE and SE measure the regularity of time series data, with SE offering improvements over AE by being less dependent on data length and more robust against noise.[15,16] These indices are particularly useful for analyzing short or noisy datasets, such as precipitation records. AE and SE were computed by comparing sequences of data points within the precipitation time series and determining the likelihood that similar

Table 1 Entropy indices used in this study.

S.No.	Entropy index	Formula	Application	References
1.	Information Entropy (IE)	$IE(X) = -\sum_{i=1}^{n} p_i \log(p_i)$	Measures the average uncertainty in a set of possible outcomes.	12
2.	Fuzzy Entropy (FE)	$FE(X) = -\sum_{i=1}^{n} p_i^m \log(p_i)$ (where $m > 0$)	Extends traditional entropy to fuzzy sets, accounting for uncertainty in membership.	14
3.	Approximate Entropy (AE)	$AE(m, r) = \frac{1}{N-m} \sum_{i=1}^{N-m} \log\left(\frac{C_i(m,r)}{C_{i+1}(m,r)}\right)$	Quantifies the regularity of a time series, measuring its complexity.	15
4.	Sample Entropy (SE)	$SE(m, r) = -\log\left(\frac{A}{B}\right)$	Similar to Approximate Entropy but avoids self-matching, providing a more robust measure.	16
5.	Wavelet Entropy (WE)	$WE = -\sum_{i=1}^{n} p_i \log(p_i)$ (using wavelet coefficients)	Analyzes the complexity of signals by decomposing them into different frequency components.	26
6.	Permutation Entropy (PE)	$PE = -\sum_{i=1}^{n} p_i \log(p_i)$ (based on ordinal patterns)	Measures the complexity of time series based on the ordering of values rather than their actual values.	13

X: random variable; p_i: probability of the ith outcome occurring; n: total number of possible outcomes; m: Length of the sequence or embedding dimension in Approximate Entropy and Sample Entropy; r: Tolerance threshold used in Approximate Entropy and Sample Entropy to determine similarity; N: Total number of data points in the time series; $C_i(m, r)$: Number of patterns of length m that match within tolerance r starting from position i; A, B: Counts of patterns used in the calculation of Sample Entropy.

patterns will repeat. Lower values indicate more regular and predictable patterns, while higher values suggest greater complexity and irregularity in precipitation.

3.1.4 Wavelet entropy (WE)

WE combines wavelet transformation with entropy measurement, allowing for a multi-scale analysis of precipitation data.[26] This index captures both the temporal and frequency components of variability, making it suitable for detecting periodic patterns or oscillations in the data. WE was calculated by decomposing the precipitation time series into different frequency bands using wavelet transformation and then applying entropy calculations to the wavelet coefficients. This approach provides insights into how precipitation complexity varies across different time scales.

3.1.5 Permutation entropy (PE)

PE measures the complexity of the time series based on the ordering of data points.[13] Unlike traditional entropy measures that rely on the magnitude of values, PE focuses on the ordinal structure of the data, making it robust against noise and outliers. In this study, PE was used to assess the temporal complexity of precipitation by comparing the relative positions of consecutive data points within the time series. Higher PE values indicate more complex and less predictable sequences.

3.2 Optimal entropy index selection

The six entropy indices provide different perspectives on precipitation complexity, but selecting the optimal index requires evaluating their reliability and stability. Reliability refers to the ability of an index to consistently distinguish between different levels of complexity, while stability refers to its robustness against small changes in the data.[21] In this study, the reliability and stability of each entropy index were assessed using the Theory of Degree of Distinction and the Theory of Sum of Serial Numbers, respectively.

3.2.1 Theory of degree of distinction

This theory was applied to evaluate the reliability of each entropy index by comparing its ability to differentiate between precipitation regimes across the 54 grid points.[27] The index that provided the greatest distinction between regions with different precipitation characteristics was considered the most reliable. The degree of distinction was calculated by ranking the entropy values for each grid point and computing the distance between consecutive ranks. The index with the largest average distance was deemed the most reliable.

3.2.2 *Theory of sum of serial numbers*

This is a method used to evaluate the stability of an evaluation system by analyzing the orderings derived from different entropy indices. This theory operates by comparing the rank positions of items across multiple evaluation rounds.[28] When applied to precipitation complexity, each entropy index generates a ranking of data points based on their complexity. The stability of these rankings is then assessed by calculating the correlation between the actual ranking of items and the expected 'reasonable' ranking.

The process starts by assigning serial numbers to objects based on their evaluated complexity, then summing the serial numbers across different evaluations. The closer the sum of the serial numbers for an object matches its expected rank, the more stable the entropy index is. High correlations between the different rankings indicate that the entropy index consistently identifies objects in a stable order, making it a reliable tool for complexity measurement.

3.3 K-means clustering for precipitation classification

K-means is an unsupervised machine learning algorithm that groups data points into clusters based on their similarity. In this case, the algorithm was used to group grid points in the river basin with similar precipitation characteristics, helping to identify distinct precipitation patterns and regimes across the basin. The K-means algorithm operates by partitioning the dataset into a predefined number of clusters (K). Each data point is assigned to the nearest cluster based on its features, and the algorithm iteratively adjusts the cluster centroids to minimize the within-cluster variance.[29] For this study, the features used for clustering included the entropy values from the six indices, as well as other relevant precipitation characteristics such as the average and maximum monthly rainfall values.

3.3.1 *Determining the optimal number of clusters (K)*

Herein, the number of clusters is determined using the Elbow Method,[30] which involves plotting the within-cluster variance as a function of 'K' and identifying the point at which the rate of decrease slows significantly (the 'elbow' point). This method balances the trade-off between having too few clusters (which may oversimplify the data) and having too many clusters (which may lead to overfitting).

3.3.2 Cluster validation

The quality of the clustering was validated using Silhouette Analysis,[31] which measures how similar each data point is to its own cluster compared to other clusters. Silhouette analysis evaluates the quality of clustering by measuring how well each data point fits within its own cluster compared to the nearest neighbouring cluster. The silhouette score ranges from, 1–1, with 1 indicating well-separated clusters, zero suggesting points lie between the clusters, and negative values implying misclassification. An average silhouette score of ≥ 0.7 indicates strong clustering, while scores between 0.5 and 0.7 suggest reasonable structure, and scores ≤ 0.5 point to weak clustering or overlap.

3.4 Spatiotemporal analysis of precipitation complexity

The integration of entropy indices and K-means clustering allowed for a detailed spatiotemporal analysis of precipitation complexity in the Brahmani River Basin. The spatial analysis revealed distinct regions with varying levels of precipitation complexity, while the temporal analysis highlighted the trends over the 66 years of study period.

4. Results and discussion

4.1 Optimal entropy index selection

The complexity of monthly precipitation across 54 grid points in the Brahmani Basin from 1954 to 2019 was analyzed using six entropy indices: IE, FE, AE, SE, WE, and PE. The analysis results, illustrated in Fig. 3, show that each entropy index produced different complexity values across the 54 gridded precipitation time series, indicating a range of spatial patterns in precipitation behaviour. These variations suggest that each entropy measure captures distinct aspects of precipitation complexity, potentially reflecting unique regional factors such as elevation, land use, and climate variability. To identify the most suitable index for accurately characterizing precipitation complexity in the Brahmani Basin, the reliability and stability of each entropy index were evaluated as discussed below.

4.1.1 Theory of degree of distinction analysis

This study employed the theory of degree of distinction to evaluate the reliability of various entropy indices in assessing complexity within monthly precipitation data for the Brahmani Basin. The six entropy

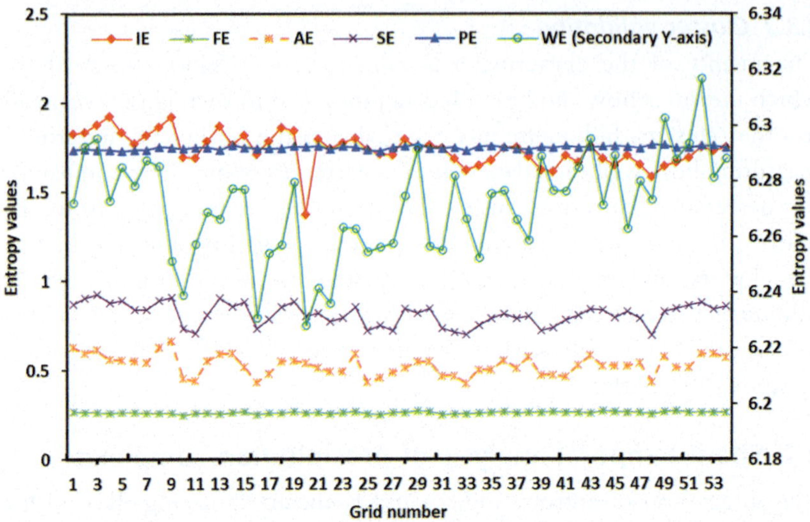

Fig. 3 Complexity of the 54 gridded monthly precipitation series in the Brahmani basin over the period 1954–2019 using the Information Entropy (IE), Fuzzy Entropy (FE), Approximate Entropy (AE), Sample Entropy (SE), Wavelet Entropy (WE), and Permutation Entropy (PE).

indices, viz. IE, FE, AE, SE, WE, and PE were analyzed and ranked according to their reliability scores. The rankings, based on the normalized values as shown in Table 2 for reliability, were obtained as follows: SE (0.43), FE (0.15), WE (0.13), IE (0.12), AE (0.10), and PE (0.08). Among these, SE was identified as the most reliable index for measuring precipitation complexity, while PE scored the lowest on this criterion. Normalizing the results was essential to create a fair basis for comparison across entropy indices, each of which operates with distinct units and scales. By applying normalization, the values were standardized, allowing for a consistent scale to evaluate and compare each index's reliability and stability in complexity measurement. The normalized reliability values indicated that the Sample Entropy is the most consistent measure of complexity across precipitation datasets, making it the preferred index for reliability.

4.1.2 Theory of sum of serial numbers

Beyond reliability, this study also examined the stability of complexity measures using the theory of the sum of serial numbers. For this, random sampling was carried out to stabilize the correlation coefficient across indices, effectively testing whether the rank order of the indices would

Table 2 Assessed reliability and stability measures of complexity in the monthly gridded precipitation time series in the Brahmani River Basin using different entropy indices.

	Information entropy	Fuzzy entropy	Approximate entropy	Sample entropy	Wavelet entropy	Permutation entropy
Reliability	33.76	44.51	28.54	125.25	36.57	22.46
Normalized reliability	0.12	0.15	0.10	0.43	0.13	0.08
Stability	108	162	216	189	189	270
Normalized stability	0.10	0.14	0.19	0.17	0.17	0.24
Normalized average	0.11	0.15	0.14	0.30	0.15	0.16

stabilize as sampling iterations increased. Using correlation analysis, stability rankings were assessed based on the normalized stability values in Table 2 with the results showing that PE scored the highest in stability (0.24), followed by AE (0.19), SE (0.17), WE (0.17), FE (0.14), and IE (0.10). Although PE exhibited high stability in complexity estimation, its lower reliability score compared to SE suggested that its use in precipitation complexity analysis may be limited.

Based on the reliability and stability measures, SE performed the best in capturing the complexity of monthly precipitation time series for the Brahmani Basin, indicating its suitability for robust analysis. PE, despite its high stability, is not optimal due to low reliability. Other indices, such as WE and FE, displayed moderate stability but were overshadowed by SE in reliability performance. Thus, SE emerged as the most balanced and effective entropy index for assessing complexity (Table 2).

4.2 Spatial and temporal distribution of precipitation complexity

Based on the spatial distribution of SE values (Fig. 4) for the monthly precipitation series in the Brahmani Basin from 1954 to 2019, significant regional variations in complexity were observed. The SE values ranged from 0.98 to 3.53, with the lower SE values (0.98–1.67) concentrated in specific regions, while higher values (2.53–3.53) were primarily located in other parts of the basin. The grids with lower SE values, indicating less complexity, are predominantly found in the northern part of the basin. In contrast, regions with higher SE values, which represent increased complexity in the monthly precipitation series, were more common in the southern and central parts of the basin. These findings suggest that the complexity of precipitation patterns is higher in the southern and central areas of the Brahmani Basin, with the lower complexity observed in the northern regions. This spatial variation in SE values reflects the diverse climatic and geographic conditions across the basin.

The temporal variation of the complexity of monthly precipitation series was also analyzed using the Sample Entropy (SE) index for monthly maximum precipitation (MMAP) and monthly average precipitation (MAP) in the Brahmani Basin (Fig. 5). Temporal analysis revealed an upward trend in SE values over the 66-year study period. The periodic spikes in complexity during extreme events, such as floods and droughts, underscore the sensitivity of precipitation patterns to both natural climatic variability and anthropogenic influences. (Fig. 6).

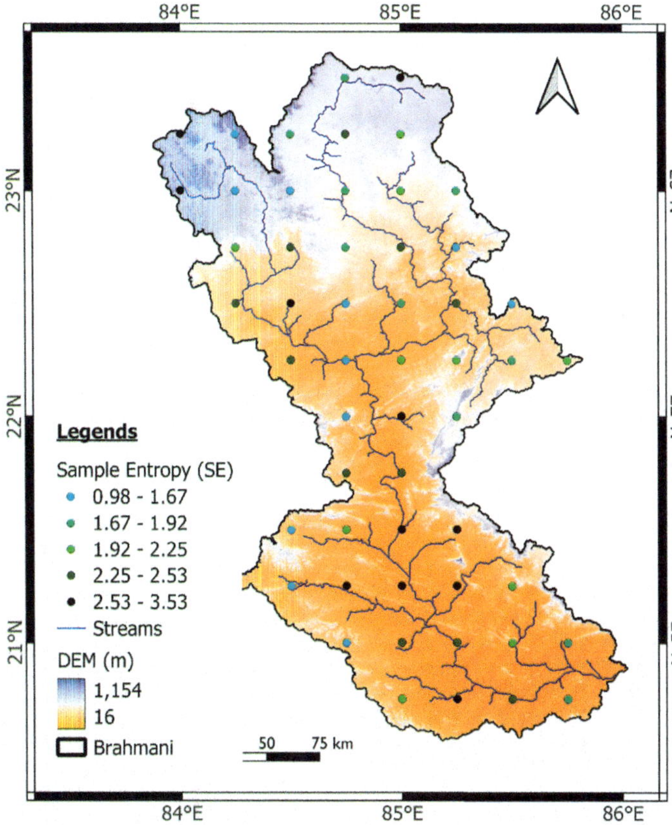

Fig. 4 Spatial distribution of the sample entropy (SE) values of the 54 gridded monthly precipitation time series in Brahmani Basin over the period 1954–2019. Map lines delineate study areas and do not necessarily depict accepted national boundaries.

4.3 Classification of precipitation regimes using K-means clustering

K-means clustering (Fig. 7) was applied to classify precipitation regimes within the basin based on entropy values and other precipitation metrics. Using the Elbow Method (Fig. 6), the optimal number of clusters was identified as K = 4. The Elbow Method graph plotted the within-cluster sum of squares (WCSS) against different values of K, showing a distinct 'elbow' at K = 4, where the WCSS reduction rate noticeably decreased. This point, indicating minimal gains from additional clusters, provided a balanced clustering model that effectively captures the complexity distribution in the basin without overfitting.

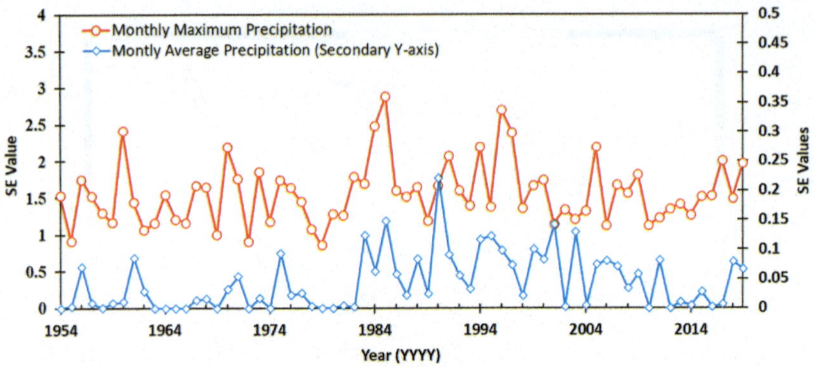

Fig. 5 Temporal variation of complexity of the monthly maximum precipitation series, and monthly average precipitation series in Brahmani Basin using the sample entropy (SE) index.

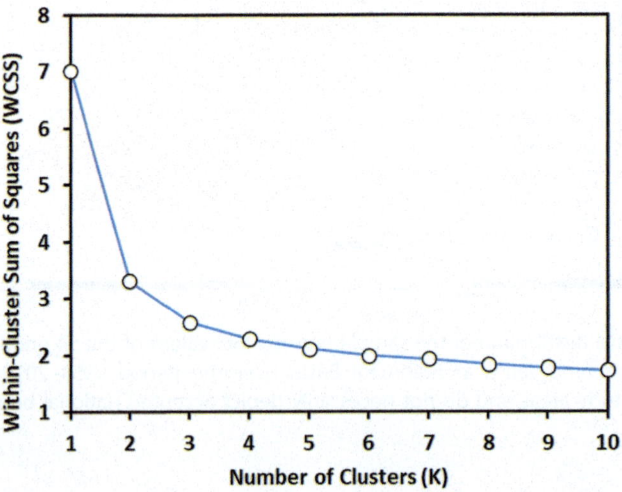

Fig. 6 Selection of optimum number of cluster (K) using Elbow method.

4.3.1 Cluster validation via silhouette analysis

The quality of clustering was validated using Silhouette Analysis, yielding an average silhouette score of 0.73. This high silhouette score confirms that the four clusters are well-separated and cohesively defined, with clear internal consistency within each cluster and sufficient separation between clusters. This score further validates that the chosen K-means clustering model accurately captures the distinct precipitation regimes within the basin. The results from the combined methods demonstrate that Sample Entropy (SE) is the optimal

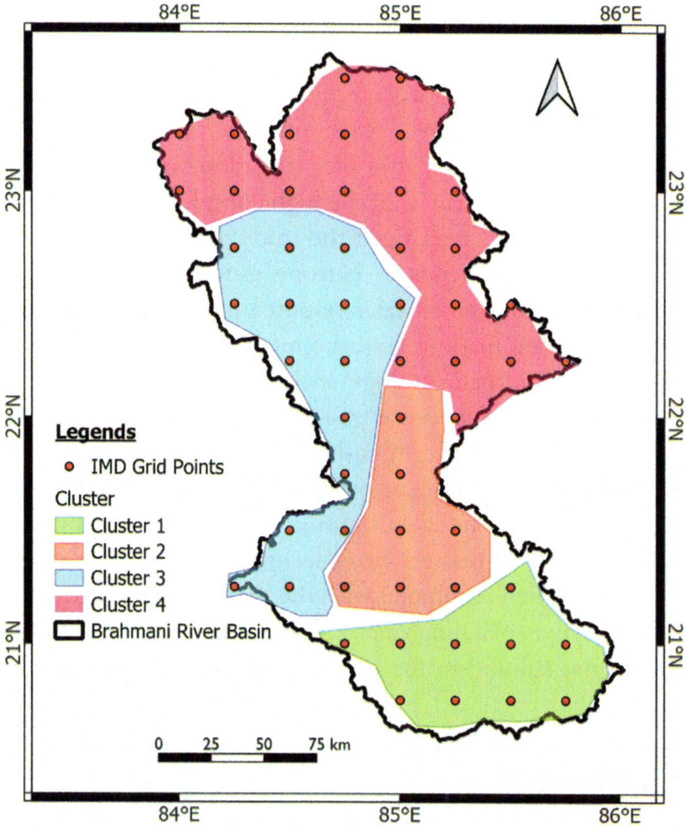

Fig. 7 Clustering of grid points using K-means clustering in the selected basin. Map lines delineate study areas and do not necessarily depict accepted national boundaries.

choice for assessing precipitation complexity in the Brahmani River Basin, providing stable and distinct complexity values that effectively differentiate between the basin's varied geographic and anthropogenic regions. The spatial analysis and K-means clustering identified distinct precipitation regimes bringing out actionable insights for water resource management. These findings provide a strong foundation for targeted water management policies, flood and drought mitigation strategies, and climate resilience planning tailored to the unique precipitation dynamics of the Brahmani River Basin.

4.4 Implications of precipitation complexity for climate-smart water management

The higher Sample Entropy estimates across the BRB reveal an increased complexity of precipitation patterns, which has profound implications on climate-smart water management. As the rainfall regimes are becoming unstable and less predictable over time, it becomes difficult to rely only on the historical rainfall trends for water infrastructure planning and its management. Therefore, adaptive water management strategies, such as decentralized water storage, flexible irrigation scheduling, and early warning systems become crucial in the high complexity zones, especially in the southern and central regions of the study area as found in this study. Additionally, the rising trends in entropy estimates over the six-decade period suggest that long-term water resource policies must evolve beyond the average-based estimations to account for increasing spatiotemporal variability in the hydroclimatology including the extreme events. The anthropogenic factors of land-use transformation, deforestation, and rapid urban growth would further intensify this complexity, highlighting the need for integrated management practices that address both climatic and anthropogenic factors. The use of entropy-based metrics, particularly SE, offers a practical tool to reassess water security risks and redesign decision-making systems. When combined with the machine learning techniques, it allows for proactive planning and supports the development of climate resilient strategies tailored to the basin's evolving hydrological conditions.

5. Conclusion

This study provides a comprehensive analysis of precipitation complexity in the Brahmani River Basin using various entropy indices, viz. Information Entropy (IE), Fuzzy Entropy (FE), Approximate Entropy (AE), Sample Entropy (SE), Wavelet Entropy (WE), and Permutation Entropy (PE). Sample Entropy (SE) was found to be the most effective metric for capturing the variability and complexity in the spatial distribution of monthly precipitation patterns, with SE values ranging from approximately 0.1–2.5 for monthly maximum precipitation (MMAP) and 0–0.5 for monthly average precipitation (MAP). These values indicate an increase in complexity over time, with the MAP series displaying higher complexity than the MMAP series, especially in the central and southern regions of the basin. This trend points to an increasingly intricate precipitation process, complicating accurate forecasting and effective water resource management.

The temporal analysis reveals an upward trend in SE values, indicating growing unpredictability in precipitation, which could impact future water resource allocation and climate adaptation strategies. The K-means clustering for capturing the spatial variability resulted in a silhouette score of 0.73, indicates that the clusters represent distinct and meaningful precipitation complexity patterns. The resulting clusters enable the identification of regions with similar complexity characteristics, allowing for tailored water management and climate resilience measures across the Basin.

Conclusively, this study underscores the utility of entropy-based methods, particularly Sample Entropy, in capturing the increasing complexity of gridded precipitation within the Brahmani Basin. The findings provide essential insights for policymakers and water resource managers in formulating adaptive strategies suited to the region's changing climatic conditions. Future research could be extended by integrating additional climate variables and exploring advanced machine-learning techniques for understanding the precipitation dynamics in response to evolving climate patterns.

Declaration of generative AI and AI-assisted technologies in the writing process

During the preparation of this work, the authors used Perplexity.ai in order to improve readability of the sentences and correct grammatical error. After using this tool/service, the authors reviewed and edited the content as needed and take full responsibility for the content of the published article.

References

1. Milly PCD, Wetherald RT, Dunne KA, Delworth TL. Increasing risk of great floods in a changing climate. *Nature* 2002;**415**:514–517.
2. Faiz MA, Liu D, Fu Q, Qama MU, Dong S, Khan MI, et al. Complexity and trends analysis of hydrometeorological time series for a river streamflow: a case study of Songhua River Basin, China. *River Res Appl* 2018;**34**:101–111.
3. Mohanty A, Sahoo B, Kale RV. A hybrid model enhancing streamflow forecasts in paddy land use-dominated catchments with numerical weather prediction model-based meteorological forcings. *J Hydrol* 2024;**635**:131225.
4. Mohanty A, Sahoo B, Kale RV. Real-time forecasting of coastal flood inundations under regulated reservoir and storm-tide influences. *Adv Water Resour* 2025:104920.
5. Roushangar K, Alizadeh F. Entropy-based analysis and regionalization of annual precipitation variability in Iran during 1960–2010 using ensemble empirical mode decomposition. *J Hydroinf* 2018;**20**(1–2):468–485.
6. Xavier SFA, Silva Jale JD, Stosic T, Santos CAC, Singh VP. An application of sample entropy to precipitation in Paraı´ba State, Brazil. *Theor Appl Climatol* 2019;**136**: 429–440.
7. Guntu RK, Maheswaran R, Agarwal A, Sivakumar B. Spatiotemporal variability of Indian rainfall using multiscale entropy. *J Hydrol* 2020:124916.
8. Soofi E. Information theoretic regression methods. In *Advances in econometrics-applying maximum entropy to econometrics problems;* Fomby, T., Carter Hill, R., Eds.; Jai Press Inc.: London, 1997.

9. Kalayci S, Kahya E. Assessment of streamflow variability modes in Turkey: 1964–1994. *J Hydrol* 2006;**324**:163–177.
10. Maurer EP, Lettenmaier DP, Mantua NJ. Variability and potential sources of predictability of North American runoff. *Water Resour Res* 2004;**40**:W09306.
11. Koutsoyiannis D. Uncertainty, entropy, scaling and hydrological stochastics. 1. Marginal distributional properties of hydrological processes and state scaling. *Hydrol Sci J* 2005;**50**:381–404.
12. Shannon CE. A mathematical theory of communication. *Bell Syst Tech J* 1948;**27**(3):379–423.
13. Bandt C, Pompe B. Permutation entropy: a natural complexity measure for time series. *Phys Rev Lett* 2002;**88**(17):174102.
14. Luca AD, Termini S. A definition of a nonprobabilistic entropy in the setting of fuzzy sets theory. *Inf Control* 1972;**20**:301–312.
15. Pincus SM. Approximate entropy as a measure of system complexity. *Proc Natl Acad Sci* 1991;**88**(6):2297–2301.
16. Richman JS, Moorman JR. Physiological time-series analysis using approximate entropy and sample entropy. *Am J Physiol Heart Circ Physiol* 2000;**278**(6): H2039–H2049.
17. Costa M, Goldberger AL, Peng CK. Multiscale entropy analysis of complex physiologic time series. *Phys Rev Lett* 2002;**89**(6):068102.
18. Zhang ZL, Xiang ZT, Chen YF, Xu JY. Fuzzy permutation entropy derived from a novel distance between segments of time series. *AIMS Math* 2020;**5**(6):6244–6260.
19. Wang J, Feng JM, Yan ZW. Impact of extensive urbanization on summertime rainfall in the Beijing area and the role of local precipitation recycling. *J Gerontol Ser A Biol Med Sci* 2018;**123**(7):3323–3340.
20. Dong L, Meng L. Application of sample entropy on measuring precipitation series complexity in Jiansanjiang Branch Bureau of China. *Nat Env Pollut Technol* 2013;**12**(2): 249–254.
21. Du L, Li X, Yang M, Sivakumar B, Zhu Y, Pan X, et al. Assessment of spatiotemporal variability of precipitation using entropy indexes: a case study of Beijing, China. *Stoch Environ Res Risk Assess* 2022:1–15.
22. Carvalho MJ, Melo-Gonçalves P, Teixeira JC, Rocha A. Regionalization of Europe based on a K-means cluster analysis of the climate change of temperatures and precipitation. *Phys Chem Earth, Parts A/B/C* 2016;**94**:22–28.
23. Dubey AD. Artificial neural network models for rainfall prediction in Pondicherry. *Int J Computer Appl* 2015;**120**(3).
24. Rawat K, Sahoo DP, Sahoo B. Entropy-based streamflow estimation in Data-Scarce rivers using MODIS-landsat fusion imagery. *J Hydrologic Eng* 2025. https://doi.org/10.1061/JHYEFF/HEENG-6387.
25. Mohanty A, Sahoo B, Kale RV. A coupled optimized hedging rule-based reservoir operation and hydrodynamic model framework for riverine flood risk management. *Water Res* 2025;**279**:123443.
26. Rosso OA, Blanco S, Yordanova J, Kolev V, Figliola A, Schurman M, et al. Wavelet entropy: a new tool for analysis of short duration brain electrical signals. *J Neurosci Methods* 2001;**105**(1):65–75.
27. Liu D, Liu C, Fu Q, Li TX, Khan MI, Cui S, et al. ELM evaluation model of regional groundwater quality based on the crow search algorithm. *Ecol Ind* 2017;**81**:302–314.
28. Zhang LL, Li H, Liu D, Fu Q, Li M, Faiz MA, et al. Identification and application of the most suitable entropy model for precipitation complexity measurement. *Atmos Res* 2019;**221**:88–97.
29. Hartigan JA, Wong MA. A k-means clustering algorithm. *Appl Stat* 1979;**28**(1): 100–108.

30. Syakur MA, Khotimah BK, Rochman EMS, Satoto BD. *Integration k-means clustering method and elbow method for identification of the best customer profile cluster. IOP conference series: materials science and engineering;* IOP Publishing, 2018; pp. 012017.
31. Rousseeuw PJ. Silhouettes: a graphical aid to the interpretation and validation of cluster analysis. *J Comput Appl Math* 1987;**20**:53–65.
32. Srivastava A, Maity R, Desai VR. Synthesising causal loop between environmental and compound droughts: A systems-driven framework for climate-resilient governance in monsoon-sensitive agroecosystems. *Next Sustain.* 2025;**6**:100159. https://doi.org/10.1016/j.nxsust.2025.100159.
33. Srivastava A, Maity R. Unveiling an Environmental Drought Index and its applicability in the perspective of drought recognition amidst climate change. *J. Hydrol.* 2023;**627**:130462. https://doi.org/10.1016/j.jhydrol.2023.130462.

SECTION B

Artificial intelligence and machine learning frontiers: From forecasting to large language models

CHAPTER FIVE

Enhancing seasonal rainfall forecasting using machine learning approaches

Mohamed Khalid AlOmar[a], Mohammed Majeed Hameed[b,*], and Adil Masood[c,*]

[a]Department of Civil Engineering, College of Engineering, University of Al-Maarif, Ramadi City, Iraq
[b]Upper Euphrates Centre for Sustainable Development Research, University of Anbar, Ramadi City, Iraq
[c]Department of Natural and Applied Sciences, TERI School of Advanced Studies, New Delhi, India
*Corresponding authors. e-mail address: mohmmag1@gmail.com; adil.masood@terisas.ac.in

Contents

Abstract

Rainfall is a critical weather phenomenon, and its accurate forecasting is crucial for water resource management and planning. In this study, several ML models — such as ANN, SVR and MLR have been developed to forecast seasonal rainfall over Al-Ramadi city, western Iraq. Two rainy seasons were selected in the study and the prediction accuracy of both models was assessed using different statistical parameters and comparison graphs such as Taylor diagrams and scatter plots. The results showed that the ANN provides more accurate results in forecasting seasonal precipitation than MLR, and SVR models. Furthermore, the ANN achieved the highest correlation

Advances in Chemical Pollution, Environmental Management and Protection, Volume 15
ISSN 2468-9289, https://doi.org/10.1016/bs.apmp.2025.11.006

coefficient (*R*) for the first season (October to December) and the second season (January to March), with *R* values of 0.734 and 0.716, respectively, which are higher than those of SVR (0.651 and 0.439) and MLR (0.543 and 0.607). For both seasons, the obtained results from ANN are stable and accurate but the performance of the MLR model varies dramatically from season to season. Finally, this paper provides promising insight into the usage of ANN in forecasting seasonal rainfall.

Keywords: Seasonal rainfall, Forecasting model, ANN, Hydrological cycle.

1. Introduction

Extreme rainfall events have become increasingly common in recent years, threatening human safety and socioeconomic well-being.[1] Playing a fundamental role in water resource management, rainfall is also a highly influential variable in forecasting water balance. Due to its considerable temporal variability, rainfall stands out as one of the more intricate parameters to simulate and comprehend in the hydrological cycle.[2] The reliable simulation of precipitation processes is vital for managing water resources. Numerous variables significantly impact precipitation formation, which is complex and associated with uncertainty.[3] Generally, there are two approaches to modeling rainfall: data-driven and process-driven.[4] The first approach, based on data-driven methods, can operate without considering the physical mechanisms of the hydrological conditions responsible for precipitation formation. This modeling approach relies on mathematical analysis of historical time series data[5] for a given catchment or station to determine the causal connections between input and output variables. As a result, models based on this approach demonstrate enhanced operational practicality.

Recent developments in AI and big data have opened new avenues for applying machine learning (ML) in rainfall forecasting. However, predicting the rainfall amount proves to be an exceptionally challenging procedure due to the complexity of the atmospheric conditions involved. Despite notable advancements in climate modeling, predicting rainfall continues to be a significant challenge within hydrology. Numerous efforts have explored numerical weather methods for rainfall forecasting.[6–8] The main constraint with numerical models is the lack of input data which proves to be extremely difficult to overcome. The computational requirements needed to execute these techniques remain very high. Different AI models were used by researchers to solve the challenging problems in numerical models for accurate prediction generation. One of these approaches i.e., ANN is widely used in rainfall forecasting due to its excellent performance.[9–12] The ANN has a good capacity to solve non-linear relations between inputs and outputs.[13] Another ML-based model is support vector regression (SVR), which has a strong reputation for modeling rainfall across different climate regions.[14–16]

The research evaluates the forecasting abilities of ANN and SVR models together with MLR to predict seasonal precipitation in western Iraq. In this study, Al-Ramadi station is selected, as it is located in Ramadi city, which falls within an arid and semi-arid climate zone, where rainfall variability differs from year to year. Additionally, the case study is affected by climatic variability, and no prior research has analyzed the fluctuations in rainfall in this area. The data used in this work includes precipitation records from 1981 to 2021. Two rainy seasons are considered in this study: the first from October to December and the second from January to March.

2. Case study

Ramadi city is one of the most important Iraqi cities located in the Anbar Governorate, western Iraq (see Fig. 1). This city is the center of Anbar Governorate and is bordered on the north by the Euphrates River. This river is the main and only source that supplies the population with fresh water. In

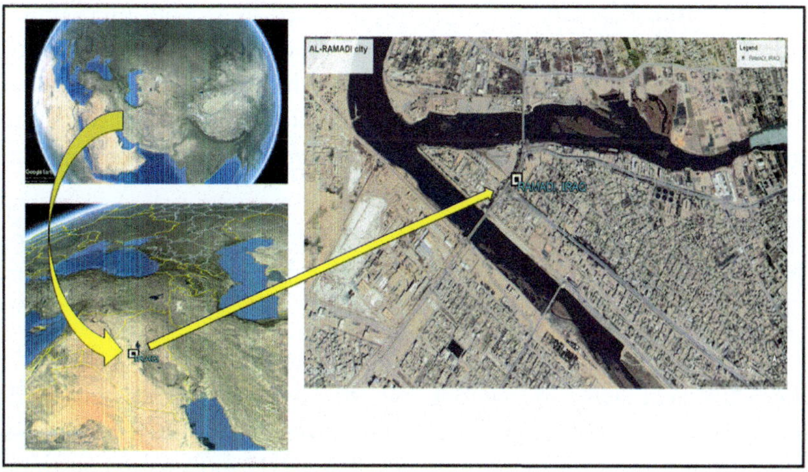

Fig. 1 The location of Al-Ramadi city.

Table 1 Statistical description of the main season rainfall.

Season	Month	Mean (mm/d)	Maximum (mm/d)	Standard deviation (mm/d)	Minimum (mm/d)
	October	0.22	1.32	0.27	0
S1	November	0.49	2.46	0.55	0
	December	0.46	1.55	0.35	0.01
Overall	0.39	1.78	0.39	0	
	January	0.55	1.54	0.03	0.38
S2	February	0.57	2.11	0.45	0.04
	March	0.49	1.62	0.43	0
Overall	0.54	1.76	0.3	0.14	

recent years, the Euphrates River has witnessed a significant decrease in its flows due to the several dams that have been built on it in Turkey and Syria. Moreover, the lack of rainfall has a pronounced effect on the decline in the water levels of the Euphrates River. In Iraq, the hydrological year begins in October,[17] and therefore,the principal focus of this work is directed towards the two wet seasons. Table 1 shows the statistical description of the selected seasons. According to the given information, the intensity of rainfall is concentrated in February (0.57 mm/d), followed by January (0.55 mm/d). Furthermore, the second season (S2) has received much average rainfall from the first season (S1). Therefore, the rainy months in that region are January, February, and March. The monthly data collected from open access sources cover a long period from 1981 to 2021 (https://power.larc.nasa.gov/data-access-viewer/).

3. Methodology

3.1 ANN

The ANN approaches are famous and effective machine learning models that simulate the biological brain[18] and are frequently used to address a variety of difficult, large-scale problems. The ANN models are constructed based on three layers namely an input layer, a hidden layer, and an output

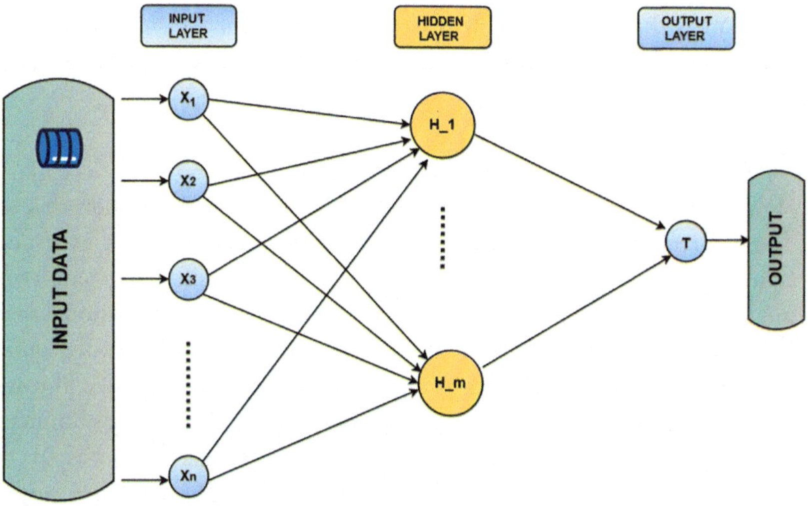

Fig. 2 Structure of ANN model.

layer. The input layer is linked to the hidden layer(s) to construct the main structure of the ANN, where the data processing stage is achieved through weighting connections. The hidden layer of ANN includes several neurons and each one is connected to all the available neurons in the output layer (see Fig. 2). The mathematical model of ANN can be expressed in the following formula:[19]

$$y(x) = F\left(\sum_{i=1}^{l} \omega_i(t).\ x_i(t) + b\right) \tag{1}$$

Where y(x) is the target, F is the transfer function, $x_i(t)$ is the input variables in the discrete time-space, ω_i is the weights and b is the bias.

3.2 MLR

Multiple linear regression (MLR) is a statistical approach that uses two or more independent variables to predict a dependent variable. in this study, this model is used to validate the efficiency of the nonlinear approach (ANN) in forecasting mean seasonal rainfall over Al-Ramadi city. However, the mathematical expression of MLR can be seen in the following formula:

$$Y_i = \alpha X1_i + \beta X2_i + \ldots \lambda Xn_i + \varepsilon \tag{2}$$

Where Yi is the target variable while $X1_i$, $X2_i$, Xn_i are the independent variables, and the ε is an unsystematic error. The coefficients(α, β, λ) can be calculated via the least square error method.[20,21]

3.3 SVR

SVR is a type of machine learning model developed by researchers that can be used for prediction and time series problems in the field of hydrology and water resources. The structure of this model consists of several components. Initially, the input data is converted to a higher-dimensional space using a kernel function. SVR supports several kernels, such as polynomial, linear, and radial basis functions. The transformed data obtained through kernel functions is essential for training the model. The optimal boundary of SVR can be determined by maximizing the margin between the boundary and the data records. The parameters of SVR are determined using a cost function that penalizes deviations from the target values.[22] The kernel function and the regularization parameter (C) are hyperparameters of the model that need to be optimized to improve performance. One of the main advantages of SVR is its ability to handle complex relationships between the inputs and their corresponding outputs by applying suitable kernel functions.

3.4 Model development and statistical parameters

This study is applied to forecast the mean seasonal rainfall over Al-Ramadi city, Iraq. Three lagged variables are selected as inputs and provided to MLR, SVR and ANN models to forecast one-season step ahead. It is important to highlight that this paper uses models to forecast rainfall for two rainy seasons. The first season, as shown in Table 1, includes October, November, and December, while the second season covers the precipitation in January, February, and March. Also, seventy percent of the data are used for training and the rest are used for testing. All the rainfall points are normalized (zero to one) to enhance the prediction capacity. Fig. 3 shows the process of model development.

To assess the efficiency of MLR and ANN in the prediction of rainfall over Al-Ramadi city, some metrics (Root mean square error ($RMSE$), mean absolute error (MAE), correlation of determination (R^2), and index of agreement (IA) have been used as below:[23,24]

$$RMSE = \sqrt{\frac{1}{N}\sum_{i=1}^{N}\left(R_{i\,obs} - R_{i\,pre}\right)^2} \tag{3}$$

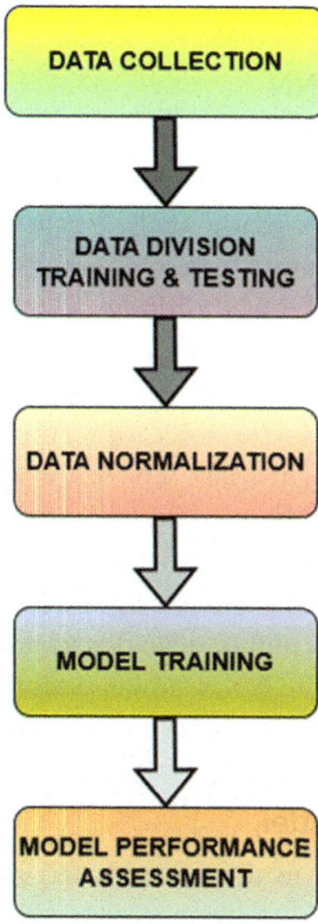

Fig. 3 Model development steps.

$$MAE = \frac{1}{N}\sum_{i=1}^{N} |(R_{i\,obs} - R_{i\,pre})| \quad (4)$$

$$IA = 1 - \frac{\sum_{i=1}^{n} (R_{obs_i} - R_{pred_i})^2}{\sum_{i=1}^{n} (|R_{pred_i} - \overline{R}_{obs}| + |R_{obs_i} - \overline{R}_{obs}|)^2} \quad (5)$$

$$R = \frac{\sum_{i=1}^{N} (R_{i\,obs} - \overline{R}_{i\,obs})(R_{i\,pre} - \overline{R}_{i\,pre})}{\sqrt{\sum_{i=1}^{N} (R_{i\,Obs} - \overline{R}_{i\,Obs})^2}\sqrt{\sum_{i=1}^{N} (R_{i\,pre} - \overline{R}_{i\,pre})^2}} \quad (6)$$

Where, N is the total number of observations, $R_{i\,obs}$, $R_{i\,pre}$ are actual and forecasted values and finally $\overline{R}_{i\,obs}$, $\overline{R}_{i\,pre}$ are the mean of actual and forecasted values.

4. Results and discussion

4.1 Results of ANN

The optimal ANN model for predicting rainfall rates in two distinct seasons, as presented here, was achieved through an iterative development process (i.e., Oct-Dec, Jan-Mar). Table 2 provides a summary of the statistical analysis results obtained from the training and testing phases. These results make it clear that the ANN model with $RMSE = 0.226$ and 0.153, $IA = 0.692$ and 0.846, $R = 0.623$ and 0.733 for training and $RMSE = 0.209$ and 0.396, $IA = 0.780$ and 0.673, $R = 0.734$ and 0.716 for testing has performed satisfactorily for forecasting the rainfall rate during both season1 (Oct-Dec) and 2 (Jan-Mar). The scatter diagram in Fig. 4 presents a visual comparison of the observed and simulated rainfall rates during the testing phase. From the figure it is revealed that the ANN model displays a tight clustering of predicted points along the ideal regression line for both seasons, with only a few points deviating significantly (Fig. 4C and F). According to the results, the ANN model performs effectively during both rainy seasons at the Ramadi city station.

4.2 Result of SVR model

SVR has been developed in this research as a supervised model for forecasting rainfall over the considered seasons. The forecasting accuracy, as shown in Table 2, indicates that the model demonstrates varied performance across the two seasons. The statistics during the training phase for

Table 2 Statistical analysis-based results for the ANN and MLR approaches during both seasons.

Season	Models	Training			Testing		
		RMSE	*IA*	*R*	*RMSE*	*IA*	*R*
Season 1	ANN	0.226	0.692	0.623	0.209	0.780	0.734
	MLR	0.248	0.561	0.438	0.286	0.467	0.543
	SVR	0.259	0.585	0.433	0.214	0.753	0.651
Season 2	ANN	0.153	0.846	0.733	0.396	0.681	0.716
	MLR	0.206	0.198	0.184	0.209	0.475	0.607
	SVR	0.200	0.334	0.264	0.226	0.410	0.439

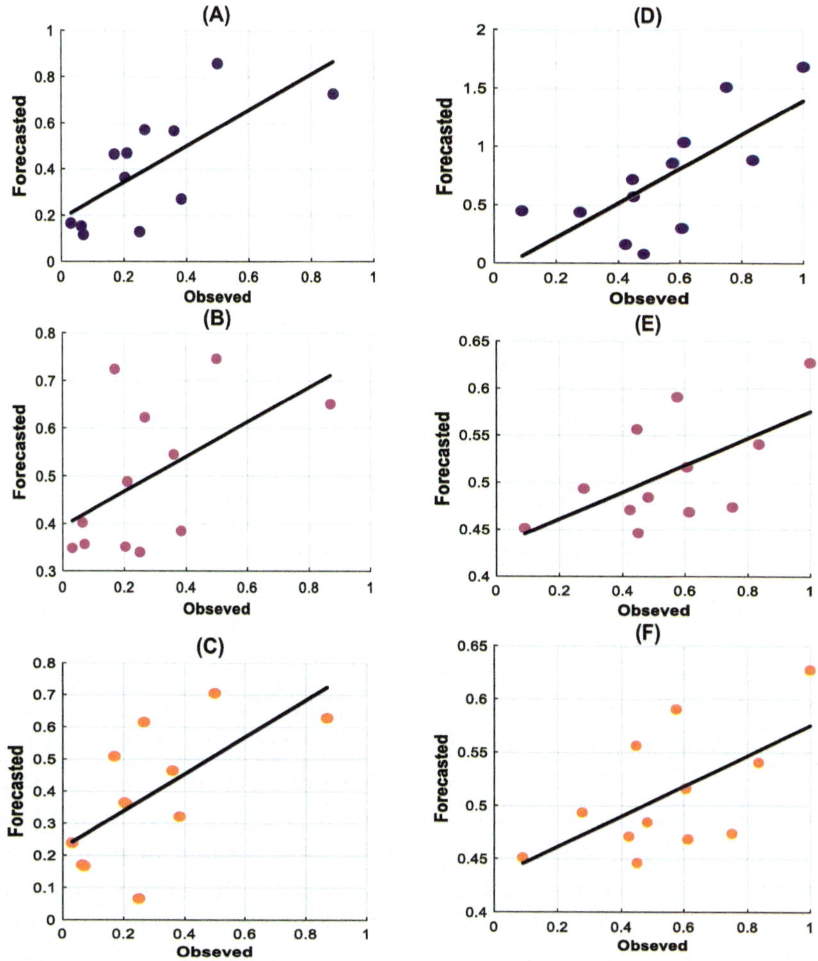

Fig. 4 Testing performance of the models across the two seasons. Panels A, B, and C show the results for SVR, MLR, and ANN in Season 1, while panels D, E, and F present the same models for Season 2.

season 1 (season 2) show *RMSE* values of 0.259 (0.200), *IA* of 0.585 (0.334), and *R* of 0.433 (0.264). In the testing phase, the variation increases significantly across both rainy seasons. For example, the *R* value for the first season is 0.651, while it declines significantly to 0.439 for the second season. In the regression plots (Fig. 3A and 4), it is evident that the model performs significantly worse during the second season, while it exhibits strong performance in the first rainy season.

4.3 Result of MLR model

The MLR model developed in this work forecasts precipitation for both rainy seasons (i.e., January-March and October-December). The model performs significantly better for the first season (October to December) compared to the second season (January-March). The scatter diagram between the observed and the predicted rate of rainfall has been shown in Fig. 4 for the testing stage. The plot demonstrates that the predicted values for the MLR model show a loose clustering along the ideal regression line (Fig. 3B and E). The results of the statistical analysis for both the training and testing phases are provided in Table 2. It is evident from these figures that the MLR model with an $RMSE = 0.248$ and 0.206, $IA = 0.561$ and 0.198, $R = 0.438$ and 0.184 for training and $RMSE = 0.286$ and 0.209, $IA = 0.467$ and 0.418, $R = 0.543$ and 0.607 for testing has presented a below par performance in predicting the rainfall rate during both season1 and 2.

4.4 Comparable models

The comparative analysis for the ANN, SVR and MLR models was carried out based on scatter plots and the Taylor diagram, as shown in Figs. 4 and 5. According to the scatter plot, the predicted rainfall rate values estimated by the ANN model for both seasons were closer to the line of perfect agreement in comparison to the MLR, and SVR models. Additionally, ANN model has a distinct advantage, as seen in Fig. 4, demonstrating consistent and stable performance in generating accurate forecasts for both rainy seasons. In contrast, the other models exhibited significant variation in performance during these two seasons. Similarly, from the Taylor plot (Fig. 5A and B) it can be ascertained that the ANN model with higher correlation and lower error outperforms the MLR model and SVR in predicting the rainfall rate during both seasons.

4.5 Discussion

Water resources in western Iraq, particularly the city of Ramadi, have been recently affected by fluctuating seasonal rainfall. These fluctuations in the density of the rainfall may be a result of climate change and its volatility, which has significantly impacted water availability and decreased water levels in dams, particularly Haditha Dam. Therefore, forecasting rainfall in this water-scarce region can enhance optimal water management. The results obtained in this study clearly indicated the superiority of the ANN model, providing highly accurate predictions for the two rainy seasons studied in this research. Furthermore, the used model (ANN) has a greater ability than

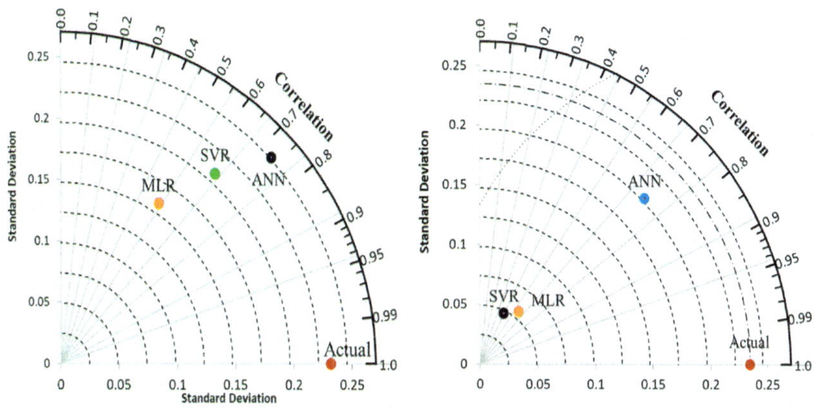

Fig. 5 Taylor diagram depicting the rainfall rate predicting the performance of ANN, SVR and MLR models. Panal a for Season 1 and b Season 2 based on the testing dataset.

other models (SVR, and MLR) to model complex nonlinear relationships, thus providing more accurate rainfall forecasts. Also, the ANN model may enhance water management, develop plans for drought preparation, and mitigating its potential impact on the agricultural sector. The changing behavior of rainfall due to climate change makes future forecasts difficult, but the proposed ANN model could adapt to these changes and generate more accurate forecasts, which reflects its great potential to respond to changing environmental conditions more effectively than other models. In a broader context, this work underscores the critical intersection of climate resilience and water security. As global warming intensifies hydrological extremes—-from prolonged droughts to erratic rainfall—regions like western Iraq face mounting pressures on freshwater resources. Reliable predictive tools such as the ANN model presented here are not merely technical advancements; they are essential instruments for building adaptive water governance, safeguarding agricultural livelihoods, and ensuring sustainable water access in an increasingly uncertain climate. Investing in such climate-informed forecasting is a vital step toward securing water futures in vulnerable arid and semi-arid regions worldwide.[25–29]

5. Conclusion

The study evaluated three data mining procedures ANN, SVR, and MLR for their performance capabilities in forecasting mean seasonal rainfall measurements during the two seasons in Ramadi city. The models were developed using forty years (1981–2021) of historical observation data. This study demonstrated that the ANN-based model surpasses other ML models in accurately forecasting rainfall values. The ANN modeling technique achieved superior forecasting precision within the first rainy season (October to December) compared to both SVR and MLR methods. ANN yielded the best predictions for that rainy season with an R value of 0.73, followed by SVR with 0.651 and MLR with 0.543. During the second rainy season (January to March), the ANN model demonstrated superior forecasting performance (R = 0.716) compared to MLR (R = 0.607) and SVR (R = 0.439). Overall,the results showed that ANN demonstrated better stability during the rainy seasons since other models (MLR and SVR) performed poorly under similar circumstances. This study demonstrates that ANN provides an accurate solution for complex problems especially involving rainfall rate forecasting.

Acknowledgment

The authors would like to thank the Al-Maarif University for supporting this research.

Declaration of generative AI and AI-assisted technologies in the writing process

The authors confirm that no generative AI was used in the creation of the content of this paper, nor in the generation of the findings and results presented herein.

References

1. Jiao X, He Z. A novel coupled rainfall prediction model based on stepwise decomposition technique. *Sci Rep* 2024;**14**(1):10853. https://doi.org/10.1038/s41598-024-61855-0.
2. Mandal T, Jothiprakash V. Short-term rainfall prediction using ANN and MT techniques. *ISH J Hydraulic Eng* 2012;**18**(1):20–26. https://doi.org/10.1080/09715010.2012.661629.
3. Chen X, Zhang L, Gippel CJ, Shan L, Chen S, Yang W. Uncertainty of flood forecasting based on radar rainfall data assimilation. *Adv Meteorol* 2016;**2016**(1):2710457. https://doi.org/10.1155/2016/2710457.
4. He R, Zhang L, Chew AWZ. Data-driven multi-step prediction and analysis of monthly rainfall using explainable deep learning. *Expert Syst Appl* 2024;**235**:121160. https://doi.org/10.1016/j.eswa.2023.121160.
5. Barrera-Animas AY, Oyedele LO, Bilal M, Akinosho TD, Delgado JMD, Akanbi LA. Rainfall prediction: a comparative analysis of modern machine learning algorithms for time-series forecasting. *Mach Learn Appl* 2022;**7**:100204. https://doi.org/10.1016/j.mlwa.2021.100204.
6. Tippett MK, Da Silva A. Model error estimation for the CPTEC Eta model. *Climanalise, Divulgacao, Cientifica* 1999;**1**:16.
7. Bustamante JFF. Evaluation of April 1999 rainfall forecasts over South America using the Eta model. *Climanálise* 1999.

8. Chou SC, Justi da Silva MGA. Objective evaluation of Eta model precipitation forecasts over South America. *Rev Climanálise* 1999;**1**(1):1–17.
9. Akiner ME. Long-term rainfall information forecast by utilizing constrained amount of observation through artificial neural network approach. *Adv Meteorol* 2021;**2021**: 5524611. https://doi.org/10.1155/2021/5524611.
10. Allawi MF, Abdulhameed UH, Adham A, Sayl KN, Sulaiman SO, Ramal MM, et al. Monthly rainfall forecasting modelling based on advanced machine learning methods: tropical region as case study. *Eng Appl Comput Fluid Mech* 2023;**17**(1):2243090. https://doi.org/10.1080/19942060.2023.2243090.
11. Samantaray S, Tripathy O, Sahoo A, Ghose DK. Rainfall forecasting through ANN and SVM in Bolangir Watershed, India. *Smart intelligent computing and applications;* Springer, 2020, pp. 767–774.
12. Gholami V, Sahour H. Simulation of rainfall-runoff process using an artificial neural network (ANN) and field plots data. *Theor Appl Climatol* 2022;**147**(1):87–98.
13. Masood A, Ahmad K. A review on emerging artificial intelligence (AI) techniques for air pollution forecasting: fundamentals, application and performance. *J Clean Prod* 2021;**322**:129072. https://doi.org/10.1016/j.jclepro.2021.129072.
14. Jamshid Mousavi S, Jafari F. PSO-SVR rainfall forecast-assisted real-time optimal operation of urban drainage systems. *EGU Gen Assembly Conf Abstr* 2023 EGU-4251).
15. Mohammadi J, Vafaeinejad A, Behzadi S, Aghamohammadi H, Hemmasi A. Integration of SVR and GPR algorithms with wavelet in modeling monthly drought forecasting. *Eng J* 2023;**16**(56):95–108.
16. Yulianto F, Mahmudy WF, Soebroto AA. Comparison of regression, support vector regression (SVR), and SVR-particle swarm optimization (PSO) for rainfall forecasting. *J Inf Technol Computer Sci* 2020;**5**(3):235–247.
17. Hameed MM, AlOmar MK, Al-Saadi AAA, AlSaadi MA. Inflow forecasting using regularized extreme learning machine: haditha reservoir chosen as case study. *Stoch Environ Res Risk Assess* 2022;**36**(12):4201–4221. https://doi.org/10.1007/s00477-022-02254-7.
18. Mamata RC, Ramlia A, Yazidb MRM, Kasab A, Razalib SFM, Bastamc MN. Slope stability prediction of road embankment using artificial neural network combined with genetic algorithm. *J Kejuruter* 2022;**34**(1):165–173.
19. Onyelowe KC, Fazal EJ, Michael EO, Ifeanyichukwu CO, George UA, Chidozie I. Artificial intelligence prediction model for swelling potential of soil and quicklime activated rice husk ash blend for sustainable construction. *J Kejuruter* 2021;**33**(4): 845–852.
20. Hamzah FB, Mohamad Hamzah F, Mohd Razali SF, El-Shafie A. Multiple imputations by chained equations for recovering missing daily streamflow observations: a case study of Langat River basin in Malaysia. *Hydrological Sci J* 2022;**67**(1):137–149. https://doi.org/10.1080/02626667.2021.2001471.
21. Hameed MM, Alomar MK, Mohd Razali SF, Kareem Khalaf MA, Baniya WJ, Sharafati A, et al. Application of artificial intelligence models for evapotranspiration prediction along the Southern Coast of Turkey. *Complexity* 2021;**2021**:1–20. https://doi.org/10.1155/2021/8850243.
22. Wani OA, Mahdi SS, Yeasin M, Kumar SS, Gagnon AS, Danish F, et al. Predicting rainfall using machine learning, deep learning, and time series models across an altitudinal gradient in the North-Western Himalayas. *Sci Rep* 2024;**14**(1):27876. https://doi.org/10.1038/s41598-024-77687-x.
23. Hameed MM, Khaleel F, AlOmar MK, Mohd Razali SF, AlSaadi MA. Optimising the selection of input variables to increase the predicting accuracy of shear strength for deep beams. *Complexity* 2022;**2022**:6532763. https://doi.org/10.1155/2022/6532763.

24. Masood A, Ahmad K. Prediction of PM2. 5 concentrations using soft computing techniques for the megacity Delhi, India. *Stoch Environ Res Risk Assess* 2022:1–14.
25. Maity R, Srivastava A, Sarkar S, Khan MI. Revolutionizing the future of hydrological science: Impact of machine learning and deep learning amidst emerging explainable AI and transfer learning. *Appl. Comput. Geosci.* 2024;**24**:100206. https://doi.org/10.1016/j.acags.2024.100206.
26. Srivastava A, Maity R, Desai VR. Assessing Global-Scale Synergy Between Adaptation, Mitigation, and Sustainable Development for Projected Climate Change. In *Ecological Footprints of Climate Change . Springer Climate;* Chatterjee U, Akanwa AO, Kumar S, Singh SK, Dutta Roy A, Eds.; Springer: Cham, 2022, https://doi.org/10.1007/978-3-031-15501-7_2
27. Masood A, Srivastava A, Hameed MM, Tanweer MS, Ahmad K, Razali SFM. Machine learning for water quality soft sensing in wastewater treatment: State of the art and future directions. In *Innovative and hybrid technologies for wastewater treatment and recycling*; Alam P, Khan AH, Américo-Pinheiro JHP, Ahmad K, Eds.; 1st ed. CRC Press, 2024.
28. Hameed MM, Masood A, Srivastava A, et al. Investigating a hybrid extreme learning machine coupled with Dingo Optimization Algorithm for modeling liquefaction triggering in sand-silt mixtures. *Sci Rep* 2024;**14**:10799. https://doi.org/10.1038/s41598-024-61059-6.
29. Masood A, Hameed MM, Srivastava A, et al. Improving $PM_{2.5}$ prediction in New Delhi using a hybrid extreme learning machine coupled with snake optimization algorithm. *Sci Rep* 2023;**13**:21057. https://doi.org/10.1038/s41598-023-47492-z.

CHAPTER SIX

Enhancing water quality assessment through machine learning: Exploring machine learning applications for accurate water quality assessment

Noureddine Daif[a], Salah Difi[b], Khaled Merabet[c], and Salim Heddam[d,*]

[a]Laboratory of Optimizing Agricultural Production in Subhumid Zones (LOPAZS), Agronomy Department, Faculty of Science, 20 August 1955 University of Skikda: Universite du 20 Août 1955 de Skikda, Route El Hadaik, BP 26, Skikda, Algeria
[b]Vegetal Chemistry-Water-Energy Laboratory (LCV2E), Department of Hydraulic, Civil Engineering and Architecture Faculty, University of Hassiba Benbouali, Chlef, Algeria
[c]Hydraulics Division, Agronomy Department, Faculty of Science, University 20 Août 1955, Route El Hadaik, BP 26, Skikda, Algeria
[d]Agronomy Department, Faculty of Science, University 20 Août 1955 Skikda, Route El Hadaik, BP 26, Skikda, Algeria
*Corresponding author. e-mail address: heddamsalim@yahoo.fr; s.heddam@univ-skikda.dz

Contents

Advances in Chemical Pollution, Environmental Management and Protection, Volume 15
ISSN 2468-9289, https://doi.org/10.1016/bs.apmp.2025.11.003

Abstract

In the present chapter, water quality assessment was conducted using advanced machine learning (ML) models. More precisely, four deep learning models were developed and compared for better prediction of dissolved oxygen concentration in river, namely, LSTM, BiLSTM, GRU and BiGRU. The four deep learning models were developed using data collected at two USGS stations, and in total five input water quality variables were selected, i.e., daily mean discharge (Q), daily mean gage height (GH), daily mean pH, daily mean specific conductance (SC), and daily mean turbidity (TU). The models were compared based on various input combination and using splitting ratio of 70 % for training and 30 % for validation. Models' performances were evaluated using various numerical indices, i.e., the RMSE, the MAE, the coefficient of correlation (R), Kling Gupta Efficiency (KGE), Willmott's Index of Agreement (WI) and Nash-Sutcliffe efficiency (NSE). According to the obtained results from the two stations, the best numerical performances were found to be ≈ 0.878, ≈ 0.760, ≈ 0.937, ≈ 0.876, ≈ 1.005, and ≈ 0.672, in terms of R, NSE, WI, KGE, RMSE and MAE, respectively, obtained using the BiGRU. Further analysis of the obtained results revealed that, the best performances was achieved by combining the five-water quality variable as input.

Keywords: Modelling, deep learning, DO, LSTM, GRU, BiGRU, BiLSTM.

1. Introduction

Dissolved oxygen (DO) in freshwater is considered as critical factor for various metabolisms process, and its accurate monitoring and control is a great importance.[1] DO is used as an indicator of water quality in various lakes, rivers and streams, and also for controlling hypoxia.[2] Predicting DO has been reported largely in the literature and it was reported that, developing accurate and concise modelling frameworks in highly nonlinear and complex in response to the large number of variables that affect the concentration of DO, among them, water quality variables and climatic conditions, are considered to be the most important factors.[3] Traditional methods for DO measuring are time consuming and not available at large scale, making continuous data availability very limited.[4] According to the literature review discussed hereafter, various machine learning models were proposed and applied for predicting DO.

Alizamir et al.[5] compared between various ML models namely, multilayer perceptron neural network (MLPNN), recurrent neural network (RNN), convolutional neural network (CNN), gated recurrent unit (GRU), long short-term memory (LSTM), seasonal autoregressive integrated moving average with exogenous variables (SARIMAX), and bayesian model averaging (BMA), for modelling DO concentration measured at two USGS stations. From the obtained results, the BMA was found to be more accurate compared to the MLPNN, GRU, CNN, RNN and SARIMAX, exhibiting an RMSE, MAE and R^2 of 0.272 mg/L, 0.216 mg/L, and 0.975, respectively. Zhang et al.[6] applied random forest regression (RFR), support Vector regression (SVR) and the MLPNN, with and without wavelet decomposition (WT). the WT-RFR was found to be the best model with RMSE, MAE and NSE values of 1.250 mg/L, 0. 950 mg/L, and 0.880, respectively. Pant et al.[7] introduced a new model for DO modelling called novel spatial-temporal attention model combined with kernel mean squared error loss function (STA-KMSE). The performances of the STA-KMSE were compared with those of the SVR, MLPNN, CNN, attention CNN (A-CNN), LSTM with previous days (LSTM-PD), A-LSTM(PD), LSTM, A-LSTM. The STA-KMSE was found to be more accurate and it surpassed all other models in terms of numerical performances and predictive accuracy. In another study, five ML models were developed and compared for predicting DO in reservoir.[8] They compared between K-Nearest Neighbors (KNN), MLPNN, LSTM, kernel ridge regression (KRR), SVR, and an ensemble-random forest regression (RFR). It was found that, the RFR was more accurate compare to all models exhibiting high R^2 (≈0.980) and NSE (≈0.980) values and very low RMSE (≈0.270) and MAE (≈0.160) values. Wang et al.[9] applied various deep learning with and without variational model decomposition (VMD) for modelling river DO concentration. They use the LSTM, GRU, A-GRU, RNN, VMD-AGRU and VMD-MAGRU. High numerical were obtained using the VMD-MAGRU with NSE value nearly equal to ≈0.980. Wu et al.[10] compared between knowledge-guided CatBoost (KGCatBoost), CatBoost and XGBoost for modelling DO concentration measured in 15 river sections of the Yangtze River Basin in Yunnan Province, China. Obtained results demonstrated that, KGCatBoost was the most accurate model having an NSE value of 0.874 and guarantying an improvement rate of 0.34 % and 3.07 % compared to the CatBoost and XGBoost, respectively. Finally, Hu et al.[11] applied the LSTM and the hybrid convolutional neural networks (CNN-LSTM) for modelling daily minimum DO concentration at the outlet of the Oyster River watershed in New Hampshire, USA. The CNN-LSTM was found to be more accurate with R^2 value of 0.865 compared to the value of 0.839 obtained using the LSTM model.

Therefore, in the present chapter, we propose an application of four deep learning models for modelling daily DO at two USGS stations (i.e., USGS 11509500 and USGS 11510700). The proposed models are respectively: (*i*) the long short-term memory (LSTM), (*ii*) the gated recurrent unit (GRU), (iii) the Bidirectional LSTM (BiLSTM), and (*iv*) the Bidirectional GRU (BiGRU).

2. Materials and methods

2.1 Study area and data

In the present chapter, data collected from two stations operated by the USGS: United States Geological Survey. The two stations are (Fig. 1): USGS 11509500 Klamath River at Keno, Oregon State, Klamath County, USA, (Latitude 42°08′00″, Longitude 121°57′40″), and USGS 11510700 Klamath River Blw John C. Boyle Pwrplnt, NR Keno, Oregon State, Klamath County, USA, (Latitude 42°05′05″, Longitude 122°04′20″). For the USGS 11509500, data covers the period ranging from 01 November 2018–22 December 2024, with a total of 2211 patterns. For USGS 11510700, data covers the period ranging from 08 November 2018–13 December 2024, with a total of 2211 patterns. Data was divided into training (70 %) and validation (30 %). For each station, data corresponds to six water quality variables measured at daily time scale namely: (*i*) daily mean discharge (Q: cu. ft/s), (*ii*) daily mean gage height (GH: feet), (*iii*) daily mean water pH, (*iv*) daily mean water turbidity (TU: FNU), (*v*) daily

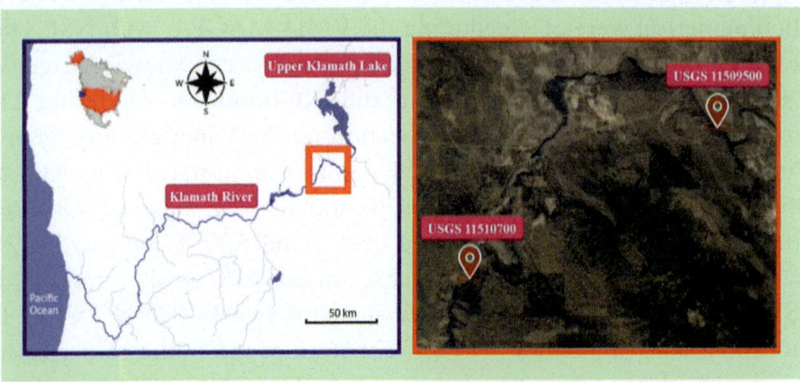

Fig. 1 Map showing the location of the two weather stations. Map lines delineate study areas and do not necessarily depict accepted national boundaries.

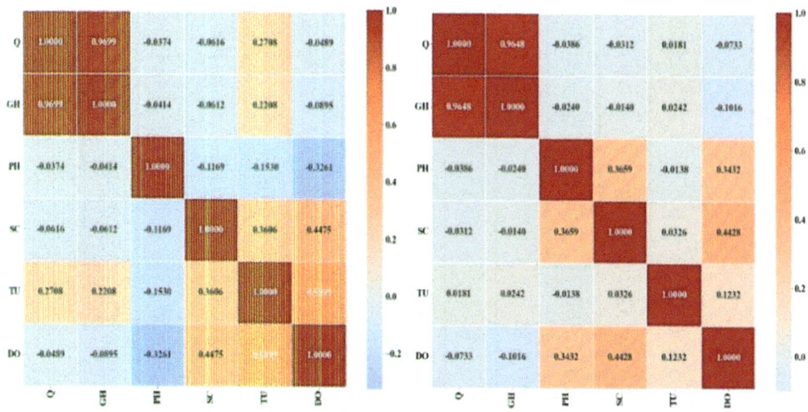

Fig. 2 The correlation heatmaps between water quality variables for all dataset: USGS 11509500 left panel and USGS 11510700 right panel.

mean specific conductance (SC: uS/cm), and daily mean dissolved oxygen concentration (DO: mg/L). The *Q*, *GH*, pH, TU and SC are used as the input variables, while DO is the predicted variable. The correlation heatmaps between water quality variables for all dataset is depicted in Fig. 2. The statistical description of all variables is reported in Table 1 for the two stations. In the present chapter, we have selected four input combination having different inputs variables (Table 2).

2.2 Accuracy performance indices

2.2.1 Root mean square error (RMSE)

The RMSE measures the difference between predicted and observed values. It is the square root of the mean square error, with **0** indicating a perfect fit, though rarely achieved. Lower RMSE signifies higher accuracy. RMSE is useful for comparing models on the same dataset but not across different datasets due to its dependence on value scale. It is sensitive to outliers, disproportionately weighting large errors.[12] The RMSE of the forecast values $\widehat{\boldsymbol{Pr}}$ for instants $\boldsymbol{t}$ of the dependent variable $\boldsymbol{M}$ of a regression, with variables observed $\boldsymbol{T}$ times, for $\boldsymbol{T}$ different predictions is calculated as the square root of the mean of the squared errors:

$$RMSE = \sqrt{\sum_{t=1}^{T} \frac{(\widehat{\mathrm{Pr}}_t - M_t)^2}{T}} \quad (1)$$

Table 1 Summary statistics of water quality variables.

Variables	Subset	Unit	X_{mean}	X_{max}	X_{min}	S_x	C_v	R
USGS 11509500 Klamath River at Keno, Oregon State, Klamath County, (USA)								
Q	Training	*Cu ft/s*	816.311	6280.000	256.000	546.639	0.670	-0.062
	Validation	*Cu ft/s*	797.870	7690.000	260.000	515.232	0.646	-0.019
	All data	*Cu ft/s*	810.781	7690.000	256.000	537.365	0.663	-0.049
GH	Training	*ft*	4.773	10.460	3.410	0.804	0.168	-0.105
	Validation	*ft*	4.750	11.370	3.430	0.740	0.156	-0.053
	All data	*ft*	4.766	11.370	3.410	0.785	0.165	-0.090
pH	Training	/	7.989	9.300	7.300	0.414	0.052	-0.302
	Validation	/	8.004	9.300	7.300	0.440	0.055	-0.378
	All data	/	7.993	9.300	7.300	0.422	0.053	-0.326
SC	Training	*uS/cm*	148.734	367.000	112.000	36.120	0.243	0.445
	Validation	*uS/cm*	148.097	374.000	113.000	38.423	0.259	0.453
	All data	*uS/cm*	148.543	374.000	112.000	36.818	0.248	0.447

TU	Training	*FNU*	11.046	47.700	1.700	5.679	0.514	0.588
	Validation	*FNU*	10.852	69.900	1.400	6.189	0.570	0.594
	All data	*FNU*	10.988	69.900	1.400	5.836	0.531	0.590
DO	Training	*mg/L*	9.546	13.200	4.900	2.022	0.212	1.000
	Validation	*mg/L*	9.359	13.000	5.100	2.054	0.219	1.000
	All data	*mg/L*	9.490	13.200	4.900	2.033	0.214	1.000
USGS 11510700 Klamath River Blw John C. Boyle Pwrplnt, NR Keno, Oregon State, Klamath County, (USA)								
Q	Training	*Cu ft/s*	1042.377	5990.000	345.000	544.252	0.522	-0.100
	Validation	*Cu ft/s*	1024.196	8210.000	359.000	564.020	0.551	-0.014
	All data	*Cu ft/s*	1036.925	8210.000	345.000	550.189	0.531	-0.073
GH	Training	*ft*	3.893	7.450	2.870	0.551	0.141	-0.128
	Validation	*ft*	3.875	8.360	2.900	0.527	0.136	-0.040
	All data	*ft*	3.887	8.360	2.870	0.544	0.140	-0.102
pH	Training	/	7.934	8.800	7.200	0.168	0.021	0.345
	Validation	/	7.933	8.800	7.300	0.172	0.022	0.340
	All data	/	7.934	8.800	7.200	0.169	0.021	0.343

(continued)

Table 1 Summary statistics of water quality variables. (*cont'd*)

Variables	Subset	Unit	X_{mean}	X_{max}	X_{min}	S_x	C_v	R
SC	Training	*uS/cm*	144.702	311.000	117.000	24.180	0.167	0.443
	Validation	*uS/cm*	143.608	333.000	114.000	24.077	0.168	0.441
	All data	*uS/cm*	144.374	333.000	114.000	24.149	0.167	0.443
TU	Training	*FNU*	10.044	655.000	0.900	27.726	2.760	0.126
	Validation	*FNU*	10.120	462.000	1.000	24.806	2.451	0.116
	All data	*FNU*	10.067	655.000	0.900	26.878	2.670	0.123
DO	Training	*mg/L*	9.985	12.600	7.400	1.304	0.131	1.000
	Validation	*mg/L*	9.900	12.600	7.700	1.312	0.133	1.000
	All data	*mg/L*	9.959	12.600	7.400	1.307	0.131	1.000

Table 2 The input combinations of different models.

	Input variables combination	Output
Scenario1	Q, GH, pH, SC, TU	DO
Scenario2	GH, pH, SC, TU	DO
Scenario3	SC, TU, pH	DO
Scenario4	SC, TU	DO

2.2.2 The mean absolute error (MAE)

The MAE is defined as the mean value of the absolute difference between predicted and measured values. A comparatively low MAE indicates good model performance.[13] The measured and predicted values are in perfect agreement when MAE = 0. The MAE is calculated using Eq. (2):

$$MAE = \frac{\sum_{t=1}^{T} |\widehat{\mathrm{Pr}}_t - M_t|}{T} \tag{2}$$

2.2.3 Pearson's correlation coefficient (R)

Pearson's correlation coefficient (R) is frequently the most widely used tool for assessing the performance of forecasting models in time series modeling.[14] The R coefficient is determined by calculating any linear regression between measured or observed values and calculated values. The term is given by Eq. (3). The correlation coefficient is always defined between −1 and 1.

$$R = \left[\frac{\sum (M_t - \bar{M}_t)(\widehat{\mathrm{Pr}}_t - \bar{\widehat{\mathrm{Pr}}}_t)}{\sqrt{\sum (M_t - \bar{M}_t)^2 - \sum (\widehat{\mathrm{Pr}}_t - \bar{\widehat{\mathrm{Pr}}}_t)^2}} \right] \tag{3}$$

Where $\bar{M}_t$ the mean measured values, $\bar{\widehat{Pr}}_t$ the mean of predicted values.

2.2.4 The Nash-Sutcliffe (NSE)

The Nash-Sutcliffe (NSE) performance criteria, proposed by Nash and Sutcliffe,[15] is a widely used criterion for optimal hydrological model calibration and performance evaluation.[16] This measure is based on the sum of squared deviations between observed (measured) and simulated (predicted) values. It has been specifically developed to facilitate meaningful comparisons between events of varying magnitude, ensuring that model performance is assessed consistently, whatever the scale of the observed data.

$$NSE = 1 - \frac{\sum_{t=1}^{T}[M_t - \widehat{\Pr}_t]^2}{\sum_{t=1}^{T}[M_t - \bar{M}_t\ ]^2} \tag{4}$$

2.2.5 The Kling-Gupta efficiency (KGE)

The KGE is an acclaimed measure of adequacy in the hydrological sciences, commonly used to assess the agreement between simulated and observed data,[17] developed by Harald Kling and Hoshin Vijai Gupta.[18] The KGE can be calculated as follow.

$$KGE = 1 - \sqrt{(R-1)^2 + (\alpha - 1)^2 + (\beta - 1)^2} \tag{5}$$

Here: $\boldsymbol{R}$ is the Pearson correlation coefficient, $\boldsymbol{\alpha} = \frac{\sigma_P}{\sigma_M}$, representing the variability ratio, $\boldsymbol{\sigma_P}$ is the standard deviation of the simulated time series, and $\boldsymbol{\sigma_M}$ is the standard deviation of the observed time series. $\boldsymbol{\beta} = \frac{\bar{\widehat{Pr}}_t}{\bar{M}_t}$, representing the bias ratio, where $\bar{\widehat{\boldsymbol{Pr}}}_t$ is the mean of the simulated time series, and $\bar{\boldsymbol{M}}_t$ is the mean of the observed time series.

2.2.6 Willmott index of agreement (WI)

The index of agreement (WI), introduced by Willmott in Willmott,[19] is a standardized measure for evaluating the performance of prediction models in hydrology. This index, denoted "WI", varies between 0 and 1, where 1 indicates a perfect match between forecasts and observations, and 0 a total absence of agreement. It quantifies forecast error by relating the root-mean-square error (RMSE) to the potential error, enabling an overall assessment of performance. The index takes into account both systematic and random deviations between measured and predicted data, providing a comprehensive view of model accuracy and reliability.[20] A major strength of this index is its ability to identify additive and proportional discrepancies between the means and variances of predicted and measured data. This helps to detect systematic biases and variabilities.

$$WI = 1 - \frac{\sum_{t=1}^{T}[\widehat{\Pr}_t - M_t]^2}{\sum_{t=1}^{T}(|\widehat{\Pr}_t - \bar{M}_t| + |M - \bar{M}_t|)^2} \tag{6}$$

3. Machines learning models

3.1 Long short-term memory (LSTM)

The long short-term memory (LSTM) model (Fig. 3A) is a specialized architecture of recurrent neural networks (RNNs) designed to address the challenges of traditional RNNs, particularly the issues of vanishing and exploding gradients. At its core, the LSTM employs a memory cell capable of maintaining information across extended time sequences. The model relies on three key gates to regulate the information flow: the forget gate determines which data should be discarded, the input gate decides which new information to integrate, and the output gate selects which information to utilize for generating the current output. These components collectively allow LSTM models to effectively capture long-term dependencies and model intricate time-series relationships in data.[21,22]

3.2 Bidirectional long short-term memory (BiLSTM)

Bidirectional long short-term memory (BiLSTM) is an enhanced LSTM model, designed to enrich time series analysis with a bidirectional approach (Fig. 3B). Unlike conventional LSTM, BiLSTM comprises two distinct networks: one processes data in chronological order, while the other analyzes it in reverse. This dual orientation simultaneously captures temporal dependencies from the past and the future, offering a more comprehensive interpretation of context. This architecture is particularly

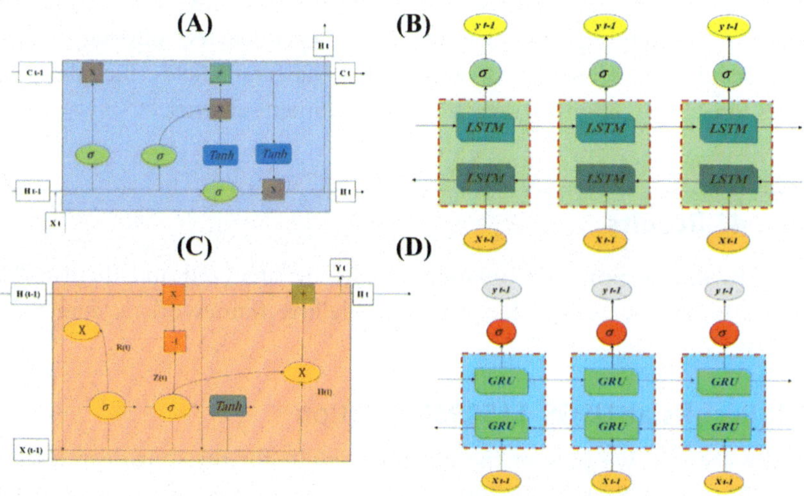

Fig. 3 Deep learning models architecture: (A) LSTM, (B) BiLSTM, (C) GRU, (D) BiGRU.

effective in applications such as speech recognition, where each word depends on its environment, or machine translation, which requires a global understanding of the sentence to faithfully render its meaning.[23,24]

3.3 Gated recurrent unit (GRU)

The gated recurrent unit (GRU) is a streamlined variant of the LSTM, designed to model temporal dependencies in time series efficiently (Fig. 3C). It uses two main gates: the updating gate, which determines what new information should be added or replaced, and the resetting gate, which decides to what extent past information should be forgotten. This simpler structure, with no separate memory cell, means that the GRU is less complex and faster to train, while maintaining performance comparable to that of the LSTM. This efficiency makes it particularly suitable for applications where computing resources are limited, such as mobile devices or embedded systems, while offering high accuracy in time-series forecasting tasks.[23,25,26]

3.4 Bidirectional gated recurrent unit (BiGRU)

The BiGRU (Bidirectional Gated Recurrent Unit) is an improved version of the GRU that processes sequential data in two directions: from past to future and from future to past (Fig. 3D). This architecture is based on two separate GRUs, combining their outputs to produce an enriched contextual representation. The BiGRU uses two main gates: the update gate, which controls what new information to integrate, and the reset gate, which decides what past information to forget. Thanks to its bidirectionality, BiGRU is particularly effective at capturing complex temporal dependencies, offering a better understanding of sequences. It is used in a variety of fields, including time series forecasting, anomaly detection and sentiment analysis.[27,28]

4. Results

In this section, the obtained results are analyzed and discussed for each station separately. The overall modelling framework is depicted in Fig. 4.

4.1 Results at USGS 11509500 station

In this section, we present the obtained results for the four deep learning models at the USGS 11509500 station. The four models, i.e., the LSTM, GRU, BiLSTM and BiGRU were calibrated according to four input

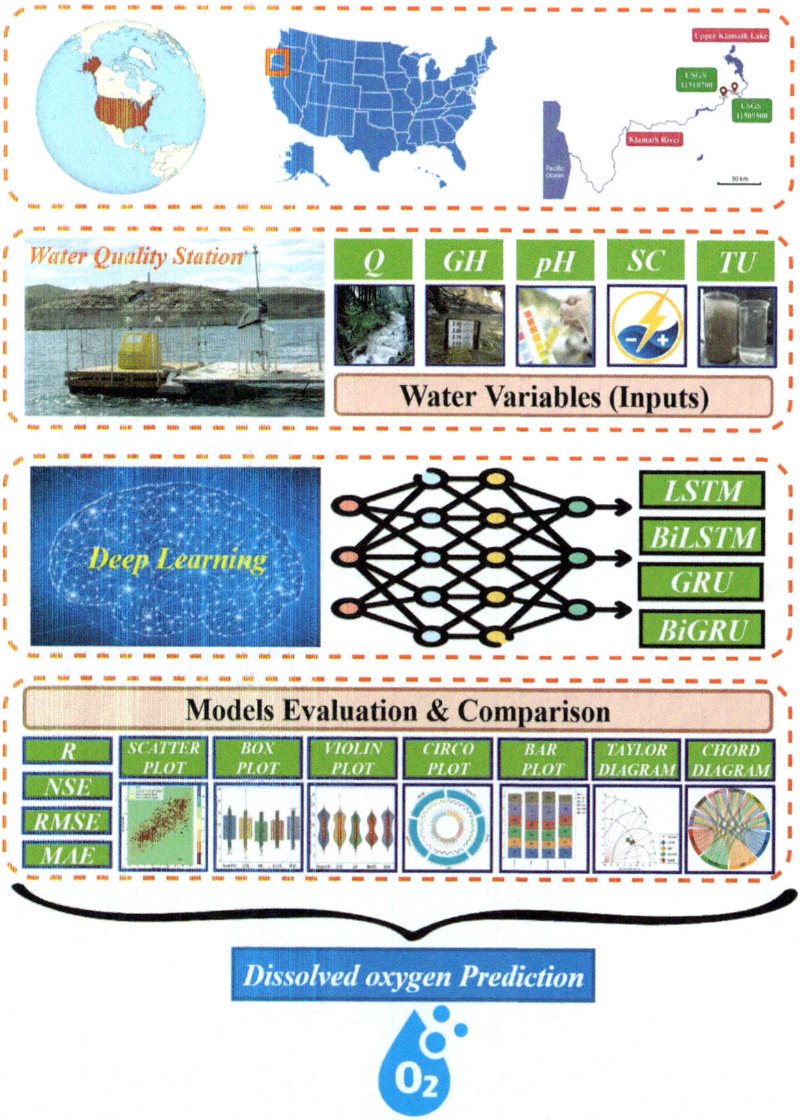

Fig. 4 Flowchart of the overall modelling farmwork.

combination using different input variables. Results are reported in Table 3 and discussed hereafter. We compare the performances of the model's based on various numerical performances and only the results obtained during the validation stage are presented and analyzed. Fig. 5 presents the scatterplot of measured and predicted DO concentration obtained using the

Table 3 Results of modelling DO at USGS 11509500.

	Training						Validation					
	R	NSE	WI	KGE	RMSE	MAE	R	NSE	WI	KGE	RMSE	MAE
LSTM1	0.860	0.726	0.918	0.821	1.058	0.805	0.863	0.734	0.920	0.826	1.059	0.815
LSTM2	**0.863**	**0.738**	**0.925**	**0.843**	**1.034**	**0.753**	**0.870**	**0.750**	**0.929**	**0.853**	**1.025**	**0.751**
LSTM3	0.836	0.698	0.902	0.752	1.111	0.839	0.856	0.731	0.913	0.763	1.064	0.801
LSTM4	0.794	0.598	0.870	0.736	1.282	0.992	0.813	0.640	0.883	0.747	1.232	0.945
GRU1	**0.898**	**0.798**	**0.940**	**0.849**	**0.909**	**0.645**	**0.814**	**0.594**	**0.882**	**0.794**	**1.309**	**0.967**
GRU2	0.863	0.721	0.917	0.826	1.068	0.769	0.728	0.399	0.829	0.716	1.591	1.137
GRU3	0.852	0.719	0.913	0.799	1.071	0.817	0.760	0.502	0.844	0.717	1.448	1.071
GRU4	0.809	0.649	0.887	0.747	1.198	0.894	0.756	0.483	0.836	0.705	1.475	1.113
BiLSTM1	**0.814**	**0.662**	**0.893**	**0.754**	**1.175**	**0.914**	**0.798**	**0.589**	**0.874**	**0.770**	**1.317**	**1.029**
BiLSTM2	0.806	0.646	0.886	0.744	1.203	0.930	0.773	0.544	0.854	0.729	1.386	1.068
BiLSTM3	0.767	0.589	0.856	0.674	1.296	1.042	0.750	0.513	0.840	0.702	1.432	1.105
BiLSTM4	0.725	0.505	0.785	0.505	1.422	1.204	0.729	0.486	0.766	0.480	1.471	1.193
BiGRU1	**0.904**	**0.808**	**0.950**	**0.904**	**0.887**	**0.617**	**0.878**	**0.760**	**0.937**	**0.876**	**1.005**	**0.672**
BiGRU2	0.866	0.742	0.930	0.861	1.027	0.715	0.862	0.735	0.927	0.852	1.057	0.736
BiGRU3	0.868	0.751	0.929	0.843	1.009	0.741	0.857	0.731	0.921	0.820	1.065	0.780
BiGRU4	0.811	0.632	0.898	0.808	1.226	0.907	0.826	0.668	0.907	0.819	1.183	0.864

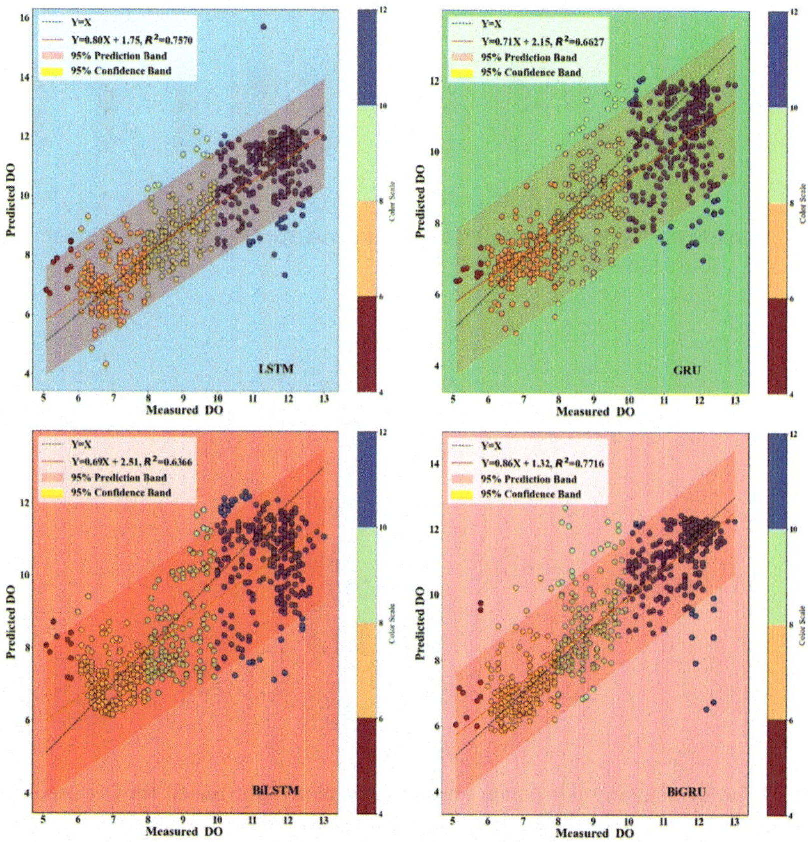

Fig. 5 Scatterplot of measured versus predicted DO using the best models: USGS 11509500 & validation dataset.

best models for the validation stage. Form Fig. 5 we can conclude that, the BiLSTM was the poorest one compared to the other having a large scattered data, while the BiGRU was the best models. The inferiority of the BiLSTM and the superiority of the BiGRU was further confirmed by the boxplot and violinplot depicted in Fig. 6. In Fig. 7, the Taylor diagram and the ImShow were plotted providing a clear insight about the model's performances. Comparing the numerical performances in Table 3 we can conclude that, generally, there is a larger error calculated between measured and predicted DO for the fourth input combination. In fact, the worst results appear obtained using the GRU2 having an RMSE and MAE values of ≈1.591 and ≈1.137, respectively, which systematically correspond

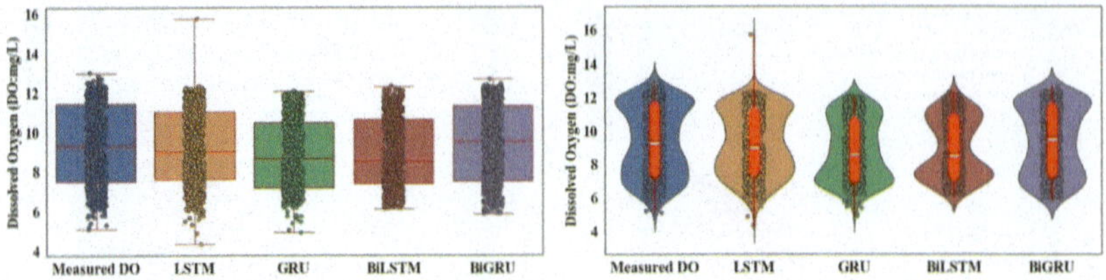

Fig. 6 Boxplot (left panel) and Violinplot (right panel) for DO prediction using the best models: USGS 11509500 & validation dataset.

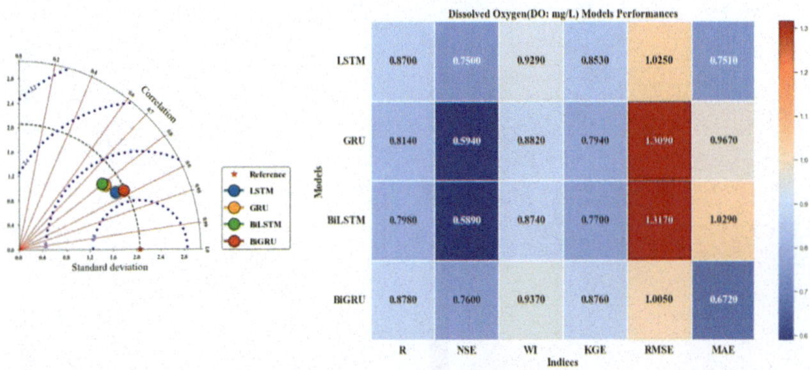

Fig. 7 Taylor diagram (left panel) and ImShow plot (right panel) for DO prediction using the best models: USGS 11509500 & validation dataset.

to low R (≈0.728), NSE (≈0.399), WI (≈0.829) and KGE (≈0.716), respectively. Furthermore, the best accuracy was obtained using the BiGRU with high R (≈0.878), NSE (≈0.760), WI (≈0.937) and KGE (≈0.876) and the lower RMSE (≈1.005) and MAE (≈0.672) values, respectively. The BiGRU was followed by the LSTM2 with high R (≈0.870), NSE (≈0.750), WI (≈0.929) and KGE (≈0.853), RMSE (≈1.025) and MAE (≈0.751) values, respectively.

Comparison between the models having various inputs combinations, we can conclude that, the best performances were obtained using the five inputs variables, i.e., Q, GH, pH, SC and TU. However, moving from one combination to another show slightly decreasing in the models' performances. This can be interpretated by the fact that, the prediction of DO is mainly governed by two variables: SC, TU, while the three other variables have slightly contribution and negligible effect on DO. Furthermore, the GRU models were the poorest one showing very low NSE values ranging from ≈0.399 to ≈0.594, respectively.

4.2 Results at USGS 11510700 station

Results obtained at USGS 11510700 station are depicted in Table 4. Scatterplot of calculated versus predicted DO are plotted in Fig. 8. According to Fig. 8, the BiGRU was the most accurate model showing less scattered data and the high R^2 value of 0.7904. the superiority of the BiGRU model was further confirmed by the boxplot and violinplot depicted in Fig. 9, in which we can see that, the BIGRU was the most similar model compared to the measured data, while the BiLSTM was the poorest one. Finally, th superiority of the BiGRU model was further confirmed by Taylor diagram depicted in Fig. 10. From the numerical results reported in Table 4, some important conclusions can be drawn. First, the BiGRU can achieve accurate daily DO prediction results for the four-input combination, from BiGRU1 to BiGRU4, exhibiting the high numerical indices with the high R, NSE, WI and KGE values, ranging from ≈0.814 to ≈0.889, ≈0.635 to ≈0.767, ≈0.895 to ≈0.942, and ≈0.800 to ≈0.881, and the lowest RMSE ad MAE values, ranging from ≈0.634 to ≈0.793, and from ≈0.462 to ≈0.583, respectively, which clearly indicates that the developed BiGRU model have strong generalization ability compared to the other models. Second, the best prediction accuracy was obtained by BiGRU2 for which the Q was removed from the input variables, which indicates that it can be more beneficial to use less input variables for improving the models' performances and to enhance the accuracy of DO prediction. Third, the BiLSTM models have lower prediction accuracy compared to the all other from the first to the fourth input combination. We can see that the R, NSE, WI and KGE values obtained using the BiLSTM, were ranged from ≈0.576 to ≈0.741, ≈0.259 to ≈0.441, ≈0.604 to ≈0.815, and ≈0.286 to ≈0.599, while the RMSE and MAE they are ranged from ≈0.980 to ≈1.128 and from ≈0.759 to ≈0.980, respectively. Fourth, the prediction accuracy of the LSTM models is close to that of the BiLSTM models with very negligible difference. Fifth, we can see that, among all models developed in the present study, the BiGRU2 improves the performances of the LSTM, GRU, and BiLSTM by ≈32.409 % and ≈32.555 %, ≈29.556 % and ≈34.094 %, ≈35.306 % and ≈39.130 %, in terms of RMSE and MAE reduction, respectively.

Table 4 Results of modelling DO at USGS 11510700.

	Training						Validation					
	R	NSE	WI	KGE	RMSE	MAE	R	NSE	WI	KGE	RMSE	MAE
LSTM1	**0.701**	**0.443**	**0.826**	**0.672**	**0.973**	**0.714**	**0.723**	**0.488**	**0.840**	**0.692**	**0.938**	**0.685**
LSTM2	0.669	0.348	0.799	0.640	1.053	0.779	0.696	0.407	0.816	0.662	1.010	0.736
LSTM3	0.627	0.345	0.768	0.545	1.055	0.815	0.661	0.408	0.787	0.562	1.009	0.774
LSTM4	0.597	0.351	0.716	0.429	1.050	0.881	0.609	0.370	0.723	0.434	1.041	0.862
GRU1	0.669	0.430	0.774	0.522	0.984	0.801	0.653	0.406	0.762	0.502	1.010	0.831
GRU2	0.752	0.493	0.855	0.739	0.928	0.712	0.705	0.164	0.795	0.696	1.199	0.901
GRU3	**0.796**	**0.600**	**0.866**	**0.685**	**0.824**	**0.653**	**0.812**	**0.529**	**0.860**	**0.717**	**0.900**	**0.701**
GRU4	0.778	0.600	0.860	0.659	0.825	0.656	0.806	0.605	0.867	0.670	0.824	0.647
BiLSTM1	**0.749**	**0.559**	**0.850**	**0.660**	**0.865**	**0.665**	**0.741**	**0.441**	**0.815**	**0.599**	**0.980**	**0.759**
BiLSTM2	0.749	0.560	0.849	0.656	0.864	0.666	0.743	0.457	0.807	0.565	0.967	0.760
BiLSTM3	0.657	0.399	0.716	0.406	1.010	0.862	0.684	0.331	0.689	0.370	1.073	0.925
BiLSTM4	0.560	0.285	0.621	0.310	1.102	0.955	0.576	0.259	0.604	0.286	1.128	0.980
BiGRU1	0.895	0.795	0.945	0.885	0.590	0.419	0.871	0.746	0.933	0.866	0.661	0.454
BiGRU2	**0.897**	**0.783**	**0.946**	**0.889**	**0.607**	**0.449**	**0.889**	**0.767**	**0.942**	**0.881**	**0.634**	**0.462**
BiGRU3	0.859	0.736	0.922	0.820	0.669	0.505	0.855	0.728	0.921	0.824	0.684	0.509
BiGRU4	0.806	0.618	0.891	0.793	0.805	0.591	0.814	0.635	0.895	0.800	0.793	0.583

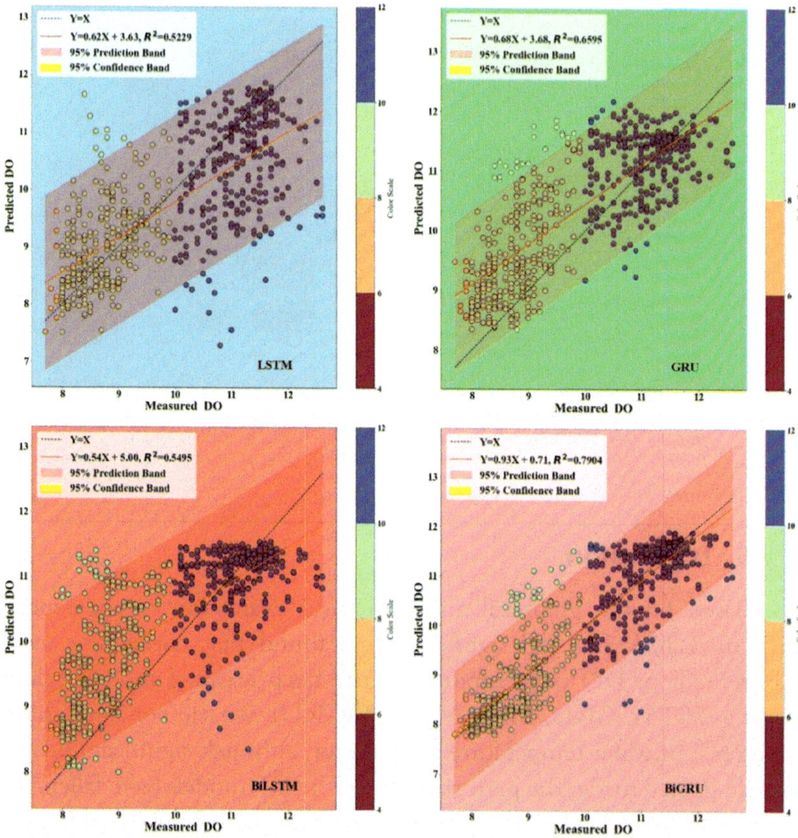

Fig. 8 Scatterplot of measured versus predicted DO using the best models: USGS 11510700 & validation dataset.

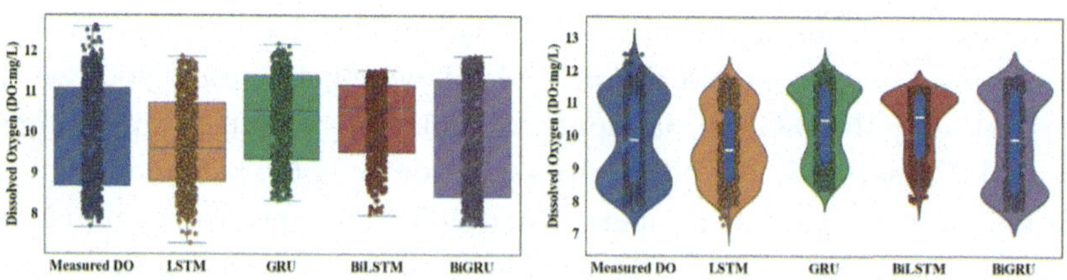

Fig. 9 Boxplot (left panel) and Violinplot (right panel) for DO prediction using the best models: USGS 11510700 & validation dataset.

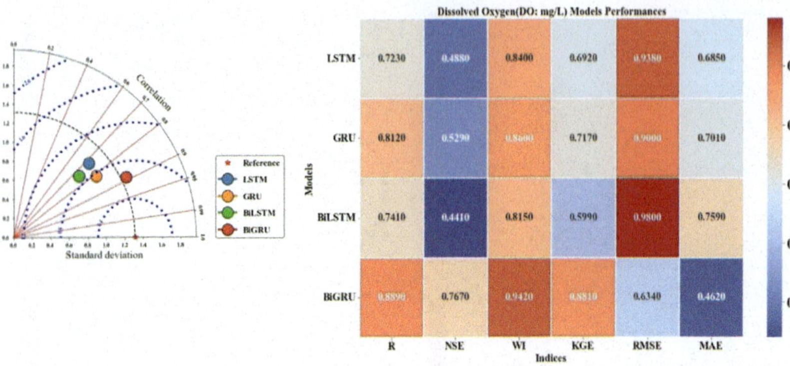

Fig. 10 Taylor diagram (left panel) and ImShow plot (right panel) for DO prediction using the best models: USGS 11510700 & validation dataset.

5. Discussion

Based on the graphical and numerical evaluation results reported above in Tables 3–4 and figures, it was found that comparing in situ measurement of DO concentration to the deep learning models calculations (i.e. LSTM, GRU, BiLSTM, and BiGRU), we can conclude that, in the absence of water temperature as predictor, although significant progress has been gained made, the performances of the DL models have fallen short of expectations. On the other hand, considering all input combinations, models based on combination 1 (Q, GH, pH, SC, TU) and combination 2 (GH, pH, SC, TU) gave the best performances in terms of correlations and errors metrics, while only one exception for GRU3 (pH, SC, TU) at the USGS 11510700, for which the best performances were obtained with R, NSE, RMSE and MAE of approximately 0.812, 0.529, 0.900 and 0.701, respectively. While none of the proposed models have guaranteed high correlations coefficient and all calculated R were below the 0.900, we can see that using the BiGRU, all inputs combination guaranteed an R ranging from 0.826 to 0.878, and NSE values superior to 0.660. as demonstrated in various previous published papers, water temperature plays a key role in the accuracy of DO prediction, we need to further analyze its impact on DO prediction. More specifically, the inclusion of various other water quality variables can probably contribute to the improvement of the prediction accuracy. Furthermore, under the present and projected impacts of climate change, rising water temperatures, altered hydrological regimes, and increased frequency of extreme weather events are likely to further complicate dissolved oxygen dynamics, emphasizing the urgent need to develop adaptive and climate-informed modeling frameworks that can robustly capture these evolving environmental interactions [29–31].

6. Conclusion

In the present chapter, four deep learning models (DL) were calibrated and applied for modelling dissolved oxygen concentration measured at two USGS stations. The DL models were respectively, the LSTM, GRU, BiLSTM and BiGRU. In detail, the performances of the models were evaluated using various numerical performances and by testing various input combinations. From the obtained results, it was found that, the BiGRU model could provide reliable and accurate DO prediction compared to the other models. The results indicate that the DO at different sites can be predicted by the developed DL models. Furthermore, it was found that, the experimental results indicate that the water TU and SC were the most significant variables compared to the Q, GH and pH. Overall, the obtained results are promising and it will be very interesting if the models were compared by the inclusion of other water quality variables that can help in better improving the models' performances.

References

1. Araujo AN, San Andres CF, Nguyen KQ, Corby TL, Rhodes MA, García J, et al. Effects of minimum dissolved oxygen setpoints for aeration in semi-intensive pond production of Pacific White shrimp (Litopenaeus vannamei). *Aquaculture* 2025;**594**:741376. https://doi.org/10.1016/j.aquaculture.2024.741376.
2. Simonetti I, Lubello C, Cappietti L. On the use of hydrodynamic modelling and random forest classifiers for the prediction of hypoxia in coastal lagoons. *Sci Total Environ* 2024;**951**:175424. https://doi.org/10.1016/j.scitotenv.2024.175424.
3. Shi P, Kuang L, Yuan L, Wang Q, Li G, Yuan Y, et al. Dissolved oxygen prediction using regularized extreme learning machine with clustering mechanism in a black bass aquaculture pond. *Aquacultural Eng* 2024;**105**:102408. https://doi.org/10.1016/j.aquaeng.2024.102408.
4. Wong WK, Fronitasari D, Juwono FH, Kong JT. An evolutionary approach to dissolved oxygen mathematical modeling: a case study of the Klamath River. *Aquacultural Eng* 2024;**106**:102428. https://doi.org/10.1016/j.aquaeng.2024.102428.
5. Alizamir M, Moradveisi K, Ahmed KO, Bahrami J, Kim S, Heddam S. An efficient data fusion model based on Bayesian model averaging for robust water quality prediction using deep learning strategies. *Expert Syst Appl* 2025;**261**:125499. https://doi.org/10.1016/j.eswa.2024.125499.

6. Zhang F, Shi X, Zhao S, Hao R, Sun B, Li G, et al. Simulation and explanatory analysis of dissolved oxygen dynamics in Lake Ulansuhai, China. *J Hydrology: Regional Stud* 2025;**57**:102109. https://doi.org/10.1016/j.ejrh.2024.102109.
7. Pant N, Toshniwal D, Gurjar BR. Short-term forecasting of dissolved oxygen based on spatial-temporal attention mechanism and kernel-based loss function. *J Water Process Eng* 2025;**69**:106677. https://doi.org/10.1016/j.jwpe.2024.106677.
8. Zhang P, Liu X, Dai H, Shi C, Xie R, Song G, et al. A multi-model ensemble approach for reservoir dissolved oxygen forecasting based on feature screening and machine learning. *Ecol Indic* 2024;**166**:112413. https://doi.org/10.1016/j.ecolind.2024.112413.
9. Wang Z, Wang Q, Liu Z, Wu T. A deep learning interpretable model for river dissolved oxygen multi-step and interval prediction based on multi-source data fusion. *J Hydrol* 2024;**629**:130637. https://doi.org/10.1016/j.jhydrol.2024.130637.
10. Wu J, Wang Z, Dong J, Yao Z, Chen X, Fan H. Multi-step ahead dissolved oxygen concentration prediction based on knowledge guided ensemble learning and explainable artificial intelligence. *J Hydrol* 2024;**636**:131297. https://doi.org/10.1016/j.jhydrol.2024.131297.
11. Hu Y, Liu C, Wollheim WM. Prediction of riverine daily minimum dissolved oxygen concentrations using hybrid deep learning and routine hydrometeorological data. *Sci Total Environ* 2024;**918**:170383. https://doi.org/10.1016/j.scitotenv.2024.170383.
12. Willmott CJ, Matsuura K. Advantages of the mean absolute error (MAE) over the root mean square error (RMSE) in assessing average model performance. *Clim Res* 2005;**30**(1):79–82. https://doi.org/10.3354/cr030079.
13. Chai T, Draxler RR. Root mean square error (RMSE) or mean absolute error (MAE). *Geosci. Model Dev Discuss* 2014;**7**(1):1525–1534. https://doi.org/10.5194/gmdd-7-1525-2014.
14. Difi S, Elmeddahi Y, Hebal A, Singh VP, Heddam S, Kim S, et al. Monthly streamflow prediction using hybrid extreme learning machine optimized by bat algorithm: a case study of Cheliff watershed, Algeria. *Hydrol Sci J* 2022;**68**(2):189–208. https://doi.org/10.1080/02626667.2022.2149334.
15. Nash JE, Sutcliffe JV. River flow forecasting through conceptual models part I-A discussion of principles. *J Hydrol* 1970;**10**(3):282–290. https://doi.org/10.1016/0022-1694(70)90255-6.
16. Difi S, Heddam S, Zerouali B, Kim S, Elmeddahi Y, Bailek N, et al. Improved daily streamflow forecasting for semi-arid environments using hybrid machine learning and multi-scale analysis techniques. *J Hydroinform* 2024; jh2024263. https://doi.org/10.2166/hydro.2024.263.
17. Knoben WJ, Freer JE, Woods RA. Inherent benchmark or not? Comparing Nash-Sutcliffe and Kling-Gupta efficiency scores. *Hydrol Earth Syst Sci* 2019;**23**(10):4323–4331. https://doi.org/10.5194/hess-23-4323-2019.
18. Gupta HV, Kling H, Yilmaz KK, Martinez GF. Decomposition of the mean squared error and NSE performance criteria: implications for improving hydrological modelling. *J Hydrol* 2009;**377**(1-2):80–91. https://doi.org/10.1016/j.jhydrol.2009.08.003.
19. Willmott CJ. On the validation of models. *Phys Geogr* 1981;**2**(2):184–194. https://doi.org/10.1080/02723646.1981.10642213.
20. Nourani M, Alali N, Samadianfard S, Band SS, Chau KW, Shu CM. Comparison of machine learning techniques for predicting porosity of chalk. *J Pet Sci Eng* 2022;**209**:109853. https://doi.org/10.1016/j.petrol.2021.109853.
21. Li Z, Yang J, Wang J, Shi L, Stein S. Integrating LSTM and BERT for long-sequence data analysis in intelligent tutoring systems. *arXiv Preprint* 2024;**2405**:05136. arXiv:2405.05136. https://doi.org/10.48550/arXiv.2405.05136.

22. Xu K, Cheng Y, Long S, Guo J, Xiao J, Sun M. Advancing financial risk prediction through optimized LSTM model performance and comparative analysis. *arXiv Preprint* 2024:arXiv.2405.20603. https://doi.org/10.48550/arXiv.2405.20603.
23. Ouyang J, Zuo Z, Wang Q, Duan Q, Qin L. Enhancing short-term power forecasting of PV clusters: A statistical upscaling and CBAM-BiLSTM approach. *J Renew Sustain Energy* 2024;**16**(6). https://doi.org/10.1063/5.0216421.
24. Zhu L, Zhu W, Xiang W, Meng X, Zhang C. An improved EEMD-PE-BiLSTM model for forecasting traffic flow. *Int J Intell Syst Technol Appl* 2024;**22**(4):445–468. https://doi.org/10.1504/IJISTA.2024.143265.
25. Xu W, Chen J, Corzo G, Xu CY, Zhang XJ, Xiong L, et al. Coupling deep learning and physically based hydrological models for monthly streamflow predictions. *Water Resour Res* 2024;**60**(2):e2023WR035618. https://doi.org/10.1029/2023WR035618.
26. Tao L, Nan Y, Cui Z, Wang L, Yang D. An explainable Bayesian gated recurrent unit model for multi-step streamflow forecasting. *J Hydrology: Regional Stud* 2025;**57**: 102141. https://doi.org/10.1016/j.ejrh.2024.102141.
27. Liu K, Li N, Liang S. Enhancing runoff prediction accuracy with superflex-Gru hybrid models in the Qinghai Lake Basin, Northeast Tibetan Plateau, China. *Preprint* 2024. https://doi.org/10.2139/ssrn.4890375.
28. Duan Y, Liu Y, Wang Y, Ren S, Wang Y. Improved BiGRU model and its application in stock price forecasting. *Electronics* 2023;**12**(12):2718. https://doi.org/10.3390/electronics12122718.
29. Maity R, Srivastava A, Sarkar S, Khan MI. Revolutionizing the future of hydrological science: Impact of machine learning and deep learning amidst emerging explainable AI and transfer learning. *Appl. Comput. Geosci.* 2024;**24**:100206. https://doi.org/10.1016/j.acags.2024.100206.
30. Srivastava A, Maity R, Desai VR. Assessing Global-Scale Synergy Between Adaptation, Mitigation, and Sustainable Development for Projected Climate Change. In *Ecological Footprints of Climate Change. Springer Climate;* Chatterjee U, Akanwa AO, Kumar S, Singh SK, Dutta Roy A, Eds.; Springer: Cham, 2022, https://doi.org/10.1007/978-3-031-15501-7_2
31. Masood A, Srivastava A, Hameed MM, Tanweer MS, Ahmad K, Razali SFM. Machine learning for water quality soft sensing in wastewater treatment: State of the art and future directions. In *Innovative and hybrid technologies for wastewater treatment and recycling;* Alam P, Khan AH, Américo-Pinheiro JHP, Ahmad K, Eds.; 1st ed. CRC Press, 2024.

CHAPTER SEVEN

Exploring the efficacy of different variants of long short-term memory networks in daily streamflow forecasting

Amina Khatun[a,*], Ashish Kumar[b], and Chandranath Chatterjee[b]
[a]Department of Natural Resource Management, College of Horticulture & Farming System Research, Assam Agricultural University, Nalbari, India
[b]Agricultural and Food Engineering Department, Indian Institute of Technology Kharagpur, Kharagpur, India
*Corresponding author. e-mail address: aminakhatun9286@gmail.com

Contents

Abstract

Floods pose a significant global threat annually, demanding precise and timely flood forecasts to mitigate potential losses. This study develops three deep-learning based streamflow forecasting models for prediction of daily streamflows at 1–3 day lead times. The developed deep learning models are three variants of state-of-the-art data-driven flood forecasting models, namely simple Long Short-Term Memory (LSTM), Stacked LSTM and Bidirectional LSTM (BiLSTM). The Hirakud Reservoir catchment in eastern India (Mahanadi River basin) is considered as the test area. The models were forced with the time-lagged upstream gauge discharges, sub-catchment averages of rainfall and temperature, and inflows to the reservoir as inputs. Considering a constant time lag of 4-days, the streamflow forecasting ability of these models were tested. Comparison of model performance reveals that the BiLSTM

Advances in Chemical Pollution, Environmental Management and Protection, Volume 15
ISSN 2468-9289, https://doi.org/10.1016/bs.apmp.2025.11.005

model outperformed the other two models with NSE values of 0.94 and 0.85 for lead-times of 1–2 days, respectively. Moreover, the Stacked LSTM model showed comparatively accurate results at 1- and 2-day lead-times with NSEs equal to 0.92 and 0.84, respectively. The overall findings indicate that the simple LSTM model is able to provide accurate forecasts at 1-day lead-time only (NSE = 0.92). The peak discharges predicted by all the three models during the testing period were compared, revealing that the data-driven models show good performance up to 2-day lead-times for most of the cases. The study's implications extend to improve flood management strategies in similar regions, highlighting the critical role of accurate streamflow predictions in reducing flood-related risks and improving disaster preparedness.

Keywords: Flood forecasting, Deep learning, Long short-term memory, BiLSTM, Stacked LSTM, Mahanadi.

1. Introduction

Flood stands as a destructive natural force, causing substantial loss of life and property. It contributes about 30 % of all losses from natural calamities.[1] As the Earth's temperature rises due to climate change, flooding events are becoming more frequent and severe.[2,3] As per predictions, frequent occurrence of intense floods can be anticipated by 2050 leading to substantial loss of about US $ 1 trillion.[4,5] Various researchers have documented the heightened occurrence of extreme flooding events over the past decade (e.g.,[6]). As reservoirs play a significant role in flood management, it becomes vital to effectively manage their overflow.[7] This management is critical to prevent dam failures and mitigate downstream flooding in large river basins. Thus, accurate inflow forecasts with ample lead-time play a pivotal role in developing early warning systems to avert floods and ensure the judicious release of water from reservoirs.[8]

Flood forecasting models fall into two categories: physically-based and empirical methods.[9] Physically-based methods employ sophisticated mathematical tools, aiming to replicate the intricate physical processes of runoff generation in a catchment.[10,11,48] Conversely, empirical methods establish a direct input-output link, bypassing the requirement of an in-depth knowledge of the physical catchment dynamics. Instead, they rely on the empirical relationship between rainfall and runoff data.[12,13] The Machine Learning (ML) or Neural Network (NN) models are data-driven models, capable of capturing the nonlinearity in the dataset reasonably well.[14,47]

Recent successes in deploying ML and Deep Learning (DL) models for simulating streamflow processes underscore their efficacy in precise runoff forecasting.[15,16,49] The most basic ML models, Artificial Neural Network (ANN), have gained popularity in hydrological applications during the 1990s.[50] They were widely used for flood forecasting (e.g.,[17,18]). However, ANN fails to deal with sequential data, especially in time-series prediction. To overcome this challenge, Recurrent Neural Networks (RNNs) are developed. Unlike standard feedforward NNs, RNNs are equipped with feedback connections which facilitates the well-capturing of temporal dependencies within the data. In flood forecasting, RNNs can effectively discern patterns and relationships between hydrological variables over time, facilitating accurate predictions of future streamflows based on historical data.[19,20] However, the problem of vanishing gradient limits their long-term dependence learning ability. Further, Long Short-Term Memory (LSTM)[21] address the issue of vanishing gradients through memory cells and gating mechanisms, enabling them to capture complex temporal patterns more effectively. Their adaptability to changing environmental conditions makes them suitable for flood forecasting tasks, where dynamic adjustments to predictions are crucial for effective management and mitigation strategies.

By stacking multiple LSTM layers, Stacked LSTM networks can learn the complex input-output patterns from different datasets, enabling accurate predictions, especially at higher lead times (e.g.,[22]). Research indicates that stacking LSTM layers enhances the capacity of the model in detecting subtle variations in streamflow and improve the accuracy of flood predictions, especially at higher lead times (e.g.,[23,24]). Additionally, Stacked LSTM architectures demonstrate superior performance in simulating discharge fluctuations across different basins, surpassing the predictive capabilities of basin-wise trained individual LSTM models (e.g.,[22,25]). Bidirectional LSTM (BiLSTM) networks, a variant of LSTM models, offer enhanced capabilities for flood forecasting by learning information in both forward and backward directions.[26] This two-way processing allows BiLSTM to leverage understanding of both historical and future observations, leading to improved accuracy in predicting streamflow levels and anticipating flood events. They can detect the complex temporal patterns and handling irregularities in hydrological time-series data.[26,27]

In the specific context of the Mahanadi basin, flood forecasting assumes paramount importance due to the basin's vulnerability to flooding and the consequent impact on the environment and local communities.[28,29] Characterized by its complex hydrological dynamics, the Mahanadi River basin, experiences recurrent flooding events driven by factors like intense rainfall, tropical cyclones and reservoir releases.[30,31] Effective forecasting of streamflows in the Mahanadi River basin necessitates the development of robust predictive models capable of capturing the intricate interactions between precipitation, runoff, and river flow dynamics. Leveraging advancements in ML techniques, researchers aim to improve the accuracy of streamflow forecasting models[32] tailored to the unique characteristics of the Mahanadi River basin.

An extensive review of literature suggests that there has been limited exploration of LSTM variants, such as Stacked LSTM and BiLSTM for flood forecasting. Moreover, the existing models struggle with accurate runoff predictions as the lead-time increases. In the present scenario, improving flood prediction methods is crucial due to the increasing frequency and severity of floods worldwide. While the physics-based models have been useful, newer data-driven techniques like LSTM, Stacked LSTM, and BiLSTM offer exciting possibilities. By focusing on the research gaps identified, flood forecasting can be made more accurate, leading to better flood preparedness. Thus, this study aims to develop a conventional LSTM model, a Stacked LSTM, and a BiLSTM model for daily short-to-medium range streamflow forecasting in the upper Mahanadi basin and compare their performance.

2. Details of the study area

The catchment of the Hirakud Reservoir (Fig. 1) in the upper Mahanadi basin lies in the eastern part of India. It constitutes around 65 % (83,400 km^2) of the geographical area of the river basin. Located between latitudes 19°90′N to 23°35′N and longitudes 80°30′E to 84°80′E, the catchment is predominantly rainfed having about 1400 mm of mean annual rainfall. A major portion of the rainfall occurs during the southwest monsoon months of June to September. The basin exhibits a tropical climate, wherein the minimum temperature ranges between 4 °C to 12 °C during December and January months. The maximum temperature of 42 °C to 45 °C is experienced during May and June months. The catchment's topography comprises plains in the central area and hills in the northern regions. Primarily designed for flood protection, the

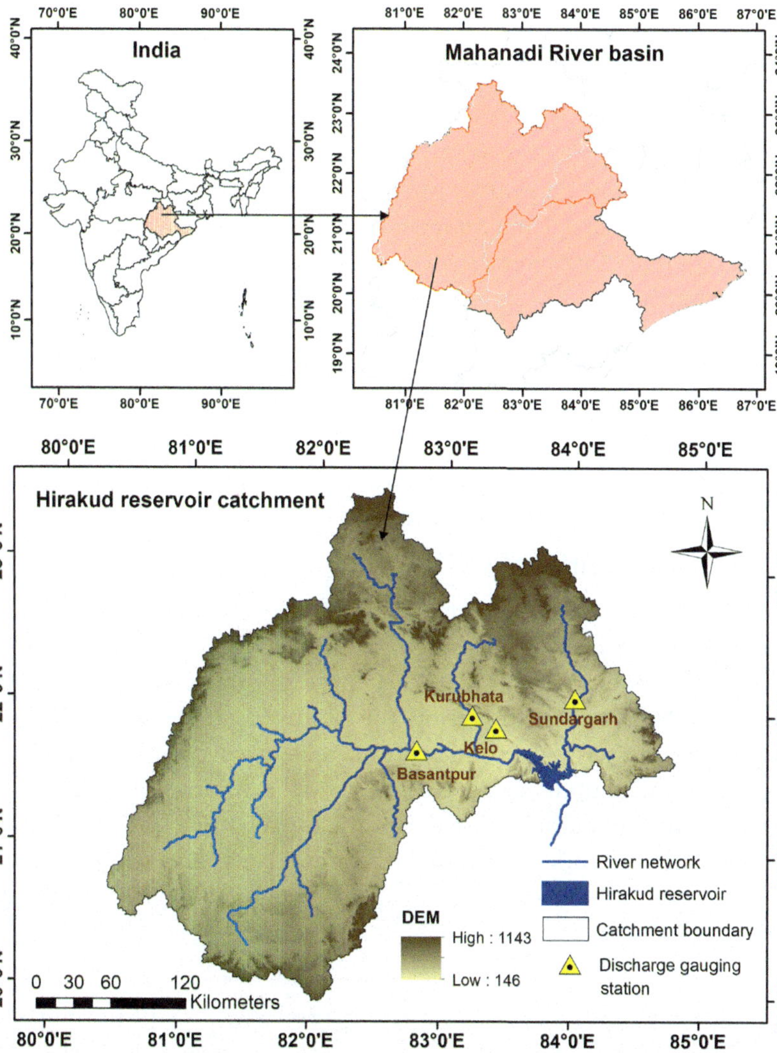

Fig. 1 Location map of the study area. Map lines delineate study areas and do not necessarily depict accepted national boundaries.

Hirakud Reservoir boasts a storage (live) capacity of about 4823 million cubic meters. Given its role in flood management in the downstream delta, this study targets the development of DL-based flood forecasting models to test their effectiveness in daily inflow forecasting for the Hirakud Reservoir.

3. Data used

The study utilizes four types of hydrometeorological datasets, namely rainfall, temperature, upstream gauge discharges, and inflow to the Hirakud reservoir. All the datasets collected are at a daily temporal resolution and for the months of June–September (monsoon season) during 2000–2014. The meteorological (rainfall and temperature) data are obtained from the India Meteorological Department (IMD), Pune at a spatial resolution of 0.25° and 1°, respectively. The discharges from four (Basantpur, Kelo, Kurubhata and Sundargarh) gauging stations are collected from India-Water Resources Information System (India-WRIS). The inflow to the reservoir are provided by the Hirakud Dam Circle (HDC), Burla.

4. Flood forecasting models

4.1 Conventional LSTM

LSTMs have demonstrated remarkable effectiveness in addressing complex problems related to hydrological time series, as evidenced by the work of Granata and Di Nunno.[33] They are characterized by cells consisting of three gates showcasing the ability to understand long-term patterns in time-series data and make accurate predictions. The input (I_t), forget (F_t) and output (O_t) gates manages the information flow in the network. An explicit hidden state is maintained at time 't' which is passed through the network and subsequently updated in each cell. Fig. 2

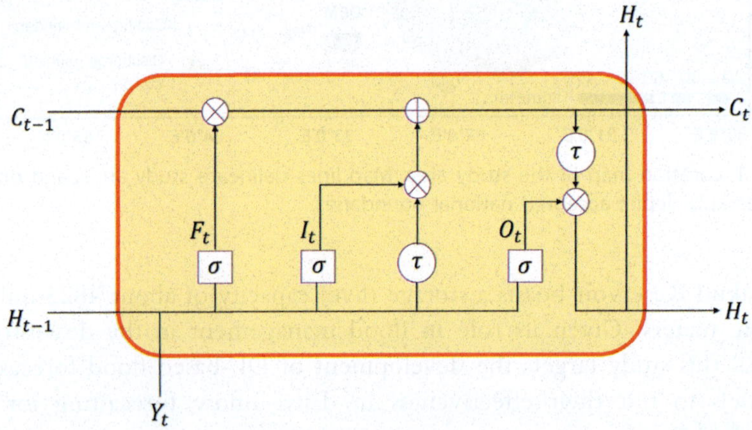

Fig. 2 Architecture of an LSTM cell showing the gates.

presents the typical LSTM cell structure, wherein '*Y*', '*H*' and '*C*' represents the current input, hidden and cell states, respectively. While I_t considers the information from the current input state to be added, the fraction of information obtained from the previous hidden state that is to be forgotten (keeping the relevant information intact) is decided by F_t. Further, O_t takes into consideration of the information from the current cell state to be provided as the output. The symbols '$\otimes$', '$\oplus$', and 'σ' indicates point-wise multiplication, point-wise addition, and sigmoid activation, respectively. At the current time step, Y_t, H_{t-1} and C_{t-1} enters the LSTM unit as input, and H_t and C_t are given out as output. Interestingly, the concept of 'LSTM cell' and 'LSTM unit' is often confusing. An 'LSTM cell' represents the LSTM block that houses the three gates regulating information flow as shown in Fig. 2. In contrast, 'LSTM unit' indicates the number of neurons in the LSTM layer containing the input and hidden states concatenated together.[8] For a detailed understanding of LSTM networks, the readers are directed towards.[34]

The conventional LSTM model developed herein contains one LSTM layer followed by one dense layer. With a ReLU activation function, the LSTM layer is equipped with 50 LSTM units. The dense layer is provided with one unit. The MSE and Adam serves as the loss function and optimizer, respectively. A batch size of 32 is provided with the number of training epochs equal to 70.

4.2 Stacked LSTM

The Stacked LSTM represents a notable advancement incorporating multiple hidden layers, each comprising multiple memory cells. While Stacked LSTM has found success in diverse applications such as machine translation and speech recognition, its adoption in hydrological problem-solving has been relatively modest compared to other domains. This study develops a deeper LSTM network architecture with the stacking of five hidden layers. Fig. 3 provides an illustrative representation of the Stacked LSTM model, where '*Y*' denotes the input vector and '*M*' denotes the output (here, α-day lead-time inflow forecast to the Hirakud Reservoir), emphasizing its hierarchical structure and potential for capturing intricate temporal patterns in hydrological data. Each layer of the developed model

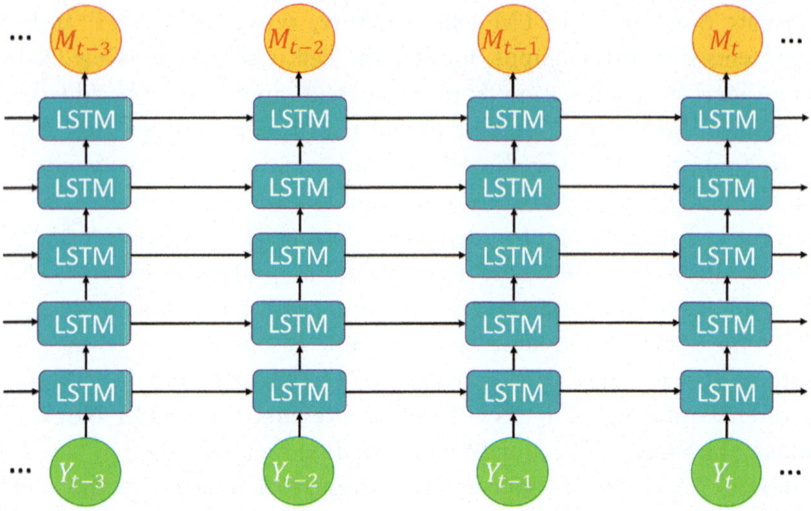

Fig. 3 Stacked LSTM network structure.

consists of 30 units. The activation function, learning rate, dropout, batch size and epochs are set to ReLU, 0.0048, 0.32, 100 and 80, respectively. The Adam optimizer is used with the Huber loss function.

4.3 BiLSTM

In contrast to conventional LSTM models that process data in a unidirectional (forward) manner, BiLSTM performs a forward-backward data processing. This advanced extension of the LSTM architecture enables the BiLSTM to generate output at a given time step (t) by combining information from two distinct hidden layers, each representing a training sequence for the forward and backward directions. As depicted in Fig. 4, the typical structure of a BiLSTM network illustrates the bidirectional flow of information, highlighting its enhanced capacity to handle temporal dependencies in both directions. Here, a BiLSTM model with four stacked layers is developed for streamflow forecasting. Each layer contains 100 units. The activation function, learning rate, dropout, batch size and epochs are set to ReLU, 0.001, 0.4, 128 and 120, respectively. The Adam optimizer is used with the Huber loss function.

The three LSTM model variants developed herein uses daily areal average values of rainfall and temperature, daily discharges of four

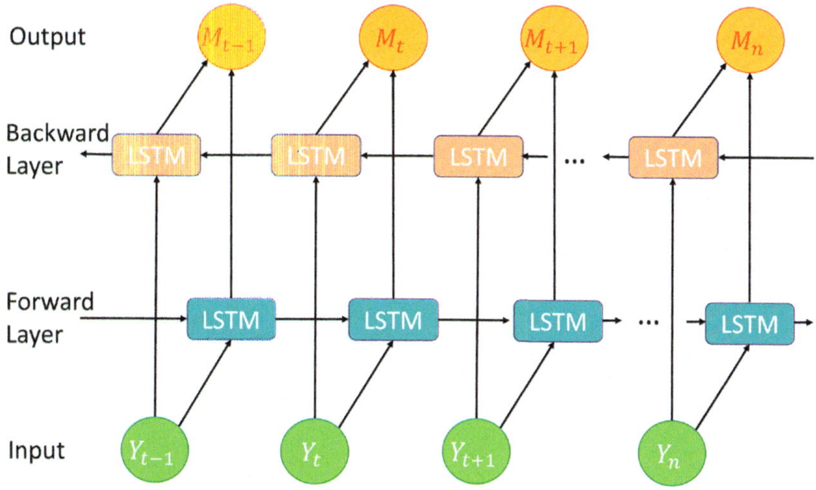

Fig. 4 BiLSTM network structure.

(Basantpur, Kelo, Kurubhata and Sundargarh) gauging stations, and daily inflow to the Hirakud Reservoir as inputs. Following Khatun et al.,[14] a constant step size of 4 is considered for each input variable. The output is 1–3 day lead-time inflow to the reservoir. Note that all the models are of sequence-to-single-output type. The dataset from 2000–2008 and 2009–2011 are used for training and validation of the models, respectively, and that from 2012–2014 is used for model testing. The model hyper-parameters are obtained by trial-and-error approach.

4.4 Model evaluation criteria

The study uses a range of error measures to compare the model performance during daily flood forecasting. The Nash Sutcliffe Efficiency (NSE) (Eq. 1) indicates the accuracy in the corresponding observed-predicted data fit. It ranges from $-\infty$ to a maximum value of 1, with an NSE >0.5 representing good model performance. However, an NSE value close to 1 is desired. The Mean Absolute Error (MAE) shows the total error in the predicted values irrespective of its direction (Eq. 2). The ratio of Root Mean Square Error (RMSE) to the standard deviation of observed data (SD_{obs}) given by RSR is a dimensionless quantity (Eq. 3). As the total incoming water volume is crucial for reservoir management, the error in total streamflow volume (E_{vol}) is also computed (Eq. 4). Finally, to capture the model's ability in predicting the peak events, the error in peak flows is measured by (Eq. 5).

$$NSE = 1 - \frac{\sum_{i=1}^{n} (O_i - C_i)^2}{\sum_{i=1}^{n} (O_i - \bar{O})^2} \quad (1)$$

$$MAE = \frac{\sum_{i=1}^{n} |O_i - C_i|}{n} \quad (2)$$

$$RSR = \frac{RMSE}{SD_{Obs}} = \sqrt{\frac{\sum_{i=1}^{n} (O_i - C_i)^2}{\sum_{i=1}^{n} (O_i - \bar{O})^2}} \quad (3)$$

$$E_{vol}\,(\%) = \frac{\sum_{i=1}^{n} C_i - \sum_{i=1}^{n} O_i}{\sum_{i=1}^{n} O_i} \times 100 \quad (4)$$

$$E_{peak}\,(\%) = \left(\frac{Q_c}{Q_o} - 1\right) \times 100 \quad (5)$$

where O_i, C_i and $\bar{O}$ are the observed streamflow, simulated streamflow and average of observed streamflow, respectively. Q_c and Q_o are the predicted and observed peak events, respectively.

5. Evaluation of different streamflow forecasting model performance

5.1 Comparison of overall time series

Fig. 5 presents the time series of predicted versus observed streamflow values at all the lead-times (1–3 days) for the testing period (2012–2014). The values of the error measures to compare the model performance are presented in Table 1. Fig. 5 and Table 1 reveals that all the three model variants are able to capture the overall time-series very well at 1-day lead time. However, comparison among the models reveals that the BiLSTM model, with an NSE = 0.94 outperforms the conventional LSTM (NSE = 0.92) and Stacked LSTM (NSE = 0.92). At 2-day lead-time, the Stacked LSTM and BiLSTM provides reasonably accurate flood predictions with an NSE = 0.84 and NSE = 0.85, respectively. Further, with the increase in lead-time to 3-day, the model performance deteriorates significantly (NSE = 0.55–0.67), but are still acceptable.[35] Furthermore, all

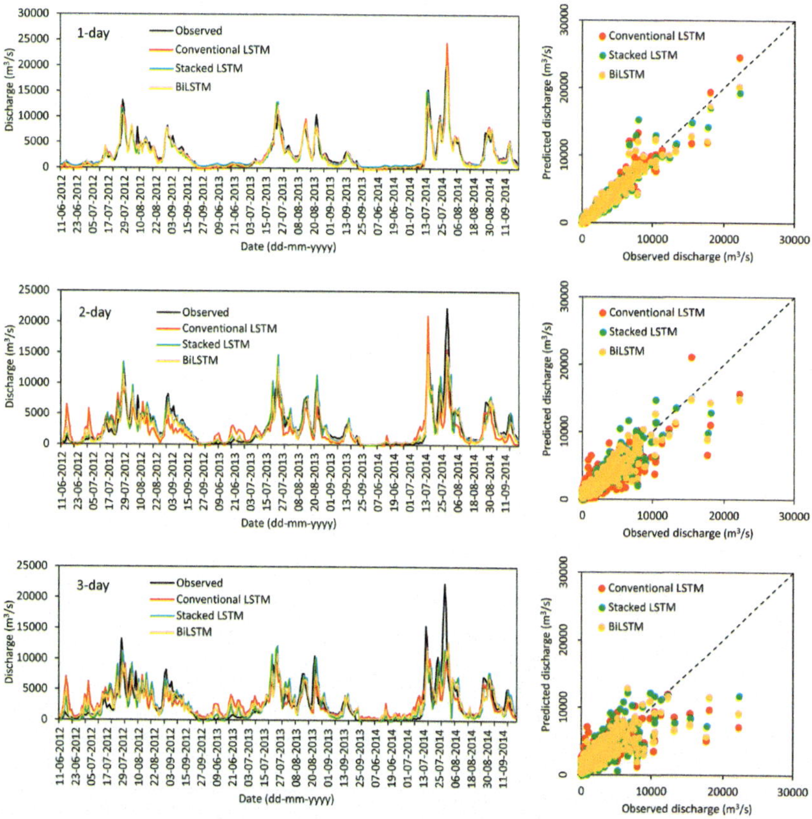

Fig. 5 Comparison of observed and predicted streamflows by the three LSTM model variants (conventional LSTM, Stacked LSTM and BiLSTM) at 1–3 days lead-times during the testing period (2012–2014).

the three model variants at 1-day, and Stacked LSTM and BiLSTM at 2-day lead-times shows 'very good' forecasting performance with an RSR value <0.5.[36] The conventional LSTM at 2-day and all the models at 3-day lead-times are not acceptable in terms of RSR (RSR > 0.7). The MAE estimates ranges from 461.69–564.95 m^3/s, 700.46–1265.73 m^3/s and 1131.30–1357.98 m^3/s at 1-, 2- and 3-day lead-time, respectively. Interestingly, model performances are acceptable in terms of MAE as the MAE estimates lies below half the standard deviation of observed data (i.e., 1559.22 m^3/s).[35] Likewise, the error in total streamflow volume prediction is also considered acceptable as the E_{vol} <20 % in all the cases.

Table 1 Error measures showing the performance of different LSTM model variants (conventional LSTM, Stacked LSTM and BiLSTM) at 1–3 days lead-time streamflow forecasting during the testing period (2012–2014).

Model	Error measures	Lead-time (days)		
		1	2	3
Conventional LSTM	NSE (–)	0.92	0.65	0.55
	RSR (–)	0.28	0.71	0.99
	MAE (m^3/s)	461.69	1265.73	1357.98
	E_{vol} (%)	−2.36	−13.98	4.69
Stacked LSTM	NSE (–)	0.92	0.84	0.67
	RSR (–)	0.3	0.43	0.73
	MAE (m^3/s)	564.95	700.46	1166.97
	E_{vol} (%)	0.85	1.96	7.81
BiLSTM	NSE (–)	0.94	0.85	0.65
	RSR (–)	0.26	0.45	0.78
	MAE (m^3/s)	467.11	743.52	1131.3
	E_{vol} (%)	−2.93	−2.43	4.8

5.2 Peak flow estimation

To test the efficacy of the model variants in capturing the individual peak flood events, the error in peak E_{peak} for each of the three peak events occurring in the testing period i.e., from 2012 to 2014 are computed. Table 2 shows the E_{peak} values for the individual events. As evidenced from Table 2, it is found that all the three models are able to predict the 2012 peak event within an error bound of 20 % ($|E_{peak}| \leq 20\%$) at 1–2 days lead time. Moreover, Stacked LSTM and BiLSTM could capture the 2013 peak flood at 1–2 days ahead, although the error in peak at 1-day lead-time is slightly higher than 20 % ($|E_{peak}| < 20.6$ m^3/s. Further, the 2014 peak event, which is one of the highest peaks that have occurred in the Mahanadi basin could be captured by the models within the acceptable range at 1-day ahead only.

Table 2 Error in peak flow computation (E_{peak}, in %) for the three typical peak events during the testing period (2012–2014).

Year	Observed peak (m^3/s)	Lead-time (Days)	Model		
			Conventional LSTM	Stacked LSTM	BiLSTM
2012	13,264	1	−12.87	−11.52	−19.3
		2	−15.11	2.36	15.23
		3	−34.97	41.94	49.47
2013	10,534	1	−26.27	−20.21	−20.6
		2	−39.97	9.13	−16.78
		3	−51.81	−18.27	−39.69
2014	22,308	1	10.41	13.53	−9.04
		2	−29.73	−33.1	−33.23
		3	−67.95	−47.51	−58.51

6. Conclusions

Due to climate change, the severity of extreme flooding events are expected to increase.[37] The extreme flooding events are the repercussions of high water levels in rivers and the reservoirs. This necessitates the hydrologic (particularly, water level and streamflow) predictions to be accurate to facilitate flood preparedness and mitigation. Thus, accurate streamflow/water level prediction plays a crucial role in optimization of water resource utilization to foster sustainable water resources development and management.[38] Appropriate model development for prediction of streamflow time series has proven to be challenging due to the involvement of numerous influencing factors and complex underlying processes (e.g.,[39]). Furthermore, during the high flow period, the randomness in water level/streamflow fluctuations and changes in flooding frequencies increases the complexity of the streamflow prediction process. While the physically-based model applications are limited by the understanding of complex underlying processes,[40] the data-driven neural network models serve as a promising tool in streamflow forecasting.[9] The neural network models provide advantages of noise tolerance, improvements in model learning ability, generalization and improved efficiency (e.g.,[41,42]). Moreover, the supremacy of these models over the physically-based models comes in their ability in capturing the nonlinear patterns and randomness reasonably well.

With the advancement of ML/DL techniques, LSTM network has emerged and gained popularity. It is a special type of RNN which has overcome the problem of vanishing gradients and shows improved ability to handle the long-term dependencies in time-series prediction.[43] Existing literature (e.g.,[44–46]) on DL models have established their superiority in capturing the hydrological processes in the complex environmental settings like climate change, glacial dynamics and land use changes. To improve the efficacy of conventional LSTM networks, different newer variants have been developed. While the conventional LSTMs have wide hydrological applications, its newer versions are still at an emerging phase. Keeping this in view, this study utilized three variants of LSTM networks, namely conventional LSTM, Stacked LSTM and BiLSTM to develop a daily flood forecasting system for the upstream catchment of the Hirakud reservoir located in the Mahanadi basin. The overall model comparison revealed that all the three model variants are equally efficient at 1-day lead time. At 2–3 day lead-times, the Stacked LSTM and BiLSTM models outperform the conventional LSTM. Beyond 2-days, the flood forecasting capability of the models deteriorates significantly. Furthermore, among the advanced LSTM variants i.e., Stacked LSTM and BiLSTM, the BiLSTM model performs slightly better as compared to the Stacked LSTM. These two models exhibits robustness in capturing the peak events up to 2-day lead times. However, they fail to predict the very high flood event (typically occurred in 2014) reasonably well at 2-days ahead. The findings of the study encourages future research on comparison of different advanced versions of ML/DL models, especially in the field of hydrology. Such studies may build insights on the efficacy and utility of ML/DL models in flood forecasting and sustainable water resources development and management.

Acknowledgement

The authors thank the India Meteorological Department (IMD), Pune, Central Water Commission (CWC), Bhubaneswar and the Hirakud Dam Circle, Burla for providing the necessary datasets to carry out the study.

References

1. Adikari Y, Yoshitani J. Global trends in water-related disasters: an insight for policymakers. World water assessment programme side publication series, insights. The United Nations, UNESCO. International Centre for Water Hazard and Risk Management (ICHARM); 2009.
2. Dikshit A, Pradhan B, Alamri AM. Long lead time drought forecasting using lagged climate variables and a stacked long short-term memory model. *Sci Total Environ* 2021;**755**:142638.
3. Shen C. A transdisciplinary review of deep learning research and its relevance for water resources scientists. *Water Resour Res* 2018;**54**(11):8558–8593.
4. Huang K, Li X, Liu X, Seto KC. Projecting global urban land expansion and heat island intensification through 2050. *Environ Res Lett* 2019;**14**(11):114037.
5. Rahman M, Chen N, Islam MM, Mahmud GI, Pourghasemi HR, Alam M, et al. Development of flood hazard map and emergency relief operation system using hydrodynamic modeling and machine learning algorithm. *J Clean Prod* 2021;**311**: 127594.
6. Najibi N, Devineni N. Recent trends in the frequency and duration of global floods. *Earth Syst Dyn* 2018;**9**(2):757–783.
7. Khatun A, Sahoo B, Chatterjee C. Two novel error-updating model frameworks for short-to-medium range streamflow forecasting using bias-corrected rainfall inputs: development and comparative assessment. *J Hydrol* 2023;**618**:129199.
8. Khatun A, Chatterjee C, Sahu G, Sahoo B. A novel smoothing-based long short-term memory framework for short-to medium-range flood forecasting. *Hydrological Sci J* 2023;**68**(3):488–506.
9. Cao Q, Zhang H, Zhu F, Hao Z, Yuan F. Multi-step-ahead flood forecasting using an improved BiLSTM-S2S model. *J Flood Risk Manag* 2022;**15**(4):e12827.
10. Deng H, Chen W, Huang G. Deep insight into daily runoff forecasting based on a CNN-LSTM model. *Nat Hazards* 2022;**113**(3):1675–1696.
11. Adnan RM, Petroselli A, Heddam S, Santos CAG, Kisi O. Comparison of different methodologies for rainfall–runoff modeling: machine learning vs conceptual approach. *Nat Hazards* 2021;**105**:2987–3011.
12. Gao S, Huang Y, Zhang S, Han J, Wang G, Zhang M, et al. Short-term runoff prediction with GRU and LSTM networks without requiring time step optimization during sample generation. *J Hydrol* 2020;**589**:125188.
13. Young CC, Liu WC. Prediction and modelling of rainfall–runoff during typhoon events using a physically-based and artificial neural network hybrid model. *Hydrological Sci J* 2015;**60**(12):2102–2116.
14. Khatun A, Nisha MN, Chatterjee S, Sridhar V. A novel insight on input variable and time lag selection in daily streamflow forecasting using deep learning models. *Environ Model & Softw* 2024;**179**:106126.
15. Ghimire S, Yaseen ZM, Farooque AA, Deo RC, Zhang J, Tao X. Streamflow prediction using an integrated methodology based on convolutional neural network and long short-term memory networks. *Sci Rep* 2021;**11**(1):17497.
16. Yaseen ZM, Mohtar WHMW, Ameen AMS, Ebtehaj I, Razali SFM, Bonakdari H, et al. Implementation of univariate paradigm for streamflow simulation using hybrid data-driven model: case study in tropical region. *IEEE Access* 2019;**7**:74471–74481.
17. Tiwari MK, Chatterjee C. Development of an accurate and reliable hourly flood forecasting model using wavelet–bootstrap–ANN (WBANN) hybrid approach. *J Hydrol* 2010;**394**(3-4):458–470.
18. Tan QF, Lei XH, Wang X, Wang H, Wen X, Ji Y, et al. An adaptive middle and long-term runoff forecast model using EEMD-ANN hybrid approach. *J Hydrol* 2018;**567**:767–780.

19. Ahmed AM, Deo RC, Feng Q, Ghahramani A, Raj N, Yin Z, et al. Deep learning hybrid model with Boruta-Random forest optimiser algorithm for streamflow forecasting with climate mode indices, rainfall, and periodicity. *J Hydrol* 2021;**599**:126350.
20. Cheng M, Fang F, Kinouchi T, Navon IM, Pain CC. Long lead-time daily and monthly streamflow forecasting using machine learning methods. *J Hydrol* 2020;**590**:125376.
21. Hochreiter S. *Long short-term memory. Neural computation;* MIT-Press, 1997.
22. Chidepudi SKR, Massei N, Jardani A, Henriot A, Allier D, Baulon L. A wavelet-assisted deep learning approach for simulating groundwater levels affected by low-frequency variability. *Sci Total Environ* 2023;**865**:161035.
23. Chandra R, Goyal S, Gupta R. Evaluation of deep learning models for multi-step ahead time series prediction. *IEEE Access* 2021;**9**:83105–83123.
24. Mehedi MAA, Khosravi M, Yazdan MMS, Shabanian H. Exploring temporal dynamics of river discharge using univariate long short-term memory (LSTM) recurrent neural network at East Branch of Delaware River. *Hydrology* 2022;**9**(11):202.
25. Gauch M, Kratzert F, Klotz D, Nearing G, Lin J, Hochreiter S. Rainfall–runoff prediction at multiple timescales with a single long short-term memory network. *Hydrol Earth Syst Sci* 2021;**25**(4):2045–2062.
26. Tran QK, Song SK. Water level forecasting based on deep learning: a use case of Trinity River-Texas-The United States. *J KIISE* 2017;**44**(6):607–612.
27. Fang L, Shao D. Application of long short-term memory (LSTM) on the prediction of rainfall-runoff in karst area. *Front Phys* 2022;**9**:790687.
28. Khatun A, Ganguli P, Bisht DS, Chatterjee C, Sahoo B. Understanding the impacts of predecessor rain events on flood hazard in a changing climate. *Hydrological Process* 2022;**36**(2):e14500.
29. Khosravi K, Pham BT, Chapi K, Shirzadi A, Shahabi H, Revhaug I, et al. A comparative assessment of decision trees algorithms for flash flood susceptibility modeling at Haraz watershed, northern Iran. *Sci Total Environ* 2018;**627**:744–755.
30. Jena PP, Chatterjee C, Kumar R, Khatun A. Flood Hazard assessment using hydrodynamic modeling under severity-frequency based changing flood regime. *Water Resour Manag* 2024:1–26.
31. Patel A, Jena PP, Khatun A, Chatterjee C. Improved Cartosat-1 based DEM for flood inundation modeling in the delta region of Mahanadi River Basin, India. *J Indian Soc Remote Sens* 2022;**50**(7):1227–1241.
32. Mosavi A, Ozturk P, Chau KW. Flood prediction using machine learning models: literature review. *Water* 2018;**10**(11):1536.
33. Granata F, Di Nunno F. Neuroforecasting of daily streamflows in the UK for short-and medium-term horizons: a novel insight. *J Hydrol* 2023;**624**:129888.
34. Kratzert F, Klotz D, Brenner C, Schulz K, Herrnegger M. Rainfall–runoff modelling using long short-term memory (LSTM) networks. *Hydrol Earth Syst Sci* 2018;**22**(11):6005–6022.
35. Moriasi DN, Arnold JG, Liew MWV, Bingner RL, Harmel RD, Veith TL, et al. Model evaluation guidelines for systematic quantification of accuracy in watershed simulations. *Trans ASABE* 2007;**50**:885–900.
36. Singh J, Knapp HV, Arnold JG, Demissie M. Hydrological modeling of the Iroquois river watershed using HSPF and SWAT. *J Am Water Resour Assoc* 2004;**41**(2):343–360.
37. Roxy MK, Ghosh S, Pathak A, Athulya R, Mujumdar M, Murtugudde R, et al. A threefold rise in widespread extreme rain events over central India. *Nat Commun* 2017;**8**(1):1–11.
38. Chenmin N, Marsani MF, Pei Shan F. Flood prediction based on feature selection and a hybrid deep learning network. *J Water Clim Change* 2024;**15**(3):1245–1261.

39. Sahoo BB, Jha R, Singh A, Kumar D. Long short-term memory (LSTM) recurrent neural network for low-flow hydrological time series forecasting. *Acta Geophysica* 2019;**67**(5):1471–1481.
40. Hunt KM, Matthews GR, Pappenberger F, Prudhomme C. Using a long short-term memory (LSTM) neural network to boost river streamflow forecasts over the western United States. *Hydrol Earth Syst Sci* 2022;**26**(21):5449–5472.
41. LeCun Y, Bengio Y, Hinton G. Deep learning. *Nature* 2015;**521**(7553):436–444.
42. Kao IF, Zhou Y, Chang LC, Chang FJ. Exploring a long short-term memory based encoder-decoder framework for multi-step-ahead flood forecasting. *J Hydrol* 2020;**583**: 124631.
43. Ma K, He D, Liu S, Ji X, Li Y, Jiang H. Novel time-lag informed deep learning framework for enhanced streamflow prediction and flood early warning in large-scale catchments. *J Hydrol* 2024;**631**:130841.
44. Bolibar J, Rabatel A, Gouttevin I, Zekollari H, Galiez C. Nonlinear sensitivity of glacier mass balance to future climate change unveiled by deep learning. *Nat Commun* 2022;**13**(1):409.
45. Singh D, Vardhan M, Sahu R, Chatterjee D, Chauhan P, Liu S. Machine-learning-and deep-learning-based streamflow prediction in a hilly catchment for future scenarios using CMIP6 GCM data. *Hydrol Earth Syst Sci* 2023;**27**(5):1047–1075.
46. Mohammadi B, Vazifehkhah S, Duan Z. A conceptual metaheuristic-based framework for improving runoff time series simulation in glacierized catchments. *Eng Appl Artif Intell* 2024;**127**:107302.
47. Khatun A, Jena PP, Sahoo B, Chatterjee C. Forecasting flood inundations in the dam-regulated Mahanadi River delta using integrated hydrologic-hydrodynamic-deep learning model. *Environ Model & Softw* 2025:191:106523.
48. Mohite AR, Khatun A, Chatterjee C, Singh R, Kumar R. Improved flood forecasting using numerical weather prediction meteorological forecasts with integrated rainfall-runoff and hydrodynamic model. *Water Resour Manag* 2025:39:4833–4851.
49. Maity R, Srivastava A, Sarkar S, Khan MI. Revolutionizing the future of hydrological science: Impact of machine learning and deep learning amidst emerging explainable AI and transfer learning. *Appl. Comput. Geosci.* 2024;**24**:100206.
50. Pande CB, Al-Ansari N, Kushwaha NL, Srivastava A, Noor R, Kumar M, et al. Forecasting of SPI and meteorological drought based on the artificial neural network and M5P model tree. *Land* 2022;**11**(11):2040.

CHAPTER EIGHT

Integrating large language models (LLMs) in climate-smart water management: Future approaches and innovations

Aman Srivastava[a,*], **Shubham Jain**[b], **and Rajib Maity**[a]
[a]Department of Civil Engineering, Indian Institute of Technology Kharagpur, Kharagpur, West Bengal, India
[b]Department of Civil Engineering, Indian Institute of Technology Bombay, Mumbai, India
*Corresponding author. e-mail address: amansrivastava1397@kgpian.iitkgp.ac.in; amansrivastava1397@gmail.com

Contents

Abstract

Climate change, urbanization, and population growth pose significant challenges to water management, necessitating more adaptive and sustainable approaches. This

Advances in Chemical Pollution, Environmental Management and Protection, Volume 15
ISSN 2468-9289, https://doi.org/10.1016/bs.apmp.2025.11.004

article explores the possibilities of integrating Large Language Models (LLMs) into climate-smart water management to improve decision-making, predictive modeling, data analysis, and stakeholder engagement. Conventional water management systems rely heavily on historical data and centralized decision-making and are generally becoming increasingly inadequate to address the complexities of climate variability, including extreme weather events. Emerging technologies, including Artificial Intelligence (AI), Machine Learning (ML), Remote Sensing (RS), Geographic Information Systems (GIS), and the Internet of Things (IoT), are already transforming conventional water management, but LLMs open up new capabilities. Their advanced Natural Language Processing (NLP) and multimodal data comprehension empower the synthesis of vast, heterogeneous datasets, uncovering understandings that traditional tools might be missing. Moreover, this article highlights key applications of LLMs across four areas: data analysis and interpretation, decision support systems, predictive modeling and forecasting, and enriching communication for stakeholder engagement. Case studies demonstrate how LLMs improve forecasting accuracy, support more informed decision-making, and advance inclusive, transparent engagement among diverse stakeholders. Challenges, such as technological limitations, data privacy concerns, and biases in model outputs, are addressed alongside ethical considerations in deploying LLMs for critical water management functions. Furthermore, the study examines future trends, including developing more efficient models, improved integration with other technologies, and the increasing role of explainable AI. It concludes by emphasizing the revolutionary potential of LLMs in making water management more resilient, equitable, and sustainable in the face of climate change.

Keywords: AI-driven water solutions, Sustainable hydrological management, Smart water systems, Digital water innovation, Resilient water infrastructure, AI-powered climate adaptation.

1. Introduction

The global water crisis, exacerbated by climate change, urbanization, and population growth, has emerged as one of the most pressing challenges of the 21st century.[1] Innovative and adaptive water management strategies have never been more critical as we grapple with increasingly unpredictable weather patterns, rising sea levels, and the growing demand for water resources.[2] Climate-smart water management, a holistic approach that integrates climate change mitigation and adaptation strategies into water resource management while promoting sustainability, resilience, and adaptability, has become a potential solution to these complex challenges.[3] This framework emphasizes the need for robust management structures, clear policy frameworks, and long-term adaptive management strategies to address the multifaceted challenges of climate change and water scarcity.[4] Climate-smart water management integrates climate resilience into traditional water management frameworks to address the multifaceted challenges posed by climate change that boosts the capacity of water systems to adapt to climatic variability and extremes, thereby promoting resilience against climate-induced stresses, such as droughts, floods, and altered precipitation patterns, aiming to ensure water security for present and future generations.[5–9] For instance, the protection and restoration of wetlands safeguard water resources and serve as a natural carbon sink, demonstrating the potential for synergies between water management and climate action.[10] Therefore, understanding and advancing climate-smart water management is critical for achieving sustainable water management in an era of unprecedented climatic change.

Despite the vital role of traditional frameworks, climate-smart water management faces significant challenges that limit its effectiveness. Traditional water management systems often operate within rigid institutional structures that are not equipped to handle the dynamic and uncertain nature of climate change.[11] These systems typically rely on historical hydrological data and static management practices, which are insufficient in the face of changing climate patterns and extreme weather events.[12] Additionally, there is often a lack of integration across sectors and scales, leading to fragmented policies and actions that fail to address the complexity of water-related climate issues.[13] In response to the challenges of managing data from various sectors, policy fragmentation, and making real-time decisions in an ever-changing environment, the integration of advanced technologies, particularly Large Language Models (LLMs), provides powerful capabilities in processing vast amounts of data, refining decision-making, and promoting public engagement, thereby holding significant potential for improving water management frameworks worldwide.[14] LLMs are a class of artificial intelligence models that have been trained on vast amounts of textual data to understand and generate human-like language. They can process and analyze large volumes of unstructured data from diverse sources, including scientific literature, policy documents, and real-time environmental data, extracting meaningful findings and supporting complex decision-making processes.[15,16]

The evolution of LLMs, exemplified by Generative Pre-trained Transformer (GPT)-3 and GPT-4 models (will be discussed in Section 3), has demonstrated significant advancements in Natural Language Processing (NLP) capabilities, empowering more sophisticated interactions with data and users.[17] The ability of LLMs to understand the context, generate coherent narratives, and even engage in complex problem-solving tasks makes them uniquely suitable to address the multifaceted challenges of climate-smart water management. From analyzing historical climate data to predicting future water availability scenarios, from summarizing research findings to generating policy recommendations, LLMs have the potential to augment human expertise and improve the efficiency of water management systems at various scales.[18] By utilizing the capabilities of LLMs, water managers and policymakers can gain a deeper understanding of climate-related risks, optimize resource allocation, develop more resilient and adaptive strategies, and support stakeholder collaboration.[19] Regarding stakeholder engagement, LLMs can improve communication by translating technical information into accessible language, nurturing greater understanding and participation among diverse groups.[20] They can also assist in identifying and analyzing policy implications, helping policymakers to design more effective and inclusive management frameworks.[18] By utilizing the power of LLMs, climate-smart water management can evolve into a more dynamic, data-driven, and participatory process, better equipped to navigate the complexities of climate change.

The convergence of climate-smart water management and LLM technologies aligns with adaptive management principles, emphasizing flexibility, learning, and collaboration in managing social-ecological systems under uncertainty.[21] As climate change continues to impact water resources, it is imperative for researchers, policymakers, and practitioners to embrace innovative solutions such as Artificial Intelligence (AI) and LLMs to strengthen management structures and outcomes.[10,22–25] However, integrating LLMs into climate-smart water management is not without challenges. Issues of data privacy, algorithmic bias, and the need for transparent and explainable AI systems must be carefully addressed to ensure equitable and ethical implementation.[26] Moreover, the successful adoption of LLMs in water management will require significant investment in infrastructure, capacity building, and the development of appropriate regulatory frameworks. Despite these challenges, the potential benefits of integrating LLMs into climate-smart water management are substantial. This chapter will delve deeper into the current state of climate-smart water management, the evolution and capabilities of LLMs, and the potential applications that could redefine the future of water management. By exploring these areas, this study aims to advance knowledge and practice in the field, encouraging a vision of water management that is resilient, inclusive, and responsive to the challenges of climate change.

2. Current state of climate-smart water management

Over the past century, traditional water management methods have been instrumental in developing water infrastructure and policies, focusing primarily on maximizing water extraction and use efficiency for economic gain.[27] However, these approaches often overlook environmental sustainability, social equity, and the complexities introduced by climate change. Integrating digital technologies into urban water management is increasingly recognized as essential for addressing the multifaceted challenges of rapid urbanization and climate change. As urban populations grow, the demand for equitable access to quality water services intensifies, necessitating innovative management frameworks that can adapt to evolving conditions.[14] Digitalization in the water sector encompasses a wide array of technologies, including Geographic Information Systems (GIS), Remote Sensing (RS), Internet of Things (IoT) devices, Machine Learning (ML), AI, and Supervisory Control and Data Acquisition (SCADA) systems, which facilitate real-time monitoring and management of water resources.[28] This section provides a thorough overview of existing water management practices by examining traditional methods and their limitations, followed by recent technological advancements in the field.

2.1 Traditional methods and their limitations

Traditional water management has historically relied on centralized, top-down management approaches emphasizing infrastructural development and regulatory control. These methods prioritize large-scale engineering projects, such as dams, canals, and irrigation systems designed to control water flow and supply.[11] Rooted in the belief that human ingenuity can dominate and manipulate natural water systems, this paradigm assumes that past hydrological patterns can reliably predict future conditions—a concept known as stationarity.[29] Policies and management strategies have often been formulated based on historical data without adequately accounting for the variability introduced by climate change.[30] Moreover, these approaches focus on maximizing water extraction and use efficiency for economic gain, frequently overlooking environmental sustainability and ecosystem needs.[27] While traditional methods have contributed to significant advancements in water supply and agricultural productivity, they are increasingly insufficient to address the complexities and uncertainties of a changing climate.

Table 1 depicts the limitations of traditional water management approaches. Reliance on historical hydrological data is undermined by altered precipitation patterns, increased frequency of extreme weather events, and shifting temperature regimes.[29] This non-stationarity renders water allocation systems based on historical data unreliable, while centralized management structures lack the flexibility required for adaptive management, leading to rigid policies that cannot swiftly respond to emerging challenges.[31–34] Traditional methods often lack the flexibility inherent in Adaptive Water Management (AWM) and struggle to implement Integrated Water Resources Management (IWRM), which advocates for coordinated and adaptive approaches.[31,35] The failure to effectively integrate interdisciplinary knowledge and technological advancements also hampers a holistic approach to water-related issues. The complexity of these issues necessitates approaches like the Water-Energy-Food (WEF) Nexus and Ecosystem-based Approaches (EbA),[36,37] yet compartmentalization into separate sectors leads to conflicting policies and missed opportunities for synergistic solutions.[31,38] This fragmentation impedes Climate Risk Assessments and Scenario Planning, essential tools for developing adaptive strategies, and the slow adoption of technologies like RS, GIS, smart water technologies, and data analytics limits real-time data gathering and proactive climate risk mitigation.[28,39]

Moreover, traditional water management often excludes local communities, indigenous peoples, and other stakeholders, undermining effectiveness, leading to conflicts, and marginalizing vulnerable populations.[43–45] The failure to adopt Participatory Water Management approaches limits the incorporation of traditional knowledge and locally relevant solutions vital for sustainable management.[40,46] Insufficient utilization of Public-Private Partnerships (PPPs) hinders investment and innovation, reducing the capacity to address emerging challenges.[47] Environmental sustainability, amidst changing climate, is inadequately addressed, as the focus on supply-side solutions and maximizing water extraction leads to overexploitation of resources and degradation of aquatic ecosystems, neglecting environmental flow needs.[41,42,48] Traditional methods often overlook approaches like Ecosystem-based Adaptation (EbA) and Water-Sensitive Urban Design (WSUD) that integrate ecological principles to improve resilience and sustainability.[37,49] Additionally, insufficient emphasis on water-efficient agriculture and irrigation practices perpetuates inefficient water use in agriculture, which is the largest consumer of fresh water globally. The lack of integration between water management and environmental conservation compromises the resilience of ecosystems vital for buffering climate impacts.[50-53]

Table 1 Limitations of traditional water management methods.

Limitation	Description	References
Assumption of Stationarity	Reliance on historical hydrological data, ignoring the impacts of climate change on water cycles and future variability.	Milly et al.[29]
Centralized Decision-Making	Top-down management structures lack the flexibility and adaptability to respond to emerging challenges and uncertainties.	Pahl-Wostl,[31] Huitema et al.[32]
Limited Stakeholder Engagement	The exclusion of local communities and stakeholders from decision-making leads to conflicts and a lack of social acceptance.	Ostrom,[33] Reed[40]
Underutilization of Technology	Failure to adopt emerging technologies for data collection, monitoring, and analysis impedes informed and proactive management.	Pulwarty and Sivakumar,[28] Buytaert et al.[39]
Environmental Neglect	Overemphasis on water extraction without considering ecological needs leads to ecosystem degradation and loss of biodiversity.	Gleick,[41] Srivastava et al.[42]
Lack of Interdisciplinary Integration	Insufficient incorporation of ecological, social, and economic perspectives, resulting in fragmented and ineffective management strategies.	Sarewitz and Pielke,[38] Pahl-Wostl[31]

Table 2 provides a comparative overview to illustrate the key differences between traditional and climate-smart water management approaches. The transition from centralized, top-down management to more integrated and adaptive approaches, such as IWRM and AWM, signifies a recognition of the interconnected nature of water systems and the need for flexibility in management strategies.[31,35] Notably, the shift towards climate risk assessments and scenario planning in decision-making processes demonstrates a proactive stance in addressing future uncertainties, contrasting sharply with the historical data-dependent approaches of traditional management. Incorporating ecosystem-based approaches and the Water-Energy-Food Nexus concept in climate-smart management underscores a growing appreciation for the intricate relationships between water resources and broader environmental and socioeconomic systems. Furthermore, adopting participatory management models and integrating smart technologies reflect a move towards more inclusive and technologically advanced management practices. These theoretical changes represent practical shifts in managing, planning, and protecting water resources. As highlighted in the table, the evolution in agricultural practices and urban water management illustrates the far-reaching implications of climate-smart approaches across various sectors. Collectively, these transformations point to a more holistic, adaptive, and forward-thinking water management paradigm that is better equipped to address the complex challenges posed by climate change and ensure sustainable water management for future generations.

2.2 Recent advances and technological integration

The limitations of traditional water management approaches have spurred significant innovation, leading to advancements in technological integration and methodological frameworks that improve climate-smart water management. Technologies such as AI, ML, RS, GIS, and the IoT have revolutionized water resource monitoring, analysis, and management.[25,64] These innovations allow precise data collection, real-time monitoring, predictive modeling, and improved stakeholder engagement, which are critical for addressing the complexities introduced by climate change.[65] For instance, AI and ML algorithms process vast datasets to identify patterns and predict trends, improving hydrological forecasts and empowering proactive planning.[66] Integrating these technologies into water management frameworks promotes more adaptive, inclusive, and resilient systems, bridging gaps left by traditional approaches and providing

Table 2 Comparison of traditional and climate-smart water management approaches.

Aspect	Traditional water management	Climate-smart water management	References
Management Approach	Centralized, top-down	Integrated Water Resources Management (IWRM), Adaptive Water Management (AWM)	Agarwal et al.,[35] Pahl-Wostl[31]
Decision-making	Based on historical data and fixed projections	Climate Risk Assessments and Scenario Planning	Dessai and Hulme[54]
Ecosystem Consideration	Often overlooked or treated separately	Ecosystem-Based Approaches (EbA)	Cohen-Shacham et al.[55,56]
Sectoral Integration	Limited, siloed approach	Water-Energy-Food (WEF) Nexus Approach	Hoff et al.[57]
Urban Water Management	Focused on grey infrastructure	Water-Sensitive Urban Design (WSUD)	Wong and Brown[58]
Technology Adoption	Slow, often resistant to change	Smart Water Technologies	Makropoulos et al.[59]
Stakeholder Engagement	Limited, often excluding marginalized groups	Participatory Water Management	Jiménez et al.[45]
Agricultural Practices	Traditional irrigation methods	Water-Efficient Agriculture and Irrigation Practices	Chartzoulakis and Bertaki,[60] Elbeltagi et al.[61]
Infrastructure Development	Primarily public sector-driven	Public-Private Partnerships (PPPs)	Ameyaw and Chan[62]
Adaptability	Low, rigid regulatory frameworks	High, flexible policies and adaptive management	Pahl-Wostl[31]
Climate Change Consideration	Often overlooked or treated as an external factor	Central to decision-making and planning processes	Milly et al.[29]
Time horizon	Short to medium-term	Long-term, considering climate projections	Haasnoot et al.[63]

innovative solutions to complex water-related challenges. To illustrate the range of technological advancements and their applications in climate-smart water management, Table 3 provides an overview of key technologies, their applications, and examples from recent research.

Building upon the capabilities of AI and ML, RS and GIS have also played pivotal roles in advancing water management by providing near-real-time data on various hydrological parameters over large spatial and temporal scales. Satellite observations and advanced image processing techniques augment our ability to track changes in water resources, assess climate change impacts, and inform decision-making processes.[96] RS technologies provide important data on precipitation, evapotranspiration, soil moisture, and surface water dynamics,[97–99] while GIS allows the integration and visualization of spatial data, supporting watershed analysis, land-use planning, and environmental impact assessments.[100] For example, the use of gravity recovery satellites to measure changes in groundwater storage has provided crucial findings into the depletion of aquifers worldwide, informing policy interventions and conservation efforts.[101] The ability to overlay climate projections, hydrological data, and socio-economic indicators facilitates in-depth and context-specific approaches to water resource management.[102] These technologies improve the assessment of water availability, identification of vulnerable areas, and infrastructure planning, leading to more informed and strategic decision-making crucial for effective climate adaptation strategies.[103] Similarly, the Internet of Things introduces new possibilities for real-time monitoring and management of water resources. IoT devices, including sensors and smart meters, collect continuous data on water quality, flow rates, and consumption patterns, allowing immediate detection of anomalies like pollution events or system malfunctions and allowing for swift corrective actions.[84] IoT technologies empower consumers by providing feedback on water usage and promoting conservation behaviors.[104]

Despite these technological innovations, challenges such as data quality, interoperability, cybersecurity, and the digital divide must be addressed to realize their full potential.[105] Many management structures lack the capacity to process and interpret the increasing volume and complexity of water-related data in a timely and meaningful way.[106] Ensuring equitable access, building technical capacity, and advancing interdisciplinary collaboration are essential to maximize the benefits of these advancements.[65] As climate change intensifies hydrological variability and uncertainty, shifting toward more adaptive, inclusive, and sustainable management models is

Table 3 Technological innovations in water management.

Technology	Applications	Contributions to climate-smart water management	Case studies/examples
Artificial Intelligence (AI) and Machine Learning (ML)	– Hydrological forecasting – Optimization of water distribution networks – Leak detection and prevention	Refines adaptive capacity through improved forecasting and efficient resource allocation	Australia: ML models for flood prediction in Mary River.[67] USA: AI for leak detection in water systems in California.[68] Egypt: ML models for forecasting the drought phenomena based on limited inputs[69] India: ML models for vector-born disease prediction amidst changing climate[70] Iraq: AI techniques for addressing difficulties in irrigation water quality prediction[71] General: Support vector machines to capture the monthly variation of streamflow time series[72]
Artificial Neural Networks (ANN)	– Modeling complex hydrological processes – Flood and drought prediction – Water quality assessment – Rainfall-runoff modeling	Improved modeling of non-linear systems and enhanced predictive capabilities.	Europe: ANN-based water quality prediction in the Danube River, improving early warning systems for pollution events[73] Iran: ANN for groundwater level prediction.[74] Canada: ANN for water quality prediction in Great Lakes.[75] India: ANN for forecasting of Standardized Precipitation Index and meteorological drought[76]

(continued)

Table 3 Technological innovations in water management. (*cont'd*)

Technology	Applications	Contributions to climate-smart water management	Case studies/examples
Remote Sensing (RS)	– Monitoring Hydrological parameters – Watershed analysis – Land-use planning	Provides detailed, large-scale data for informed decision-making and long-term planning	India: RS for drought assessment in Rajasthan[77] and land use land cover change assessment in Maharashtra[78] Bangladesh: Flood mapping using RS.[79]
Geographic Information Systems (GIS)	– Watershed analysis – Land-use planning – Resource allocation – Risk mapping – - Scenario planning	Facilitates integrated watershed management and climate risk assessment	China: GIS-based water management in Yellow River Basin.[80] USA: GIS for water quality assessment in Mississippi River.[81] India: GIS for choosing appropriate locations for built-up growth while maintaining the natural balance[82]
Internet of Things (IoT)	– Real-time water quality monitoring – Smart metering – Infrastructure management – Early warning systems	Allow rapid response to changing conditions and promote water conservation by improved transparency and user engagement	Spain: IoT sensors in Ebro River Basin.[83] South Africa: Smart meters in Cape Town for conservation.[84]
Blockchain	– Water rights management – Secure data management – Transparent transactions – Water trading platforms	Increases accountability and trust in water allocation systems	China: Blockchain-based water quality data management system in Guangzhou[85] USA: Blockchain for water rights trading in California.[86] Australia: Blockchain in Murray-Darling Basin water markets.[87]

Digital Twins	– Infrastructure management – Scenario testing – Operational optimization – Virtual modeling of water systems – Predictive maintenance	Supports resilience planning and adaptive management strategies	Greece: Digital twin of urban water systems in Athens, allowing real-time optimization and reducing operational costs by 15 %[88] Netherlands: Digital twin of Amsterdam's water system.[89]
Big Data Analytics	– Processing large datasets – Pattern recognition – Risk assessment – Trend analysis – Decision support	Data-driven findings improving understanding of complex water-climate interactions that support in evidence-based policymaking	China: Big data-driven water demand forecasting in Beijing, improving accuracy by 25 % and supporting water conservation efforts[90] Brazil: Analytics for drought monitoring in Northeast region.[91]
Cloud Computing	– Data storage, processing, and sharing – Computational power for modeling – Collaboration platforms	Facilitates knowledge sharing and cross-sectoral collaboration in water management	Global: Cloud-based hydrological modeling.[92] USA: Cloud services for water data management.[93]
Citizen Science and Mobile Apps	– Community-based monitoring – Data collection by citizens – Public engagement – Reporting of water issues	Increase data coverage and promote inclusive management and local adaptation strategies	Europe: FreshWater Watch program for water quality[94]. Kenya: Mobile apps for reporting water access issues, empowering communities, and improving water access for over 50,000 people.[95]

urgently needed. Recognizing and understanding these limitations is essential for reforming water management frameworks to improve resilience and ensure the sustainable management of water resources in a changing climate. The next frontier in climate-smart water management lies in creating synergies between cutting-edge technologies and the wealth of contextual knowledge held by communities and practitioners on the ground. This holistic approach, combining technological innovation with social and institutional adaptations, holds the key to addressing the complex water challenges of the 21st century.

3. Introduction to LLMs in water management

LLMs, developed through AI and computation power advancements, can understand complex language patterns, extract meaningful information from unstructured data, and support decision-making processes.[16] Their potential applications span various domains, including NLP, data analysis, predictive modeling, and stakeholder engagement. In the context of water management, LLMs provide opportunities to augment information management, improve communication among stakeholders, and facilitate more informed and adaptive decision-making.[107] LLMs can also help stakeholders retrieve recent and precise emission data, thus refining decision-making.[108] The evolution of LLMs has empowered them to perform complex tasks such as language translation, summarization, question-answering, and content generation with remarkable accuracy.[24] Subsequent subsections will highlight how these advancements position LLMs as beneficial tools for improving climate-smart water management.

3.1 Evolution and capabilities of LLMs

The evolution of LLM represents a significant milestone in artificial intelligence, particularly in NLP, where early language models relied on statistical methods and n-gram probabilities to predict word sequences, which had limited capabilities due to computational constraints and simplistic algorithms.[109] The introduction of neural networks and deep learning techniques marked a turning point, empowering models to learn complex language patterns and representations from large datasets.[110] Recurrent Neural Networks (RNNs) and Long Short-Term Memory (LSTM) networks improved the handling of sequential data, capturing dependencies across longer text spans.[111] However, these models faced challenges with long-term context and parallelization. The advent of the Transformer Architecture addressed these limitations by utilizing self-attention mechanisms, allowing models to process entire sequences simultaneously and capture global dependencies more effectively.[112] This breakthrough led to the development of models like Bidirectional Encoder Representations from Transformers (BERT) and Generative Pre-trained Transformers (GPT), significantly advancing the state of NLP.

Building upon the Transformer architecture, models like BERT introduced bidirectional training, allowing the model to simultaneously consider both left and right contexts during training.[16] BERT achieved state-of-the-art results on various NLP benchmarks, demonstrating a superior understanding of context and semantics. In contrast, GPT models focused on unidirectional language modeling but scaled up in terms of parameters and data, leading to impressive generative capabilities.[113] With billions of parameters, GPT-2 and GPT-3 showcased the ability to generate coherent and contextually relevant text across diverse topics, including articles, summaries, and even code.[15] These models capitalized on unsupervised pre-training on massive text corpora, followed by fine-tuning specific tasks, allowing them to generalize effectively across various applications. The release of GPT-4 marked a significant advancement in LLM capabilities with advances in reasoning, understanding, and creativity, designed to be more aligned with human values and intent.[114] While specific parameter counts were not publicly disclosed, GPT-4 demonstrated improved performance over its predecessors in tasks requiring complex reasoning and understanding. Notably, GPT-4 expanded into multimodal capabilities, processing not only text but also images, allowing it to interpret and generate responses based on visual inputs.[114] This multimodal integration represents a significant step toward models that can understand and interpret vast amounts of contextual data from diverse modalities, identifying correlations and patterns across text and visual information. For example, they can simultaneously analyze historical climate data, water usage patterns, remote sensing data, satellite imagery, and socio-economic indicators, uncovering subtle relationships that might escape human analysts.[115] This capability is crucial for developing more accurate predictive models and scenario analyses in water resource management.

In addition to GPT-4, other models like Google's Pathways Language Model (PaLM) and Claude 3.5 Sonnet have pushed the boundaries of LLM capabilities and exhibit remarkable proficiency in understanding and generating human-like text, focusing on scaling up model size and training data to improve performance[116] (https://www.anthropic.com/news/claude-3–5-sonnet, accessed October 21, 2024). Meta's LLaMA (Large Language Model Meta AI) series contributed to the open research community, providing access to LLMs with varying sizes to facilitate research and development.[117] With the advent of multimodal models like GPT-4 and Gemini Ultra, the ability to process and interpret visual data in conjunction with textual information augments the applicability of LLMs in domains where visual context is important (https://blog.google/technology/ai/google-gemini-ai/, accessed October 21, 2024). Table 4 summarizes significant models, highlighting their architectures, parameter sizes, and key advancements over recent years. This overview provides context for understanding how LLMs have evolved to achieve their current capabilities, setting the stage for their potential applications in climate-smart water management. A task that might take human experts weeks or months to complete – such as synthesizing findings from thousands of research papers on climate impacts on water resources – could potentially be accomplished by an LLM in a matter of hours, providing decision-makers with timely and detailed findings.[118]

3.2 Existing applications in related fields

The application of LLMs has expanded across various sectors, demonstrating their potential to transform data analysis, decision-making, and stakeholder engagement processes. Considering only the related fields, such as environmental science and disaster management, LLMs have been employed to process vast amounts of unstructured data, extract meaningful observations, and support critical operations.[120–122] In environmental science, LLMs have been applied to analyze climate data, research publications, and policy documents to support environmental monitoring and policy development. For example, LLMs have been used to extract information on climate change impacts from scientific literature, aiding researchers and policymakers in understanding complex environmental phenomena.[123] These applications demonstrate the potential of LLMs to process and synthesize large volumes of technical information, providing a foundation for evidence-based decision-making in water management. Similarly, in disaster management, LLMs have been employed to analyze social media feeds, news reports, and sensor data to detect and respond to natural hazards.[124,125] Models like GPT-3.5 have been used to monitor and interpret data related to earthquakes, floods, and wildfires, permitting authorities to issue timely warnings and coordinate emergency responses.[126] Models capable of processing live video feeds from

Table 4 Evolution of large language models.

Model	Year	architecture	Parameters	Key advancements	References
Generative Pre-trained Transformer (GPT)	2018	Transformer (Unidirectional)	117 million	Demonstrated effectiveness of unsupervised pre-training for Natural Language Processing (NLP) tasks.	Radford et al.[113]
Bidirectional Encoder Representations from Transformers (BERT)	2019	Transformer (Bidirectional)	340 million	Introduced bidirectional training; improved understanding of context and semantics.	Devlin[16]
GPT-2	2019	Transformer (Unidirectional)	1.5 billion	Showcased advanced text generation capabilities; raised awareness of ethical considerations.	Radford et al.[113]
Text-To-Text Transfer Transformer (T5)	2020	Transformer (Text-to-Text)	11 billion	Unified NLP tasks under a text-to-text framework; improved transfer learning across tasks.	Raffel et al.[119]
GPT-3	2020	Transformer (Unidirectional)	175 billion	Achieved few-shot learning; demonstrated remarkable generalization across diverse tasks.	Brown[15]
GPT-4	2023	Transformer (Multimodal)	Undisclosed	Advanced reasoning and understanding; introduced multimodal capabilities to process text and images.	Achiam et al.[114]
Pathways Language Model (PaLM)	2022	Transformer (Unidirectional)	540 billion	Scaled-up model focusing on improved reasoning and understanding; strong performance on benchmarks.	Chowdhery et al.[116]
Large Language Model Meta AI (LLaMA)	2023	Transformer (Unidirectional)	Up to 65 billion	Open-source models facilitate research and contribute to community advancements in LLM development.	Touvron et al.[117]

drones or surveillance cameras can detect signs of natural disasters such as wildfires, floods, or landslides.[103] Audio data analysis, such as processing emergency calls or social media audio posts, provides meaningful information during crises.[127] Integrating these capabilities augments situational awareness and enables faster, more effective emergency responses, essential for managing water-related disasters. The capability of LLMs to process real-time data and generate predictive analytics is particularly relevant for water management, where timely decision-making is essential to address rapidly changing environmental conditions. The LLMs have also been applied in the energy sector to forecast energy demand, optimize grid operations, and facilitate communication with consumers. By analyzing data from smart meters, customer interactions, and market trends, LLMs contribute to more efficient energy distribution and consumption.[128] The parallels between energy and water systems suggest that similar approaches can be adopted in water management to optimize resource allocation, improve infrastructure management, and engage with consumers.

Table 5 provides a summarized overview highlighting how these technologies have been employed, their contributions, and the lessons that can be applied to climate-smart water management. This table connects existing innovations in related fields to potential strategies for improving water management practices. One of the most prominent applications of LLMs in environmental science has been in the realm of scientific literature analysis and synthesis. The exponential growth of research publications on climate change and water resources has made it increasingly challenging for researchers and policymakers to stay abreast of the latest findings. LLMs have shown remarkable capabilities in addressing this challenge. For instance, Miao et al.[18] demonstrated using LLMs to perform automated literature reviews on climate change impacts, extracting key findings and synthesizing information from thousands of papers in a fraction of the time it would take human experts. They can also facilitate communication between policymakers, practitioners, and the public by translating technical jargon into accessible language, thereby improving transparency and stakeholder participation. Li & Ning[122] developed an LLM-powered conversational interface that allows users to query and visualize complex geospatial data through natural language commands, democratizing access to water-related spatial data, empowering stakeholders at all levels to explore and analyze information on water resources, infrastructure, and climate impacts more intuitively. LLMs have been employed to augment climate model interpretability and improve communication of complex scientific concepts to non-expert stakeholders.

Table 5 Existing applications of LLMs and multimodal models in related fields.

Field	Applications of LLMs	Contributions	Lessons for water management	References
Environmental Science	– Extracting climate impact data – Analyzing policy documents – Satellite image analysis for environmental monitoring – Processing climate models – Audio analysis of wildlife	– Refined policy development – Improved environmental monitoring	– Use of remote sensing data – Multimodal data integration – Synthesis of technical information – Evidence-based decision-making	Rillig et al.,[124] Agathokleous et al.[123]
Disaster Management	– Video analysis for real-time disaster detection – Image processing from drones – Audio monitoring of emergency communications – Analyzing social media for early warnings – Coordinating emergency responses	– Timely warnings – Effective emergency coordination – Improved situational awareness	– Real-time data processing of visual and audio data – Improved situational awareness	Kapoor et al.,[126] Li et al.[103]
Energy Management	– Forecasting energy demand – Optimizing grid operations – Customer engagement	– Efficient energy distribution – Better consumer relations	– Resource optimization – Consumer engagement strategies	Zaboli et al.[128]
Urban Planning and Infrastructure	– Image analysis for infrastructure inspection – Video Monitoring of construction sites – Audio detection of structural anomalies	– Preventive maintenance – Risk reduction	– Infrastructure monitoring – Early detection of issues	Zhou et al.,[129] Cruz-Castro et al.,[130] Dhanuka et al.[131]

4. Integration of LLMs in water resource management

Integrating LLMs into climate-smart water management presents a transformative opportunity to improve water management practices' effectiveness, adaptability, and inclusivity. LLMs present opportunities and challenges for refining water management, particularly in the context of Sustainable Development Goal 6 (SDG 6), which emphasizes ensuring access to clean water and environmental sanitation for all. LLMs can apply their computational power to analyze vast amounts of environmental data, identify pollution sources, optimize water usage, and forecast water scarcity scenarios.[132] This data-driven approach allows for the formulation of effective water management strategies and conservation measures, contributing to a more informed and proactive society. This section explores the potential applications of LLMs in climate-smart water management, focusing on four key areas: data analysis and interpretation, decision support systems, predictive modeling and forecasting, and communication and stakeholder engagement.

4.1 Data analysis and interpretation

Effective data analysis and interpretation are critical components of climate-smart water management, given the vast amounts of data generated from various sources, such as hydrological models, remote sensing, sensor networks, and policy documents.[133] LLMs can process and analyze large, diverse datasets, including unstructured textual data, such as research articles, policy documents, technical reports, and stakeholder feedback, allowing the extraction of meaningful observations that support evidence-based decision-making.[134] For instance, LLM could analyze historical water usage data alongside climate projections, satellite imagery, and demographic trends to better understand future water demand scenarios.[135] This improved analytical capability is particularly meaningful in addressing the challenges of climate change, which requires integrating multidisciplinary data to develop adaptive management strategies. Traditional data analysis methods often struggle with unstructured data due to its complexity and variability. By analyzing large-scale historical data, LLMs can identify subtle patterns and anomalies that might escape human analysts. This capability is particularly significant for the early detection of emerging water quality issues or shifts in hydrological patterns due to climate change.[135] For example, LLMs can analyze a corpus of climate research papers to identify emerging trends in water scarcity or flood risks, providing important information for strategic planning. By automating the analysis of large volumes of textual data, LLMs reduce the time and effort required for manual review, allowing decision-makers to focus on implementing effective interventions.

Furthermore, the LLMs can integrate data from various modalities, including text, images, and audio, to provide a more holistic analysis of water systems.[114] For instance, when combined with remote sensing data, LLMs can interpret satellite imagery and correlate it with textual reports on water quality or land-use changes.[135] Tools like FloodGen visually represent climate threats, empowering communities to adopt preventive strategies while shaping policy decisions through impactful visualizations (https://beta.nyc/2024/05/03/betanyc-presents-floodgen-a-advocacy-tool-that-uses-generative-ai-imagery/, accessed October 21, 2024). This multimodal analysis allows the identification of correlations and patterns that may not be apparent when examining data sources in isolation. By uncovering these findings, water managers can better understand the drivers of water resource changes and develop targeted strategies to address them. The ability to analyze and interpret diverse data types improves the responsiveness and adaptability of water management systems. For example, in environmental monitoring, LLMs have been used to analyze large environmental reports and sensor data datasets to detect pollution events and assess ecosystem health.[123] Li et al.[136] Li et al.[137] demonstrated the application of an LLM in analyzing multi-source data for the Yellow River Basin in China. The model integrated historical hydrological data, climate projections, land use changes, and policy documents to provide thorough findings into the basin's water resource dynamics. The LLM-driven analysis identified previously overlooked correlations between land use changes and water quality, informing more effective conservation policies. In the context of water management, similar approaches can be employed to monitor water quality, detect contamination, and assess the effectiveness of conservation measures. These applications underscore the potential of LLMs to revolutionize data analysis and interpretation in climate-smart water management. By improving data analysis and interpretation, LLMs lay a strong foundation for improved decision-making processes.

4.2 Decision support systems (DSS)

Decision Support Systems (DSS) are critical tools in water management, assisting managers and policymakers in making informed decisions based on data analysis, modeling, and scenario evaluation.[138] LLMs can process and interpret vast amounts of data, generate insightful summaries, and facilitate interactive dialogues with users, making DSS more user-friendly and effective. One of the primary benefits of incorporating LLMs into DSS is enhancing user interaction through natural language processing. Traditional DSS often require specialized knowledge to operate and interpret outputs, which can be a barrier for non-expert stakeholders.[139] LLMs enable the development of conversational interfaces that allow users to input queries and receive explanations in natural language, thereby increasing accessibility and user engagement.[140] For example, a water manager could ask the DSS, "What are the projected impacts of a 2 °C temperature increase on regional water availability?" or planners could ask, "What are the projected water demands for the next quarter in District X?" and receive a detailed, comprehensible response generated by the LLM. The LLM provided concise summaries and visualizations based on real-time data and predictive models. This capability democratizes access to complex analytical tools, promoting inclusive decision-making processes.

Additionally, LLMs augment the analytical capabilities of DSS by integrating qualitative data and expert knowledge into decision-making processes. Water management often involves complex socio-economic and environmental considerations not fully captured by quantitative models alone.[31] The holistic approach of LLMs ensures that decisions are informed by an extensive understanding of the factors influencing water resources, including social, cultural, and political dimensions. By bridging the gap between quantitative models and qualitative findings, LLMs boost the robustness and relevance of DSS outputs.[141] Furthermore, integrating LLMs into DSS facilitates scenario analysis and risk assessment by allowing the generation and evaluation of multiple future scenarios.[142] LLMs can assist in constructing narratives around potential climatic and socio-economic changes, helping stakeholders visualize the implications of different management strategies.[143] For instance, LLMs can generate detailed descriptions of how specific policy interventions may impact water security under various climate scenarios.[144] This supports more strategic planning and helps identify adaptive measures that fit a range of possible futures. By improving DSS through integrating user-friendly interfaces powered by LLMs, water management can become more adaptive, inclusive, and effective in addressing climate-related challenges.

4.3 Predictive modeling and forecasting

Predictive modeling and forecasting are essential to climate-smart water management, empowering stakeholders to anticipate future conditions and plan accordingly.[66] LLMs contribute to these efforts by incorporating quantitative and qualitative information, providing a more thorough prediction basis.[132] This integration improves the accuracy and relevance of forecasts, supporting proactive management strategies that mitigate risks associated with climate variability and change. One significant advantage of using LLMs in predictive modeling is their ability to process and learn from unstructured textual data, such as news reports, scientific literature, and social media feeds, allowing LLMs to capture emerging trends and signals that may not be reflected in numerical datasets.[134] For example, analysis of news articles and policy discussions can reveal shifts in water usage patterns or regulatory changes that could impact future water demand.[145] By incorporating this information, LLMs provide more nuanced forecasts for socio-economic and climatic factors.

Moreover, the LLMs with multimodal capabilities can integrate visual data from satellite imagery and remote sensing, further enriching predictive modeling efforts.[114] The LLMs can identify environmental trends that influence hydrological cycles by analyzing changes in land use, vegetation cover, and surface water extent. Combining this visual data with textual information on weather patterns and human activities allows more accurate forecasts of events such as droughts, floods, and water quality issues.[135] These improved predictive capabilities support early warning systems and help stakeholders implement timely interventions to mitigate adverse impacts. By identifying patterns and correlating diverse data sources, the LLM improved the accuracy of flood predictions.[146] This empowers early warning systems to provide timely alerts to communities, allowing for better preparedness and response to flooding events exacerbated by climate change. Water management can become more proactive and resilient by capitalizing on the predictive modeling and forecasting capabilities of LLMs. Accurate forecasts allow for better resource allocation, infrastructure planning, and risk mitigation strategies.

4.4 Communication and stakeholder engagement

Effective communication and stakeholder engagement are fundamental to successful water management, particularly in the context of climate change, where decisions often have wide-reaching impacts.[40] LLMs present significant potential to improve these aspects by facilitating the dissemination of complex information in accessible language, permitting interactive dialogues, and supporting inclusive participation.[140] Boosting communication promotes collaborative decision-making processes that consider diverse perspectives and knowledge systems. Water management involves intricate technical details that can be challenging for stakeholders without specialized knowledge. LLMs can create plain-language summaries, visualizations, and multilingual translations to make information more accessible.[147] For example, an LLM can summarize a technical report on climate projections and their implications for local water resources, providing community members with the necessary information to engage in discussions and decision-making processes. This inclusivity boosts the legitimacy and acceptance of management outcomes.

Additionally, LLMs can facilitate interactive communication platforms where stakeholders can ask questions and receive real-time informative responses.[148] These conversational agents can serve as virtual assistants, providing stakeholders customized information and clarifications on water-related issues. For instance, a farmer might inquire in their local language about best practices for water conservation under projected drought conditions, and the LLM-powered assistant can provide evidence-based recommendations.[149] Williams et al.[150] have described how the City of Helsinki has implemented AI-driven models to improve access to urban services such as parking, sports facilities, and rental housing through platforms that grant rule-based dialogue systems using pre-loaded data (https://ai.hel.fi/en/ai-register/, accessed October 21, 2024). Similarly, CitizenLab (www.citizenlab.co, accessed October 21, 2024) employs conversational agents for participatory processes like budgeting, gathering public feedback on proposals, and using AI to summarize community input. Researchers at the University of Helsinki have taken a more extensive approach with Co-Creation Radar, a suite of AI tools designed to help city officials align municipal resources with the needs of citizens. One notable service, City-Tips, utilizes AI to transform citizens' requests into actionable projects. Moreover, LLMs can assist in sentiment analysis and stakeholder feedback collection by processing social media posts, surveys, and public comments.[151] This empowers water managers to gauge public opinion, identify concerns, and understand the social dimensions of water issues.[152] This data lets decision-makers personalize communication strategies and policies to address stakeholder needs and expectations effectively. This responsiveness improves the adaptability of management systems and strengthens the relationship between authorities and the communities they serve. Improving communication and stakeholder engagement through LLMs improves the inclusivity of water management and supports more effective implementation of policies and strategies. LLMs build social capital and community resilience by promoting transparent and collaborative processes.

5. Challenges, ethical considerations, and future directions

Integrating LLMs into climate-smart water management holds significant promise for augmenting data analysis, decision-making, predictive modeling, and stakeholder engagement. However, this integration has challenges and ethical considerations that must be carefully addressed to ensure effective and responsible implementation. This section examines the potential obstacles related to technological limitations, data privacy, bias, and ethical concerns. It also explores future trends and innovations that could further advance the role of LLMs in water management.

5.1 Technological challenges and limitations

Despite the advanced capabilities of LLMs, several technological challenges and limitations can hinder their effective integration into water management. One of the primary concerns is the computational resources required to train and deploy large models. LLMs like GPT-4 demand significant processing power and memory to train on custom resources, which can be a barrier for organizations with limited technological infrastructure.[153] This limitation may exacerbate disparities between well-resourced institutions and those in developing regions, potentially widening the digital divide in water management. Addressing this challenge requires exploring more efficient models, leveraging cloud computing resources, or adopting smaller, specialized models that require less computational power. Another technological challenge is the quality and availability of data required to train LLMs effectively. Water management involves diverse datasets, including hydrological data, policy documents, and socio-economic information, which may be incomplete, inconsistent, or inaccessible.[39] LLMs rely on large volumes of high-quality data to perform accurately, and deficiencies in data can lead to suboptimal performance or erroneous outputs. Notably, the latest models of LLM often lack up-to-date information and can produce ambiguous or imprecise outputs, which may lead to miscommunication in scientific contexts.[108,154] Overcoming this limitation involves investing in data collection and management systems, standardizing data formats, and promoting data sharing among stakeholders.

Moreover, LLMs often lack interdisciplinary integration and transparency, essential for understanding complex sustainability issues that span environmental science, economics, and sociology, making it challenging for users to understand how the models arrive at their conclusions.[132] This "black box" nature of LLMs can lead to mistrust among users and stakeholders who are reluctant to rely on outputs they cannot explain. Transparency is crucial for accountability and acceptance in water management, where decisions have significant environmental and socio-economic impacts.[31] While LLMs focus on efficiency and cost-effectiveness, human decision-makers consider broader societal factors such as sustainability, equity, and the cultural significance of water sources.[132] Addressing this issue requires developing methods for model interpretability, such as explainable AI techniques that provide observations into the model's decision-making processes. Improving interpretability can build trust and facilitate the adoption of LLMs in water management applications. These technological challenges highlight the need for careful consideration and strategic planning when integrating LLMs into water management. By acknowledging these limitations, stakeholders can develop solutions to mitigate their impact, such as investing in infrastructure, improving data quality, and improving model transparency. Addressing technological challenges is a critical step toward utilizing the full potential of LLMs in climate-smart water management.

5.2 Ethical considerations and data privacy

The deployment of LLMs in water management raises several ethical considerations regarding data privacy, bias, and potential unintended consequences. One of the foremost concerns is handling sensitive data, including personal information from stakeholders and proprietary data from organizations.[155] LLMs require vast amounts of data for training, and including sensitive information can lead to privacy violations if not properly managed. In the context of water management, data such as water usage patterns, billing information, and stakeholder communications must be protected to comply with legal regulations and ethical standards.[156] Implementing strong data anonymization and encryption methods is essential to safeguard privacy while allowing LLMs to process necessary information. Bias in LLM outputs is another significant ethical concern. LLMs learn from the data they are trained on, and if that data contains biases, the models may perpetuate or even amplify those biases. In water management, biased outputs could lead to unfair resource allocation, exclusion of marginalized communities, or reinforcement of existing inequalities. For example, if historical data reflects systemic disparities in water access, an AI might recommend policies that inadvertently favor certain groups over others.[44] Ethical considerations surrounding algorithm transparency and bias must be prioritized to prevent exacerbating existing social inequalities and to ensure that all voices, particularly those of marginalized communities, are included in management processes.[132] Addressing this issue requires careful dataset curation, bias detection mechanisms, and incorporating fairness constraints into model training.[157] Ensuring that LLMs promote equitable outcomes is crucial for ethical and effective water management.

The potential for unintended consequences also necessitates ethical consideration. LLMs may generate technically accurate outputs but contextually inappropriate or misaligned with cultural values.[155] In diverse communities with varying beliefs and practices related to water use, LLM recommendations must be sensitive to local contexts. Engaging with stakeholders to understand their needs and perspectives is essential to designing LLM applications appropriately.[158] The improved capabilities of LLMs in processing sensitive data and making high-stakes recommendations necessitate strong ethical frameworks and careful consideration of potential biases and limitations.[159] Additionally, establishing oversight mechanisms to review and validate LLM outputs can prevent the implementation of misguided or harmful policies. These ethical considerations emphasize the importance of responsible AI practices in deploying LLMs for water management. By prioritizing data privacy, addressing bias, and preventing unintended consequences, stakeholders can ensure that LLMs contribute positively to climate-smart water management. Ethical deployment mitigates risks and boosts the legitimacy and acceptance of LLM-driven initiatives.

5.3 Future trends and innovations

As the world approaches the midpoint of the United Nations' 2030 Agenda for Sustainable Development, the integration of LLMs in climate-smart water management presents promising avenues for augmenting SDGs, particularly SDG 6, which focuses on clean water and sanitation for all (https://sdgs.un.org/goals/goal6, accessed October 21, 2024). The potential for LLMs to assist in identifying pollution sources, optimizing water use, and forecasting water scarcity scenarios could significantly transform how water resources are managed.[132] One significant development is advancing more efficient and specialized models, often called "small but mighty" models, which aim to deliver high performance with reduced computational requirements.[160] These models can make LLM technology more accessible to organizations with limited resources, promoting wider adoption in diverse contexts. Techniques such as knowledge distillation and model compression are being explored to create lightweight models without significant loss of capability.[161] Another promising trend is the integration of LLMs with other emerging technologies like Internet of Things (IoT) devices, blockchain, and edge computing.[162] For instance, combining LLMs with IoT sensors can augment real-time data analysis and decision-making in water management.[163] Blockchain technology can provide secure and transparent data-sharing platforms, addressing some data privacy and trust issues.[164] Edge computing allows data processing closer to the data source, reducing latency and bandwidth requirements, which is beneficial for remote or resource-constrained areas.[165] These integrations can create more powerful and responsive water management systems.

Generative AI models are increasingly facilitating access to data analysis, with examples such as GeoGPT, an adaptation of OpenAI's GPT model, and LLM-Geo. These models aim to integrate GISs with LLMs, simplifying complex geographical data analysis and allowing instant map creation for non-experts. GeoGPT, for instance, streamlines data collection, processing, and visualization using natural language commands.[166] Similarly, GeoQAMap uses an LLM to convert user questions into SPARQL queries, allowing users to navigate complex databases without specialized programming knowledge.[167] The private sector has also adopted the application of LLMs in geospatial data analysis. For example, Aino.world (www.aino.world, accessed October 21, 2024) delivers web services that support prompt-based data queries and mapping visualizations. Meanwhile, ArcGIS (www.arcgis.com, accessed October 21, 2024) has incorporated LLMs into its platform to assist users with queries, data analysis, and geospatial visualization (https://mediaspace.esri.com/media/t/1_c1ftfyju, accessed October 21, 2024). These tools are particularly beneficial for expert data scientists who can validate their analyses and understand the algorithms' limitations.

The development of multilingual and culturally adaptive LLMs is another area of innovation.[168] As water management involves stakeholders from various linguistic and cultural backgrounds, LLMs capable of understanding and generating content in multiple languages can improve communication and inclusivity.[169] Culturally adaptive models that account for local contexts and practices can provide more relevant and acceptable recommendations. This advancement supports more equitable and effective stakeholder engagement in water management. These future trends and innovations hold the potential to overcome current challenges and expand the capabilities of LLMs in climate-smart water management. By staying abreast of these developments and actively incorporating them into practice, stakeholders can improve the resilience and sustainability of water management systems in the face of ongoing climate change.

5.4 Policy and management implications

A key challenge lies in the existing management structures, which often lack the flexibility to adapt to rapidly changing climate conditions.[170] The Water Act, for instance, establishes a legal framework for adaptive management; however, reforms have been significantly politicized, leading to reduced water recovery for environmental needs due to socio-economic pressures.[171] This politicization complicates the integration of AI technologies, as stakeholder trust in management processes is paramount, and any perceived inequities can exacerbate existing tensions among water users and governments. The rapid advancement of LLM technologies and their integration into water management systems will necessitate significant changes in policy frameworks and management structures. Future water management policies must be more adaptive and responsive, capable of evolving with technological advancements and emerging environmental challenges. This may involve the development of new regulatory frameworks that can effectively oversee the use of AI in water management decisions, ensuring transparency, accountability, and ethical considerations are adequately addressed.[172] Additionally, there will be a growing need for policies that promote data sharing and standardization across different sectors and jurisdictions, facilitating the seamless integration of diverse data sources that LLMs rely on for detailed analysis. The concept of "algorithmic water rights" may emerge, where water allocation decisions are increasingly influenced by AI-driven analyses, necessitating careful consideration of equity and fairness in how these systems are designed and implemented. Furthermore, future management frameworks will likely emphasize building human and technological capacity, recognizing that effective use of advanced LLM tools in water management requires a workforce skilled in data science and environmental management. This dual focus on technological and human capital development will be crucial in ensuring that the potential of LLMs is fully realized in advancing climate-smart water management.

In this rapidly evolving landscape, it is crucial to remain mindful of the ethical implications and potential risks associated with the increased reliance on AI in water management. Future research and policy development must grapple with questions of algorithmic bias, data privacy, and the appropriate balance between human judgment and machine-driven decision-making in water management. Developing and adhering to ethical guidelines and best practices is essential for responsible AI deployment.[173] Establishing standards for data privacy, bias mitigation, and transparency ensures that LLM applications uphold ethical principles. Organizations can implement oversight committees or ethical review boards to monitor AI initiatives and address potential issues proactively, contributing to accountability and public trust.[174] There is also a need to ensure that the benefits of these technological advancements are equitably distributed, avoiding scenarios where only wealthy regions or large corporations have access to the most sophisticated water management tools. As LLMs become more integral to water management, gaining public trust and understanding of these technologies will be paramount. This may involve developing new approaches to stakeholder engagement and public education that can effectively communicate the complexities of AI-driven water management strategies to diverse audiences. By proactively addressing these challenges and thoughtfully shaping the integration of LLMs into water management frameworks, stakeholders can work towards a future where technology serves as a powerful ally in our efforts to ensure sustainable and equitable water access for all in the face of climate change.

6. Conclusions

Integrating Large Language Models (LLMs) into climate-smart water management marks a significant shift toward more adaptive, inclusive, and sustainable management of water resources. Traditional approaches, often reliant on historical data and centralized decision-making, are increasingly inadequate to address the complexities posed by climate change. LLMs provide ground-breaking potential by boosting data analysis, decision support, predictive modeling, and stakeholder engagement. Their ability to synthesize vast, multimodal datasets and generate deep understandings from quantitative and qualitative sources can revolutionize decision-making processes and improve the accuracy of climate-related forecasts. In water management, LLMs enrich stakeholder communication by translating complex technical data into accessible language, fostering inclusive dialogues, and ensuring diverse perspectives are integrated into policy and practice. This inclusive approach strengthens the social dimensions of management, promoting equity and resilience in water management. However, the integration of LLMs is not without challenges. Technical limitations, such as computational resource demands, data quality issues, and the need for model interpretability, must be addressed for successful deployment. Ethical concerns, including data privacy, algorithmic bias, and the unintended consequences of automation, also require careful consideration. Specific barriers, such as institutional resistance, financial costs, and technical expertise, must be tackled through capacity-building initiatives.

Future research should improve LLM efficiency, model transparency, and ensure culturally adaptive capabilities to make them more accessible and relevant across diverse water management contexts. Additionally, there is a need for interdisciplinary collaborations among researchers, policymakers, and practitioners to ensure the responsible deployment of these technologies. Policymakers must establish frameworks that promote the ethical use of LLMs while facilitating innovation, and practitioners should actively engage with LLM-driven tools to co-create sustainable solutions for water management. The adoption of LLMs will not only strengthen water management but will also enable systems to be more resilient, equitable, and responsive to the multifaceted impacts of climate change on water resources. By addressing both environmental and social challenges, LLMs hold the potential to contribute to the sustainable well-being of communities and ecosystems globally.

Author's contribution

All authors contributed to the study's conception and design. AS and SJ collected the data, documents, reports, and research articles and conducted the analysis, while RM supervised and guided the material preparation, data/report collection, and analysis. AS and SJ wrote the first draft of the manuscript, while RM commented on previous versions of the manuscript. All authors read and approved the final manuscript.

Conflict of interest

The authors declare that they have no known competing financial interests or personal relationships that could have appeared to influence the work reported in this study.

Declaration of generative AI and AI-assisted technologies in the writing process

While preparing this work, the authors used the 'OpenAI - Cloud Computing Platform Version 4′ to improve language. After using this service, the authors reviewed and edited the content as needed and took full responsibility for the publication's content.

References

1. IPCC. Summary for policymakers. In *Climate change 2023: synthesis report. Contribution of working groups I, II and III to the sixth assessment report of the intergovernmental panel on climate change;* Lee, H., Romero, J., Eds.; IPCC: Geneva, Switzerland, 2023; pp. 1–34. https://doi.org/10.59327/IPCC/AR6–9789291691647.001. Core Writing Team.
2. Melo Zurita MDL, Thomsen DC, Holbrook NJ, Smith TF, Lyth A, Munro PG, et al. Global water governance and climate change: identifying innovative arrangements for adaptive transformation. *Water* 2018;**10**(1):29. https://doi.org/10.3390/w10010029.
3. Özerol G, Vinke-de Kruijf J, Brisbois MC, Flores CC, Deekshit P, Girard C, et al. Comparative studies of water governance. *Ecol Soc* 2018;**23**(4).
4. Maggie White. Swed Water House; 2020. https://siwi.org/latest/why-water-governance-is-key-to-tackling-climate-change/ (Accessed 21 October 2024).
5. UN-Water. Water and climate change; 2024 Retrieved from https://www.unwater.org/water-facts/climate-change/ (Accessed 21 October 2024)
6. FAO. Climate-Smart Agriculture Sourcebook. Food and Agriculture Organization of the United Nations; 2017. https://www.fao.org/climate-smart-agriculture-sourcebook/en/ (Accessed October 21, 2024)
7. Maity R, Suman M, Verma NK. Drought prediction using a wavelet based approach to model the temporal consequences of different types of droughts. *J Hydrol* 2016;**539**:417–428. https://doi.org/10.1016/j.jhydrol.2016.05.042.
8. Srivastava A, Pandey A, Maity R. Comparative multi-index analysis of existing drought typology and environmental droughts in a climate-stressed region. *Sci Rep* 2025;**15**(1):34349. https://doi.org/10.1038/s41598-025-16859-9.
9. Srivastava A, Maity R, Desai VR. Synthesising causal loop between environmental and compound droughts: A systems-driven framework for climate-resilient governance in monsoon-sensitive agroecosystems. *Next Sustain* 2025;**6**:100159. https://doi.org/10.1016/j.nxsust.2025.100159.
10. IPCC. Summary for Policymakers. In *;* Pörtner, H.-O., Roberts, D. C., Tignor, M., Poloczanska, E. S., Mintenbeck, K., Alegría, A., Eds.; Cambridge University Press: Cambridge, UK and New York, NY, USA, 2022; pp. 3–33. https://doi.org/10.1017/9781009325844.001.
11. Gleick PH. Roadmap for sustainable water resources in southwestern North America. *Proc Natl Acad Sci* 2010;**107**(50):21300–21305.
12. Vörösmarty CJ, McIntyre PB, Gessner MO, Dudgeon D, Prusevich A, Green P, et al. Global threats to human water security and river biodiversity. *Nature* 2010;**467**(7315):555–561.
13. Biswas AK, Tortajada C. *Future water governance: problems and perspectives. Improving water policy and governance;* Routledge, 2013; pp. 1–11.
14. Wang X. *Leveraging artificial intelligence in addressing water safety challenges. Routledge handbook of urban water governance;* Routledge, 2022; pp. 316–330.
15. Brown TB. Language models are few-shot learners. *arXiv Preprint* 2020. arXiv:2005.14165.
16. Devlin J. Bert: pre-training of deep bidirectional transformers for language understanding. *arXiv Preprint* 2018. arXiv:1810.04805.
17. Min B, Ross H, Sulem E, Veyseh APB, Nguyen TH, Sainz O, et al. Recent advances in natural language processing via large pre-trained language models: a survey. *ACM Comput Surv* 2023;**56**(2):1–40. https://doi.org/10.1145/3605943.
18. Miao C, Hu J, Moradkhani H, Destouni G. Hydrological research evolution: a large language model-based analysis of 310,000 studies published globally between 1980 and 2023. *Water Resour Res* 2024;**60**(6). p.e2024WR038077.
19. Banerjee C, Bhaduri A, Saraswat C. Digitalization in Urban water governance: case study of Bengaluru and Singapore. *Front Environ Sci* 2022;**10**:816824.

20. Bhat A. *A human-centered approach designing effective large Lang Model (llm based tools Writ Softw Tutor;* McGill University: Canada, 2023.
21. Folke C, Hahn T, Olsson P, Norberg J. Adaptive governance of social-ecological systems. *Annu Rev Env Resour* 2005;**30**(1):441–473.
22. Srivastava A, Maity R. Assessing the potential of AI–ML in urban climate change adaptation and sustainable development. *Sustainability* 2023;**15**(23):16461. https://doi.org/10.3390/su152316461.
23. Srivastava A, Maity R. Unveiling an environmental drought index and its applicability in the perspective of drought recognition amidst climate change. *J Hydrol* 2023;**627**:130462. https://doi.org/10.1016/j.jhydrol.2023.130462.
24. Yang J, Jin H, Tang R, Han X, Feng Q, Jiang H, et al. Harnessing the power of llms in practice: a survey on chatgpt and beyond. *ACM Trans Knowl Discovery Data* 2024;**18**(6):1–32.
25. Maity R, Srivastava A, Sarkar S, Khan MI. Revolutionizing the future of hydrological science: impact of machine learning and deep learning amidst emerging explainable AI and transfer learning. *Appl Comput Geosci* 2024:100206. https://doi.org/10.1016/j.acags.2024.100206.
26. Mensah GB. Artificial intelligence and ethics: a comprehensive review of bias mitigation, transparency, and accountability in AI Systems. *Preprint* 2023 November, 10.
27. Falkenmark M, Folke C. The ethics of socio-ecohydrological catchment management: towards hydrosolidarity. *Hydrol Earth Syst Sci* 2002;**6**(1):1–10.
28. Pulwarty RS, Sivakumar MV. Information systems in a changing climate: early warnings and drought risk management. *Weather Clim Extremes* 2014;**3**:14–21. https://doi.org/10.1016/j.wace.2014.03.005.
29. Milly PC, Betancourt J, Falkenmark M, Hirsch RM, Kundzewicz ZW, Lettenmaier DP, et al. Stationarity is dead: whither water management?*Science* 2008;**319**(5863):573–574.
30. UN-Water. *Climate change and water;* UN-Water: Geneva, Switzerland, 2019.
31. Pahl-Wostl C. The implications of complexity for integrated resources management. *Environ Model & Softw* 2007;**22**(5):561–569. https://doi.org/10.1016/j.envsoft.2005.12.024.
32. Huitema D, Mostert E, Egas W, Moellenkamp S, Pahl-Wostl C, Yalcin R. Adaptive water governance: assessing the institutional prescriptions of adaptive (co- management from a governance perspective and defining a research agenda. *Ecol Soc* 2009;**14**(1).
33. Ostrom E. A general framework for analyzing sustainability of social-ecological systems. *Science* 2009;**325**(5939):419–422. https://doi.org/10.1126/science.1172133.
34. Grafton RQ, Pittock J, Davis R, Williams J, Fu G, Warburton M, et al. Global insights into water resources, climate change and governance. *Nat Clim Change* 2013;**3**(4):315–321.
35. Agarwal A, De Los Angeles MS, Bhatia R, Chéret I, Davila-Poblete S, Falkenmark M, et al. *Integrated water resources management;* Global water Partnership: Stockholm, 2000; pp. 1–67.
36. Ringler C, Bhaduri A, Lawford R. The nexus across water, energy, land and food (WELF: potential for improved resource use efficiency?*Curr OpEnviron Sustainability* 2013;**5**(6):617–624.
37. Urban MC, Bocedi G, Hendry AP, Mihoub JB, Pe'Er G, Singer A, et al. Improving the forecast for biodiversity under climate change. *Science* 2016;**353**(6304):aad8466. https://doi.org/10.1126/science.aad8466.
38. Sarewitz D, Pielke RA, Jr. The neglected heart of science policy: reconciling supply of and demand for science. *Environ Sci & Policy* 2007;**10**(1):5–16.

39. Buytaert W, Baez S, Bustamante M, Dewulf A. Web-based environmental simulation: bridging the gap between scientific modeling and decision-making. *Environ Sci & Technol* 2014;**46**(4):1971–1976.
40. Reed MS. Stakeholder participation for environmental management: a literature review. *Biol Conserv* 2008;**141**(10):2417–2431.
41. Gleick PH. Global freshwater resources: soft-path solutions for the 21st century. *Science* 2003;**302**(5650):1524–1528.
42. Srivastava A, Maity R, Desai VR. Assessing global-scale synergy between adaptation, mitigation, and sustainable development for projected climate change. In *Ecological footprints of climate change. Springer climate;* Chatterjee, U., Akanwa, A. O., Kumar, S., Singh, S. K., Dutta Roy, A., Eds.; Springer: Cham, 2022 https://doi.org/978-3-031-24767-5_39/978-3-031-15501-7_2.
43. Tortajada C. *Water governance: some critical issues. Improving water policy and governance;* Routledge, 2013; pp. 169–179.
44. Zeitoun M, Goulden M, Tickner D. Current and future challenges facing transboundary river basin management. *Wiley Interdiscip Reviews: Clim Change* 2013;**4**(5):331–349.
45. Jiménez A, Cortobius M, Kjellén M. Water, sanitation and hygiene and indigenous peoples: a review of the literature. *Water Int* 2014;**39**(3):277–293.
46. Fabricius C, Folke C, Cundill G, Schultz L. Powerless spectators, coping actors, and adaptive co-managers: a synthesis of the role of communities in ecosystem management. *Ecol Soc* 2007;**12**(1).
47. Kang S, Mulaphong D, Hwang E, Chang CK. Public-private partnerships in developing countries: factors for successful adoption and implementation. *Int J Public Sect Manag* 2019;**32**(4):334–351. https://doi.org/10.1108/IJPSM-01-2018-0001.
48. Srivastava A, Maity R, Desai VR. Investigating spatio-temporal trends and anomalies in long-term meteorological variables to determine if maharashtra is an emerging warming state in India. In *Civil Engineering for Multi-Hazard Risk Reduction. IACESD 2023. Lecture Notes in Civil Engineering;* Sreekeshava, K. S., Kolathayar, S., Vinod Chandra Menon, N., Eds.; Springer: Singapore, 2024 https://doi.org/10.978-3-031-24767-5_39/978-981-99-9610-0_25.
49. Kabisch N, Korn H, Stadler J, Bonn A. *Nature-based solutions to climate change adaptation in urban areas: linkages between science, policy and practice.* Springer Nature, 2017.
50. Chinnasamy P, Srivastava A. Revival of traditional cascade tanks for achieving climate resilience in drylands of South India. *Front Water* 2021;**3**:639637. https://doi.org/10.3389/frwa.2021.639637.
51. Falkenmark M, Rockström J. The new blue and green water paradigm: breaking new ground for water resources planning and management. *J Water Resour Plan Manag* 2006;**132**(3):129–132.
52. Senapati U, Srivastava A, Maity R. High-resolution agricultural drought hazard mapping using the potential of geospatial data and machine learning approaches. *Environ Monit Assess* 2025;**197**(11):1–25. https://doi.org/10.978-3-031-24767-5_39/s10661-025-14538-w.
53. Srivastava A, Jain S, Maity R, Desai VR. Demystifying artificial intelligence amidst sustainable agricultural water management. *Curr Dir Water Scarcity Res* 2022;**7**:17–35. https://doi.org/10.1016/B978-0-323-91910-4.00002-9.
54. Dessai S, Hulme M. Assessing the robustness of adaptation decisions to climate change uncertainties: a case study on water resources management in the East of England. *Glob Environ Change* 2007;**17**(1):59–72.
55. Cohen-Shacham E, Walters G, Janzen C, Maginnis S. Nature-based solutions to address global societal challenges. *IUCN: Gland, Switz* 2016;**97**:2016–2036.

56. Sahu H, Purohit P, Singh R, Mishra AP, Arunachalam K, Kumar U. Comparative assessment of soil parameters and ecological dynamics in the Western Himalayan Wetland and its surrounding periphery. *Environ Qual Manag* 2024;**34**(1):e22283. https://doi.org/10.1002/tqem.22283.
57. Hoff H, Alrahaife SA, El Hajj R, Lohr K, Mengoub FE, Farajalla N, et al. A nexus approach for the MENA region—from concept to knowledge to action. *Front Environ Sci* 2019;**7**:48. https://doi.org/10.3389/fenvs.2019.00048.
58. Wong TH, Brown RR. The water sensitive city: principles for practice. *Water Sci Technol* 2009;**60**(3):673–682.
59. Makropoulos C, Nikolopoulos D, Palmen L, Kools S, Segrave A, Vries D, et al. A resilience assessment method for urban water systems. *Urban Water J* 2018;**15**(4):316–328.
60. Chartzoulakis K, Bertaki M. Sustainable water management in agriculture under climate change. *Agriculture Agric Sci Procedia* 2015;**4**:88–98.
61. Elbeltagi A, Deng J, Li Z, et al. Forecasting vapor pressure deficit for agricultural water management using machine learning in semi-arid environments. *Agric Water Manag* 2023;**283**:108302. https://doi.org/10.1016/j.agwat.2023.108302.
62. Ameyaw EE, Chan AP. Evaluation and ranking of risk factors in public–private partnership water supply projects in developing countries using fuzzy synthetic evaluation approach. *Expert Syst Appl* 2015;**42**(12):5102–5116.
63. Haasnoot M, van't Klooster S, Van Alphen J. Designing a monitoring system to detect signals to adapt to uncertain climate change. *Glob Environ Change* 2018;**52**:273–285.
64. Jayaraman P, Nagarajan KK, Partheeban P, Krishnamurthy V. Critical review on water quality analysis using IoT and machine learning models. *Int J Inf Manag Data Insights* 2024;**4**(1):100210.
65. Sivapalan M, Blöschl G. Time scale interactions and the coevolution of humans and water. *Water Resour Res* 2015;**51**(9):6988–7022.
66. Mosavi A, Ozturk P, Chau KW. Flood prediction using machine learning models: literature review. *Water* 2018;**10**(11):1536.
67. Cui F, Salih SQ, Choubin B, Bhagat SK, Samui P, Yaseen ZM. Newly explored machine learning model for river flow time series forecasting at Mary River, Australia. *Environ Monit Assess* 2020;**192**:1–15.
68. Baird GM, Hatler D, Carpenter P. *How cost effective is machine learning/AI applied to leak detection and pipe replacement prioritization?Pipelines;* American Society of Civil Engineers: Reston, VA, 2019; pp. 274–283.
69. Elbeltagi A, Ehsan M, Sharma G, et al. Advanced stacked integration method for forecasting long-term drought severity: CNN with machine learning models. *J Hydrology: Regional Stud* 2024;**53**:101759. https://doi.org/10.1016/j.ejrh.2024.101759.
70. Kumar S, Srivastava A, Maity R. Modeling climate change impacts on vector-borne disease using machine learning models: case study of Visceral leishmaniasis (Kala-azar from Indian state of Bihar. *Expert Syst Appl* 2024;**237**:121490. https://doi.org/10.1016/j.eswa.2023.121490.
71. Al-Mukhtar M, Khadke L, Al-Musawi T, et al. Prediction of irrigation water quality indices using random committee, discretization regression, REPTree, and additive regression. *Water Resour Manag* 2024;**38**(1):343–368. https://doi.org/10.1007/s11269-023-03674-y.
72. Maity R, Bhagwat PP, Bhatnagar A. Potential of support vector regression for prediction of monthly streamflow using endogenous property. *Hydrological Processes: An Int J* 2010;**24**(7):917–923. https://doi.org/10.1002/hyp.7535.
73. Palani S, Liong SY, Tkalich P. An ANN application for water quality forecasting. *Mar Pollut Bull* 2008;**56**(9):1586–1597.

74. Chitsazan M, Rahmani G, Neyamadpour A. Groundwater level simulation using artificial neural network: a case study from Aghili plain, urban area of Gotvand, south-west Iran. *Geopersia* 2013;**3**(1):35–46.
75. Lu F, Zhang H, Liu W. Development and application of a GIS-based artificial neural network system for water quality prediction: a case study at the Lake Champlain area. *J Oceanology Limnology* 2020;**38**(6):1835–1845.
76. Pande CB, Al-Ansari N, Kushwaha NL, et al. Forecasting of SPI and meteorological drought based on the artificial neural network and M5P model tree. *Land* 2022;**11**(11):2040. https://doi.org/10.3390/land11112040.
77. Mishra D, Goswami S, Matin S, Sarup J. Analyzing the extent of drought in the Rajasthan state of India using vegetation condition index and standardized precipitation index. *Modeling Earth Syst Environ* 2022;**8**(1):601–610.
78. Pande CB, Moharir KN, Radwan N, Mohd Sidek L, Alshehri F, Pal SC, et al. Characterizing land use/land cover change dynamics by an enhanced random forest machine learning model: a Google Earth Engine implementation. *Environ Sci Europe* 2024;**36**(1):84. https://doi.org/10.1186/s12302-024-00901-0.
79. Hoque R, Nakayama D, Matsuyama H, Matsumoto J. Flood monitoring, mapping and assessing capabilities using RADARSAT remote sensing, GIS and ground data for Bangladesh. *Nat Hazards* 2011;**57**:525–548.
80. Li S, Abdelkareem M, Al-Arifi N. Mapping groundwater prospective areas using remote sensing and GIS-based data driven frequency ratio techniques and detecting land cover changes in the Yellow River Basin, China. *Land* 2023;**12**(4):771.
81. Yang X, Jin W. GIS-based spatial regression and prediction of water quality in river networks: a case study in Iowa. *J Environ Manag* 2010;**91**(10):1943–1951.
82. Liladhar Rane N, Purushottam Choudhary S, Saha A, et al. Delineation of environmentally sustainable urban settlement using GIS-based MIF and AHP techniques. *Geocarto Int* 2024;**39**(1):2335249. https://doi.org/10.1080/10106049.2024.2335249.
83. Moles S, Gozzo S, Ormad MP, Mosteo R, Gómez J, Laborda F, et al. Long-term study of antibiotic presence in Ebro River Basin (Spain: identification of the emission sources. *Water* 2022;**14**(7):1033. https://doi.org/10.3390/w14071033.
84. Adedeji KB, Hamam Y, Abe BT, Abu-Mahfouz AM. *Burst leakage-pressure dependency in water piping networks: its impact on leak openings. 2017 IEEE AFRICON;* IEEE, 2017; pp. 1502–1507.
85. Xia W, Chen X, Song C. A framework of blockchain technology in intelligent water management. *Front Environ Sci* 2022;**10**:909606.
86. Liu Y, Shang C. Application of blockchain technology in agricultural water rights trade management. *Sustainability* 2022;**14**(12):7017.
87. Satilmisoglu TK, Keskin H. *Smart water chain: immutable, distributed and decentralized water transaction ledgers. IOP conference series: earth and environmental science;* IOP Publishing, 2023; pp. 012014. No. 1.
88. Evangelou T, Gkeli M, Potsiou C. Building digital twins for smart cities: a case study in Greece. *ISPRS Ann Photogrammetry, Remote Sens Spat Inf Sci* 2022;**10**:61–68.
89. Ham Y, Kim J. Participatory sensing and digital twin city: updating virtual city models for enhanced risk-informed decision-making. *J Manag Eng* 2020;**36**(3):04020005.
90. Shuang Q, Zhao RT. Water demand prediction using machine learning methods: a case study of the Beijing–Tianjin–Hebei region in China. *Water* 2021;**13**(3):310.
91. Horita FE, de Albuquerque JP, Marchezini V, Mendiondo EM. Bridging the gap between decision-making and emerging big data sources: An application of a model-based framework to disaster management in Brazil. *Decis Support Syst* 2017;**97**:12–22.
92. Wang T, Duan J, Zhai J, Zhao J, Gao Y, Gao F, et al. Research on a cloud model intelligent computing platform for water resource management. *J Hydroinformatics* 2024. p.jh2024223.

93. Goodall JL, Robinson BF, Castronova AM. Modeling water resource systems using a service-oriented computing paradigm. *Environ Model & Softw* 2011;**26**(5):573–582.
94. Thornhill I, Chautard A, Loiselle S. Monitoring biological and chemical trends in temperate still waters using citizen science. *Water* 2018;**10**(7):839.
95. Krell NT, Giroux SA, Guido Z, Hannah C, Lopus SE, Caylor KK, et al. Smallholder farmers' use of mobile phone services in central Kenya. *Clim Dev* 2021;**13**(3):215–227.
96. Famiglietti JS, Cazenave A, Eicker A, Reager JT, Rodell M, Velicogna I. Satellites provide the big picture. *Science* 2015;**349**(6249):684–685.
97. Pal M, Maity R. Soil moisture retrieval using Quad-Polarized SAR data from Radar imaging Satellite 1 (RISAT1 through artificial intelligence-based soft computing techniques. *J Indian Soc Remote Sens* 2019;**47**(10):1671–1682. https://doi.org/10.1007/s12524-019-01015-4.
98. Pal M, Maity R. Assimilation of remote sensing based surface soil moisture to develop a spatially varying vertical soil moisture profile database for entire Indian mainland. *J Hydrol* 2021;**601**:126807. https://doi.org/10.1016/j.jhydrol.2021.126807.
99. Duan W, Maskey S, Chaffe PL, Luo P, He B, Wu Y, et al. Recent advancement in remote sensing technology for hydrology analysis and water resources management. *Remote Sens* 2021;**13**(6):1097.
100. Mishra AK, Coulibaly P. Developments in hydrometric network design: a review. *Rev Geophys* 2009;**47**(2).
101. Rodell M, Famiglietti JS, Wiese DN, Reager JT, Beaudoing HK, Landerer FW, et al. Emerging trends in global freshwater availability. *Nature* 2018;**557**(7707):651–659.
102. Xu Y, Tung YK, Li J, Niu S. Alternative risk measure for decision-making under uncertainty in water management. *Prog Nat Sci* 2009;**19**(1):115–119.
103. Li M, Wiedmann T, Liu J, Wang Y, Hu Y, Zhang Z, et al. Exploring consumption-based planetary boundary indicators: an absolute water footprinting assessment of Chinese provinces and cities. *Water Res* 2020;**184**:116163.
104. Mwansa MN. *Modelling and predicting energy usage from smart meter data and consumer behaviours in residential houses*. Liverpool John Moores University: United Kingdom, 2020.
105. Bhandari P, Creighton D, Gong J, Boyle C, Law KM. Evolution of cyber-physical-human water systems: challenges and gaps. *Technol Forecast Soc Change* 2023;**191**:122540.
106. Biesbroek R, Berrang-Ford L, Ford JD, Tanabe A, Austin SE, Lesnikowski A. Data, concepts and methods for large-n comparative climate change adaptation policy research: a systematic literature review. *Wiley Interdiscip Reviews: Clim Change* 2018;**9**(6):e548. https://doi.org/10.1002/wcc.548.
107. Krishnan SR, Nallakaruppan MK, Chengoden R, Koppu S, Iyapparaja M, Sadhasivam J, et al. Smart water resource management using Artificial Intelligence—A review. *Sustainability* 2022;**14**(20):13384. https://doi.org/10.3390/su142013384.
108. Kraus M, Bingler JA, Leippold M, Schimanski T, Senni CC, Stammbach D, et al. Enhancing large language models with climate resources. *arXiv preprint* 2023. arXiv:2304.00116.
109. Manning C, Schutze H. *Foundations of statistical natural language processing;* MIT Press, 1999.
110. Goodfellow I, Bengio Y, Courville A. *Deep learning;* MIT Press, 2016.
111. Hochreiter S. *Long short-term memory;* Neural Computation MIT-Press, 1997.
112. Vaswani A. Attention is all you need. Advances in Neural Information Processing Systems, 31st Conference on Neural Information Processing Systems (NIPS 2017, Long Beach, CA, USA; 2017.
113. Radford A, Wu J, Child R, Luan D, Amodei D, Sutskever I. Language models are unsupervised multitask learners. *OpenAI Blog* 2019;**1**(8):9.

114. Achiam J, Adler S, Agarwal S, Ahmad L, Akkaya I, Aleman FL, et al. Gpt-4 technical report. *arXiv Preprint* 2023. https://doi.org/10.48550/arXiv.2303.08774. arXiv:2303.08774.
115. Bommasani R, Hudson DA, Adeli E, Altman R, Arora S, von Arx S, et al. On the opportunities and risks of foundation models. *arXiv Preprint* 2021 arXiv:2108.07258.
116. Chowdhery A, Narang S, Devlin J, Bosma M, Mishra G, Roberts A, et al. Palm: scaling language modeling with pathways. *J Mach Learn Res* 2023;**24**(240):1–113.
117. Touvron H, Lavril T, Izacard G, Martinet X, Lachaux MA, Lacroix T, et al. Llama: open and efficient foundation language models. *arXiv Preprint* 2023;**2302**:13971. arXiv:2302.13971.
118. Thulke D, Gao Y, Pelser P, Brune R, Jalota R, Fok F, et al. Climategpt: towards ai synthesizing interdisciplinary research on climate change. *arXiv Preprint* 2024. arXiv:2401.09646.
119. Raffel C, Shazeer N, Roberts A, Lee K, Narang S, Matena M, et al. Exploring the limits of transfer learning with a unified text-to-text transformer. *J Mach Learn Res* 2020;**21**(140):1–67.
120. Chen M, Tao Z, Tang W, Qin T, Yang R, Zhu C. Enhancing emergency decision-making with knowledge graphs and large language models. *Int J Disaster Risk Reduct* 2024;**113**:104804. https://doi.org/10.1016/j.ijdrr.2024.104804.
121. Clusmann J, Kolbinger FR, Muti HS, Carrero ZI, Eckardt JN, Laleh NG, et al. The future landscape of large language models in medicine. *Commun Med* 2023;**3**(1):141.
122. Li Z, Ning H. Autonomous GIS: the next-generation AI-powered GIS. *Int J Digital Earth* 2023;**16**(2):4668–4686.
123. Agathokleous E, Saitanis CJ, Fang C, Yu Z. Use of ChatGPT: what does it mean for biology and environmental science?*Sci Total Environ* 2023;**888**:164154.
124. Zhou B, Zou L, Mostafavi A, Lin B, Yang M, Gharaibeh N, et al. VictimFinder: harvesting rescue requests in disaster response from social media with BERT. *Computers, Environ Urban Syst* 2022;**95**:101824.
125. Koshy R, Elango S. Applying social media in emergency response: an attention-based bidirectional deep learning system for location reference recognition in disaster tweets. *Appl Intell* 2024;**54**(7):5768–5793.
126. Kapoor NR, Kumar A, Kumar A, Kumar A, Arora HC. *Artificial intelligence in civil engineering: an immersive view. Artificial intelligence applications for sustainable construction;* Woodhead Publishing, 2024; pp. 1–74.
127. Song X, Zhang H, Akerkar R, Huang H, Guo S, Zhong L, et al. Big data and emergency management: concepts, methodologies, and applications. *IEEE Trans Big Data* 2020;**8**(2):397–419. https://doi.org/10.1007/978-3-031-24767-5_39.
128. Zaboli A, Kasimalla SR, Park K, Hong Y, Hong J. A comprehensive review of behind-the-meter distributed energy resources load forecasting: models, challenges, and emerging technologies. *Energies* 2024;**17**(11):2534.
129. Zhou Z, Lin Y, Jin D, Li Y. Large language model for participatory urban planning. *arXiv Preprint* 2024. arXiv:2402.17161.
130. Cruz-Castro L, Castelblanco G, Antonenko P. LLM-based system for technical writing real-time review in urban construction and technology. Proceedings of 60th Annual Associated Schools, 5; 2024, p. 130–138.
131. Dhanuka A, Khadke L, Kushwaha NL. Smart geometric design of highways using HTML programming for sustainable and climate resilient cities. In *Urban commons, future smart cities and sustainability. Springer geography;* Chatterjee, U., Bandyopadhyay, N., Setiawati, M. D., Sarkar, S., Eds.; Springer: Cham, 2023https://doi.org/10.1007/978-3-031-24767-5_39
132. Wu Q, Xu Y, Xiao T, Xiao Y, Li Y, Wang T, et al. Surveying Attitudinal Alignment Between Large Language Models Vs. Humans Towards 17 Sustainable Development Goals. *arXiv Preprint* 2024. arXiv:2404.13885.

133. Lehmann A, Giuliani G, Ray N, Rahman K, Abbaspour KC, Nativi S, et al. Reviewing innovative Earth observation solutions for filling science-policy gaps in hydrology. *J Hydrol* 2014;**518**:267–277.
134. Prabhod KJ. Integrating large language models for enhanced clinical decision support systems in modern healthcare. *J Mach Learn Healthc Decis Support* 2023;**3**(1):18–62.
135. Kadiyala LA, Mermer O, Samuel DJ, Sermet Y, Demir I. The implementation of multimodal large language models for hydrological applications: a comparative study of GPT-4 Vision, Gemini, LLaVa, and Multimodal-GPT. *Hydrology* 2024;**11**(9):148.
136. Li W, Liu C, Xu Y, Niu C, Li R, Li M, et al. An interpretable hybrid deep learning model for flood forecasting based on Transformer and LSTM. *J Hydrology: Regional Stud* 2024;**54**:101873.
137. Li N, Zahra S, Brito M, Flynn C, Görnerup O, Worou K, et al. Using LLMs to build a database of climate extreme impacts. In: Proceedings of the 1st Workshop on Natural Language Processing Meets Climate Change (ClimateNLP 2024; 2024. p. 93–110.
138. Giupponi C, Sgobbi A. Decision support systems for water resources management in developing countries: learning from experiences in Africa. *Water* 2013;**5**(2):798–818.
139. Ahmad S, Simonovic SP. Spatial system dynamics: new approach for simulation of water resources systems. *J Comput Civ Eng* 2004;**18**(4):331–340.
140. Prakash K, Rao S, Hamza R, Lukich J, Chaudhari V, Nandi A. Integrating LLMs into Database Systems Education. In Proceedings of the 3rd International Workshop on Data Systems Education: Bridging education practice with education research; 2024. p. 33–39.
141. Xu B, Wen L, Li Z, Yang Y, Wu G, Tang X, et al. Unlocking the potential: benchmarking large language models in water engineering and research. *arXiv Preprint* 2024. arXiv:2407.21045.
142. Changeux A, Montagnier S. Strategic decision-making support using large language models (llms. *Manag J Adv Res* 2024;**4**(4):102–108.
143. Cecconi F, Marconi L, Barazzetti A. *Climate change mitigation policies using GPT-4;* MISC, 2024.
144. Zhu H, Tiwari P. Climate change from large language models. *arXiv Preprint* 2023. arXiv:2312.11985.
145. Arslan M, Mahdjoubi L, Munawar S. Driving sustainable energy transitions with a multi-source RAG-LLM system. *Energy Build* 2024:114827.
146. Zhu J, Dang P, Cao Y, Lai J, Guo Y, Wang P, et al. A flood knowledge-constrained large language model interactable with GIS: enhancing public risk perception of floods. *Int J Geograp Inf Sci* 2024;**38**(4):603–625. https://doi.org/10.1080/13658816.2024.2306167.
147. Pavlova A, Gerazov B, Barreiro A. Large language models and OpenLogos: an educational case scenario. *Open Res Europe* 2024;**4**(110):110.
148. Otal HT, Stern E,Canbaz MA. Llm-assisted crisis management: building advanced llm platforms for effective emergency response and public collaboration. In: 2024 IEEE Conference on Artificial Intelligence (CAI, IEEE; 2024. p. 851–859.
149. Sapkota R, Qureshi R, Hassan SZ, Shutske J, Shoman M, Sajjad M, et al. Multi-Modal LLMs in agriculture. *A Compr Rev* 2024 10.36227/techrxiv. 172651082.24507804/v1.
150. Williams S, Beery S, Conley C, Evans ML, Garces S, Gordon E, et al. People-powered gen AI: collaborating with generative AI for civic engagement; 2024. https://mit-genai.pubpub.org/pub/6uejzuox/release/2 (Accessed 21 October 2024.
151. Laskar MTR, Fu XY, Chen C, Tn SB. Building real-world meeting summarization systems using large language models: a practical perspective. *arXiv Preprint* 2023. arXiv:2310.19233.

152. Anderson T, Sarkar S, Kelley R. Analyzing public sentiment on sustainability: a comprehensive review and application of sentiment analysis techniques. *Nat Lang Process J* 2024:100097.
153. Hadi MU, Qureshi R, Shah A, Irfan M, Zafar A, Shaikh MB, et al. A survey on large language models: applications, challenges, limitations, and practical usage. *Authorea Prepr* 2023.
154. Bulian J, Schäfer MS, Amini A, Lam H, Ciaramita M, Gaiarin B, et al. Assessing large language models on climate information. *arXiv Preprint* 2023;**2310**:02932. arXiv:2310.02932.
155. Bender EM, Gebru T, McMillan-Major A, Shmitchell S. On the dangers of stochastic parrots: can language models be too big?*Proc 2021 ACM Conf Fairness, Accountability, Transpar* 2021:610–623.
156. Regulation Protection. Regulation (EU) 2016/679 of the European parliament and of the council. Regulation (eu); 2016; 679, p. 2016.
157. Mitchell M, Wu S, Zaldivar A, Barnes P, Vasserman L, Hutchinson B, et al. Model cards for model reporting. *Proc Conf fairness, accountability, Transpar* 2019:220–229.
158. Pankajakshan R, Biswal S, Govindarajulu Y, Gressel G. Mapping LLM security landscapes: a comprehensive stakeholder risk assessment proposal. *arXiv Preprint* 2024. arXiv:2403.13309.
159. Weidinger L, Mellor J, Rauh M, Griffin C, Uesato J, Huang PS, et al. Ethical and social risks of harm from language models. *arXiv Preprint* 2021. arXiv:2112.04359.
160. Sun S, Cheng Y, Gan Z, Liu J. Patient knowledge distillation for bert model compression. *arXiv Preprint* 2019. arXiv:1908.09355.
161. Wang W, Chen W, Luo Y, Long Y, Lin Z, Zhang L, et al. Model compression and efficient inference for large language models: a survey. *arXiv Preprint* 2024. arXiv:2402.09748.
162. Ullah A, Qi G, Hussain S, Ullah I, Ali Z. The role of llms in sustainable smart cities: applications, challenges, and future directions. *arXiv Preprint* 2024. arXiv:2402.14596.
163. Ignjatović D, Karagiannis V, Chettakattu A , Havlik D, Neubauer G. Crisis management in the era of the IoT, edge computing, and LLMs. In: 2024 11th International Conference on Internet of Things: Systems, Management and Security (IOTSMS, IEEE; 2024. p. 224–231.
164. Dogo EM, Salami AF, Nwulu NI, Aigbavboa CO. Blockchain and internet of things-based technologies for intelligent water management system. *Artif Intell IoT* 2019:129–150.
165. Shi W, Cao J, Zhang Q, Li Y, Xu L. Edge computing: vision and challenges. *IEEE Internet Things J* 2016;**3**(5):637–646.
166. Zhang Y, Wei C, Wu S, He Z, Yu W. GeoGPT: understanding and processing geospatial tasks through an autonomous GPT. *arXiv Preprint* 2023. arXiv:2307.07930.
167. Feng Y, Ding L, Xiao G, Geoqamap-geographic question answering with maps leveraging LLM and open knowledge base (short paper. In: 12th International Conference on Geographic Information Science (GIScience 2023. Schloss Dagstuhl-Leibniz-Zentrum für Informatik; 2023.
168. Singh P, Patidar M, Vig L. Translating across cultures: LLMs for intralingual cultural adaptation. *arXiv Preprint* 2024. arXiv:2406.14504.
169. Conneau A. Unsupervised cross-lingual representation learning at scale. *arXiv Preprint* 2019. arXiv:1911.02116.
170. Amaru S, Chhetri NB. Climate adaptation: institutional response to environmental constraints, and the need for increased flexibility, participation, and integration of approaches. *Appl Geogr* 2013;**39**:128–139.
171. Wyborn CA, van Kerkhoff LE, Colloff MJ, Alexandra J, Olsson R. The politics of adaptive governance: water reform, climates change, and First Nations' justice in Australia's Murray-Darling Basin. *Ecol Soc* 2023;**28**(1).

172. Díaz-Rodríguez N, Del Ser J, Coeckelbergh M, de Prado ML, Herrera-Viedma E, Herrera F. Connecting the dots in trustworthy Artificial Intelligence: from AI principles, ethics, and key requirements to responsible AI systems and regulation. *Inf Fusion* 2023;**99**:101896.
173. Floridi L, Cowls J. A unified framework of five principles for AI in society. *Mach Learn city: Appl architecture urban Des* 2022:535–545.
174. Sigfrids A, Nieminen M, Leikas J, Pikkuaho P. How should public administrations foster the ethical development and use of artificial intelligence? A review of proposals for developing governance of AI. *Front Hum Dyn* 2022;**4**:858108.
175. Rillig MC, Ågerstrand M, Bi M, Gould KA, Sauerland U. Risks and benefits of large language models for the environment. *Environ Sci & Technol* 2023;**57**(9):3464–3466.

SECTION C

Geospatial technologies and remote sensing: Precision tools for water resources management

CHAPTER NINE

Prioritizing and assessing the surface water quality through geospatial techniques

Abhijeet Das*
Department of Civil Engineering, C.V. Raman Global University (CGU), Bhubaneswar, Odisha, India
*Corresponding author. e-mail address: das.abhijeetlaltu1999@gmail.com; laltu.dasabhijeet1999@gmail.com

Contents

Abstract

Generally, management of surface water quality is a crucial component of the endeavor to supply the growing demand for water. With rapid urbanization and population growth, the water quality is deteriorating nowadays. To achieve this objective, water quality has to be evaluated and tracked using cutting-edge methods. While considering the Brahmani River, Odisha, the proposed study is undertaken by incorporating Comprehensive Pollution Index (CPI), and multi-criteria decision-making (MCDM) approaches like Grey Relational Analysis (GRA) and Ordered Weighted Averaging (OWA), aimed to assess the water quality, and analyzed its suitability for

Advances in Chemical Pollution, Environmental Management and Protection, Volume 15
ISSN 2468-9289, https://doi.org/10.1016/bs.apmp.2025.11.012

drinking purposes, using 15 physicochemical parameters, obtained from different surveyed places. Water samples were collected during pre-monsoon season for a period of 6 years (2018–2024), from previously selected 7 sampling stations. From the resultant outcomes, the order of anionic abundance: $SO_4^{2-} > Cl^- > NO_3^- > F^- > PO_4^{3-}$. In this contrary, the order of cationic abundance is however, resulted as: $Ca^{2+} > Mg^{2+} > Na^+ > K^+$. Again, the data was interpreted using pollution indices such as CPI, the value ranged between 0.46 and 1.90, that resulted as qualified to polluted category. Notably, poor indicated sites indicates dissolved organic matter; anthropogenic point source contamination was the primary source of nitrogen, which was the major pollutant. Furthermore, MCDM techniques were employed in an effort to pinpoint the critical factors influencing changes in seasonal and sectional water quality. The calculated GRA values are between 0.03–0.059 in the relevant examined area, which signifies that water quality at sites P-(3), (4) and (5), placed under good category. The remaining stations namely, P-(1), (2) and (7), conferred under polluted category. Subsequently, it was found that key latent factors influencing the water quality of rivers in the research region were chemical, organic, and conventional contaminants. In addition, the OWA score, fluctuate between 0.47 and 0.93 and the results revealed that P-(1) is termed to be the most polluted detected location, followed by P-(2) and P-(7). The two approaches namely GRA and OWA demonstrated that these sites were primarily impacted by harmful and oxygen-consuming organic pollutants. The spatial distribution of parameters indicated the effect of urbanization and commercialization (small house hold industries) at P-(2), (7) and (3). Hence, the poor water quality of this selected area needs immediate government attention. The findings suggest that the most contaminated river sectors and critical surface water quality indices can be determined using CPI, GRA, and OWA as essential instruments. As a result, the combination of all approaches is effective and provides us with a clear image for evaluating the parameters that affect surface water's appropriateness for drinking.

Keywords: Brahmani river, Comprehensive pollution index, Grey relational analysis, Ordered weighted averaging, Anthropogenic.

1. Introduction

The water supply is deemed essential for human well-being and a basic prerequisite for human activity and economical utility. Surface water quality is a vital aspect of human life and, because of its delicate nature, a significant worldwide issue.[1] Although water is a vital resource, its availability for industrial, agricultural, and residential usage has been threatened by a number of issues, the most significant of which is climate unpredictability. Moving further, rivers are the most often used surface water sources because of their availability and abundance, which has caused fast human population expansion and advancement in the areas surrounding water courses.[2] Since prehistoric times, riverine systems have been essential to the development of human civilization because they offer a handy and necessary supply of fresh water for industrial, agricultural, and domestic use. Urban and industrial effluents are regarded as important suppliers of nutrients and chemicals to the aquatic ecology. Chemical pollutants in the fluvial systems, however, are mostly caused by human and natural processes such soil erosion, precipitation, and industrial and agricultural operations.[3]

In the current scenario, rivers have faced serious environmental problems as a result of pollution from widespread pesticide runoff from agriculture and effluent from factories, sewage, and other sources of urban trash, particularly in emerging nations. Poor water quality is caused by a number of reasons, including precipitation, recharge water, inland surface water, and subsurface geochemical processes. Surface water quality is a highly delicate topic that is of global environmental relevance. The entire existence of life on Earth is now seriously threatened by unplanned industrialization and urbanization brought about by today's socioeconomic breakthroughs, since they pollute the water supply.[4] In rural areas, low-quality water entering the drinking water distribution system frequently leads to corrosion and scaling. As a result, it causes a variety of challenges, such as pipe blockage, shortened equipment lifespan, and health and financial problems brought on by dissolved substances in the water.[5] The quality of surface water has drawn attention from all over the world because of the high-quality water need for agriculture and residential use. There are three categories of characteristics related to surface water quality: physicochemical, bacteriological, and trace metals. Thus, river pollution prevention requires efficient monitoring of physicochemical and microbiological parameters.[6]

Development of water quality (WQ) indices is thought to be the best instrument for evaluating the quality of water. Therefore, physicochemical characteristics that are derived from hydro-chemical metrics offer a preliminary comprehension of various geochemical mechanisms, water classification, and water facies. A mathematical equation that rates the water quality and determines whether it is suitable for human consumption is combined with a number of water quality criteria to create water quality indices.[7] Because of this, the main objective of this study is to evaluate the quality of water using statistical and geographic approaches that are in line with previous research while also creating new techniques for use in the future. Moreover, water quality indices (WQIs) are some of the more effective ways to describe water quality because they use numerical index values to support strategic planning related to water quality management programs and interpret the overall water quality at several monitoring sites at a specific time by combining original data from multiple water quality measures into a single value. Typically,[8] and[9] examined the WQI in order to evaluate and simulate the water quality for irrigation and human use. Moreover,[10] assessed the water's suitability for irrigation and drinking in central Telangana, India. It was discovered that 96 % of the samples met the necessary quality requirements, and irrigation was thus advised. Also, the WQI was implemented by[11] to evaluate groundwater for drinking and determine the problems that are influencing it in Rashdiya, Iraq. Again,[12] and[13] in separate studies also assessed the water quality primarily for irrigation, but also created geographical thematic layers and suitability maps to supplement earlier research. In this context,[14] utilized the WQI to keep an eye on the Par River's water quality in the year 2019, situated in the South Gujarat, India, and discovered that the variation of water quality was rated as poor. Thus, these studies confirm that thorough understanding of the pollution status and surface water quality is essential for long-term surface water management.

Proceeding further, another promising tool, the Comprehensive Pollution Index (CPI) is used to evaluate the effects of contaminants; its primary goal was to confirm the amount of contamination. Because of its adaptability and capacity to summarize intricate environmental data, the CPI is a useful tool in environmental research, directing initiatives to reduce pollution and maintain ecological balance. Most often, this approach is amenable to remote sensing (RS) technology monitoring. Owing to rapid improvements in remote sensing data gathering methods, many satellites or nearby hyperspectral technologies have shown to be cost-effective and practical on a broad scale for integrative evaluation of numerous water quality indices. Eminent academician namely,[15] used the WQI method to assess the quality of the drinking water in the area of Bounamoussa River, located in Algeria's eltarf region, during the winter and summer season, and noticed that both seasons had very good water quality. Moreover,[16] investigated and evaluated the water quality of the Ray River in both the wet and dry seasons. The river is situated in Vietnam, and the study achieve its objective by gathering 13 physico-chemical parameters and a WQI, and demonstrated that the dry season's WQI was significantly greater than the wet season's. Therefore, examining chemical contaminants, the viability of surface water for drinking and irrigation, and representations that have been successfully used are the main goals of this strategy.[17]

Recently, Geographical Information System (GIS) has sophisticated capabilities and indicates as a powerful tool for geographical study and computerized mapping. Resource planning and environmental conservation are made easier by its usage in the analysis and presentation of geographical data.[18] One such ability is the use of satellite image data in GIS-based assessments. However, because GIS techniques provide a comfortable workspace and a useful tool for manipulating and analysing satellite information, they have been used to evaluate the morphometric features of stream watersheds and water quality features. This is especially true for future information collection and recognition for a better awareness. Recently, a study executed by,[19] that evaluated water quality using the rivers' minimum quality index in the Lake Chaohu Basin and demonstrated that the minimum WQI consists of five necessary criteria, and that there were no discernible differences between this combination and the WQI based on 15 elements. Again,[20] assessed the Aksu River's water quality using the WQI and GIS. This river often termed to be the principal river, that replenishing Karacaorem Dam Lake, and indicated that the Lake's water quality is typically good based on the WQI map; however, the water quality in the river basin is poor to extremely harmful in the north and south. Furthermore,[21] evaluated water in Abeokuta North, Nigeria, by taking three indicators such as pH, EC, and TDS, and displayed that the river's water quality is found to be excellent. Similarly,[22] employed a WQI to evaluate the water, that flows in the place called Shatt Al-Arab River, Iraq, and the investigation is carried out by using seven indicators and finally, the results revealed that the water quality fall under the poor category.

The method used in GIS is known as Inverted Distance Weighting (IDW), that make it possible to identify potential causes or sources of the variations in water quality, have an impact on the water system, and allocate the sources; as a result, they provide a useful tool for creating strategies that are appropriate for managing the water resources.[23] As per the[24] work, who implemented WQI and GIS, and considered different physicochemical parameters, to determine whether surface water could be applied to the irrigation system in the Maikunkele Fadama region of Niger State and observed that the farm's stream water is appropriate for irrigation. Hence, the advancements in remote sensing and GIS have made it possible to monitor water quality and conduct morphometric studies of natural drains with greater accuracy and affordability. In this regard, also,[25] considered 9 parameters for the analysis of WQI determination in Al Najaf,

Iraq, and demonstrated that the WQI category was low, which affects the socioeconomic and health conditions of people. Similarly,[26] employed GIS and WQI statistical analysis to evaluate the water quality, which revealed a strong correlation between the majority of the water quality parameters.

Calculation of WQI methods have been created and enhanced in various ways to more precisely and successfully reflect the surface water quality. As even a slight alteration in weighting would affect the overall sense of water quality, choosing an appropriate set of weights for the various water quality factors is crucial in this situation. To address this arbitrary ambiguity, some researchers found that multi-objective decision-making (MCDM) analysis could be applied to analyse the weight relationship of affecting elements in order to correct, determine, and optimize and forecasting object by Grey Relational Analysis (GRA) and Ordered Weighted Averaging (OWA) method.[27,28] Researchers have assessed the potential of MCDM methods in stream restoration efforts beyond WQIs. These methods include the ability to adjust WQI ranking and address management challenges related to storage systems, demand response, performance assessment, and renewable energy sources. These techniques serve as a useful framework for calculating the Euclidean distances between several viable options. The fundamental idea is to establish the objective weight of the indicators using two fundamental ideas: contrast intensity, which is the standard deviation, expressed value difference between the various evaluation schemes of the same index, and conflict between the evaluation indicators, which depends on the relationship between the indicators. Again, this method uses two metrics to calculate the method's cumulative criterion. The first measure is defined as the positive distance, that is obtained from the average solution and, the second is referred as the negative distance, which is extracted from the average solution. Although, earlier studies have concentrated on adjusting the anticipated conflicts between the irrigation and drinking water quality indices, as well as validating surface water and its classes through the use of these optimization models.

In this regard,[29] present a case study of the Iranshahr Basin in Iran, sought to determine the basin's suitable site-specific recharge regions. They combined GIS fuzzy logic spatial modelling with the MCDM model on a single platform. Research by[30] revealed that using OWA could lead to a trustworthy analysis of the weights of various physicochemical parameters. Furthermore,[31] presented about the topic of proactive integrated water resource management utilizing MCDM, which necessitates anticipating future situations involving water resources and creating long-term, collaborative, multi-sector strategic development plans to address them. Similarly,[32] demonstrate the use of decision-making techniques to determine the Southern Johor River System's most logical and effective use in Malaysia. This river system's evaluation encompassed both qualitative and quantitative elements, represented by four main criteria and twenty supporting factors. Proving that such a method can outperform the other strategies, which are now in use, was an additional issue that this study attempted to solve.[33,34] Again, a study implemented by,[35,36] which spent several years researching the spatiotemporal variance of water quality in the Giang Province in the Vietnamese part of the Mekong River Delta, which is seen as an area experiencing agricultural intensification. Hence, since interval judgment is more accurate than fixed value modifications, this technique has been used. Due to this, the decision-making systems must not only provide an overall evaluation of the locations by taking into account physicochemical and heavy metal characteristics, but also prioritize judgments during emergency scenarios. These methods have reportedly presented comparatively accurate analyses on resolving disputes over agricultural land, allocating coastal regions, reducing the risk of flooding, utilizing water reservoirs, and reserving water resources.

The current investigation was carried out in Brahmani River Basin, Odisha. However, surface water is essential to domestic, animal, and agricultural operations for the rural inhabitants in this basin. Prior research in this basin documented degradation of the surface waters, with the majority of the studies focusing on geology, hydro-geology, and a scant examination of water suitability. Regarding the water chemistry and metal pollution in this basin, none of them offered in-depth expertise. It appears that research on surface water quality assessment for agricultural and drinking uses is necessary. Reducing biological, chemical, and physical water pollution in surface waters requires an evaluation. Hence, the present study can help scientists learn more about the state of pollution, its sources, and the related geochemical processes that affect the surface water in the chosen catchment.

Therefore, a huge data matrix from a 6-year (2018–2024) monitoring program, is treated to several methodologies in the current study, in order to fulfil the objectives: (1) by employing CPI, to gather details regarding the parallels and discrepancies among the sampling locations, and (2) determine the water quality variables that affect fluctuations in water quality both spatially and temporally, using GRA and OWA. Also, this study took into account potential pollution sources that might affect the quality of the water in the area being studied.

Broadly, the findings should aid in assessing the spatial-temporal evolution and, as a result, help managers identify the primary causes of pollution at each site along the river basin. Importantly, the findings of this study may aid in the development of optimal, scientifically grounded water health regulations by policymakers and may also contain information on possible water quality suitable for industrial, agricultural, and drinking uses in this river basin.

2. Materials and methodology

2.1 Study area

The river often referred as Brahmani River, formed at the site of confluence close to Vedvyas, in Odisha, at a height of 200 m above mean sea level. Regarded as a major seasonal river and stands for 2nd largest river in the Odisha State.[37] The river basin occupies 15 % of the total state geographical area, and its geographical coordinates lies between 21–23°00′ N and 84–87° E. The study region is defined by a 'semi-arid to arid' climate, having an average of yearly precipitation of 1400 mm and the average yearly temperature is recorded as 25.7 °C. In particular, the drainage area of the catchment is estimated to be as approximately 39,268 Km^2. From the total area, 22,516 Km^2 is coveted under the state of Odisha, nearly 15,405 Km^2 belongs to the Jharkhand State and finally, 1347 Km^2 complying inside the Chhattisgarh State. Most interestingly, the specific land uses found in the basin, comprised water, forests, cattle, grassland, row crops, commercial, industrial, and residential areas.[38] The river's base map created by GIS is displayed in Fig. 1.

2.2 Sample collection, experimental design, and its analysis

The State Pollution Control Board (SPCB), is in responsibility of monitoring the water quality along the River Brahmani, in the Eastern region of India, in the State of Odisha. The sampling primarily targeted sites, which are prone to anthropogenic pollution and dumping of domestic sewage, and was further examined for a period of 2018–2024 (6 years) during the pre-monsoon season, where the water levels are anticipated to be low. The water quality indicator's value, considered for the analysis, in the current research, is taken as average. Typically, water samples were

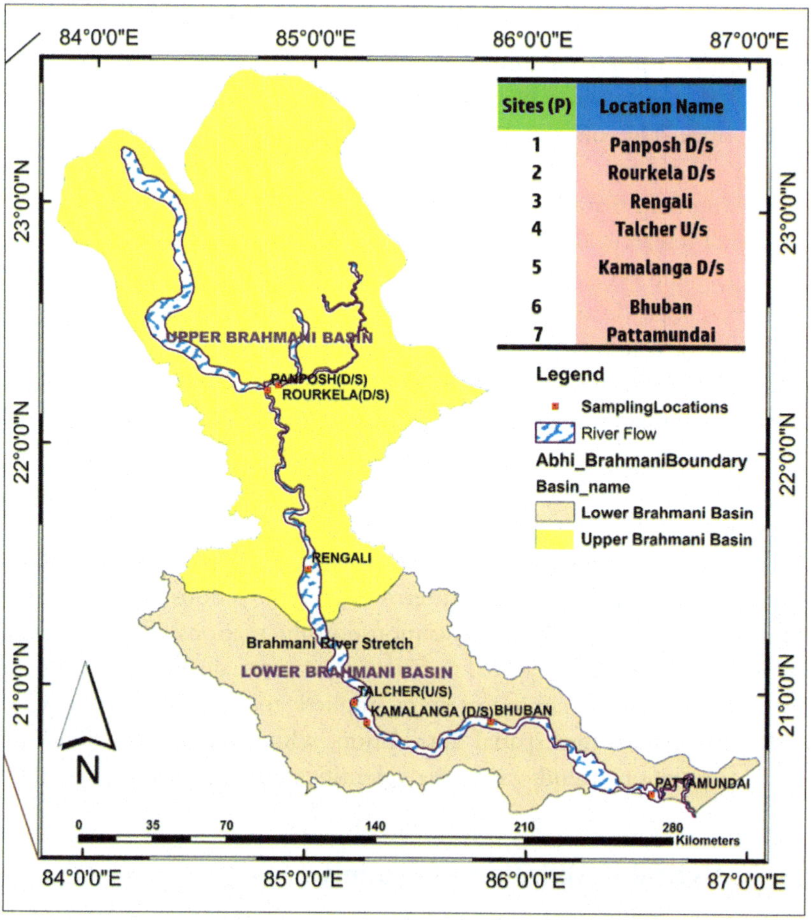

Fig. 1 Location of sampling sites along the stretch of the river. Map lines delineate study areas and do not necessarily depict accepted national boundaries.

gathered into 1000 ml polypropylene bottles, dried, and then cleaned with distilled water and an acidic solution. The samples were transferred to the laboratory of C.V. Raman Global University (CGU), Bhubaneswar, adhering to the criteria's, set by American Public Health Association (APHA), at an ideal temperature of value, that is less than 4 °C.[39] Hence, quality control was maintained in all metal assays, as per the specific instruction manuals, and the method's precision exceeded 95 % in confidence interval (CI), in conjunction with correlation coefficient, (p = 1) of the corresponding calibration curves. It is noteworthy that the calculated ionic balance errors are within 5 %.

In this current investigation, seven sampling stations were carefully selected for 15 physicochemical parameters, to depict the principal stream and its tributaries while keeping in mind the standards for choosing the sample sites. The dissolved oxygen (DO) is verified by using Winkler method.[40] Evidently, alkalinity was evaluated by EDTA titration method, with taking help from Eriochrome Black T, which acts as an indicator. In the overall research, the concentration of large anions, such as Cl^-, SO_4^{2-}, PO_4^{3-}, F^-, and NO_3^-, were examined using by ion chromatography (IC). Moreover, total dissolved solids (TDS), electrical conductivity (EC), and pH of the selected water samples, from the investigated locations, were estimated using conductivity/TDS meter, and pH meter, respectively. Again, the temperature is recorded using digital thermometer, while the major cationic concentration, which includes Na^+, K^+, Ca^{2+}, and Mg^{2+}, was obtained out by understanding the principle of flame atomic absorption spectrophotometer (FAAS).

The collected data were examined using a variety of decision-making techniques in order to find patterns and connections among various characteristics. To determine how monsoon fluctuations affect the chemistry of surface water, seasonal variations in water quality are evaluated. Furthermore, GIS techniques were applied to produce maps of the investigated parameters' spatial distribution, which allowed for the display of contamination trends and the identification of possible pollution sources[41].

2.3 Analytical description of models

Surface water quality index is a useful tool for communicating water quality because it gives policymakers and concerned citizens access to information about water quality in a straightforward, consistent, and provable unit of measurement. Therefore, it is referred to as a unique quantification that outlines the primary causes of variations in water quality. It is computed, nevertheless, utilizing factors related to people's appropriateness for the water. The literature previously stated indicates that the combined use of CPI, GRA, and OWA for decision-making is limited to a small number of research studies. Consequently, an attempt is made to incorporate both of the multi-criteria decision techniques and apply them to the resolution of material selection problems in this research work. The research area's surface water quality was estimated using a step-by-step methodology, as demonstrated in Fig. 2. Here, CPI is applied to determine the weight requirement, whereas GRA and OWA, cumulatively utilized to determine

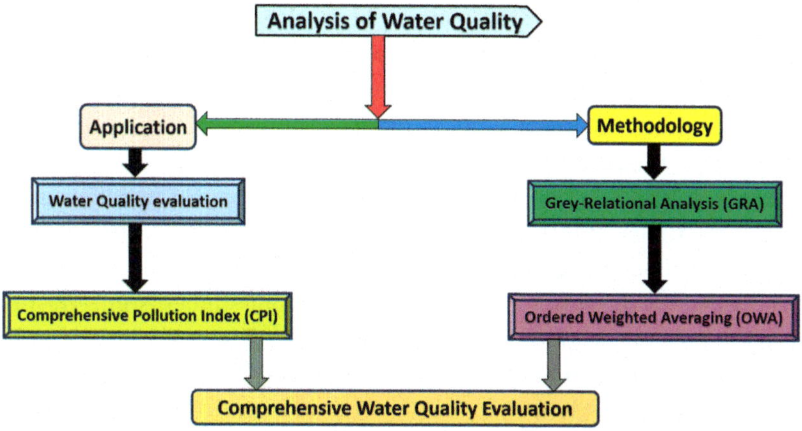

Fig. 2 Framework of the complete research process.

which option is best after ranking the possibilities. These techniques lead to a weighted and ranked computation that has been applied in numerous investigations to arrive at this kind of conclusion.

The application of water quality and MCDM methods with GIS, enabled the creation of maps, providing valuable information for drinking purposes. Diverse assessment methodologies have been employed to facilitate the precise assessment of water quality.[42] Water Quality assessment through quantification of Comprehensive Pollution Index (CPI), through various physicochemical parameters, is quite impossible to comprehend in its original form and hardly captures the true situation. This mechanism was examined in this investigation in order to verify the contamination level determined by WQI. Meanwhile, this pollution determinant is a useful instrument for evaluating and ranking the quality of water stations.[43] The specified standards are used to determine each qualitative parameter's value, which is then associated with other parameters. Its ease of use led to its adoption in numerous studies after it was created to evaluate the effects of contaminants on water quality.[44] Additionally, this index is categorized into six different classes, which is expressed as <0.20, suggests as clean; 0.21–0.40, as sub-clean; 0.41–0.80, as qualified; 0.81–1.0, as basically qualified; 1.01–2.0, as polluted and >2.0, represented as seriously polluted water quality.

Recently multi-criteria decision-analysis (MCDA) such as GRA and OWA, were used to define the possible zones for surface water. In this method, the dilemma, along with the objectives and standards of the study, arises from a hierarchical perspective. Thus, these processes adopt a pairwise comparison method, which is most often utilized and entails comparing each criterion to the others, which was done in pairs to determine which criterion was more important than the other for a specific goal. Pertinently, we need to determine the weighted criteria, presuming that each criterion is weighted.[45] However, this technique was to determine the critical physicochemical factors that influence the chemistry or water quality of surface water and to look at potential pollution sources. The effectiveness of the formulation varies throughout studies; some are more accurate, efficient, economical, and time-saving than the old procedures, which take a lot of time.[46] Utilizing OWA and GRA, the calculation of relative weights involves three steps: First, the sum of each column was calculated considering the pairwise comparison matrix. Second, the matrix was normalized by dividing each element in the matrix by its column total (the resulting matrix is referred to as the normalized pairwise comparison matrix). Third, the average of the elements in each row of the normalized matrix was calculated by dividing the sum of normalized scores for each row by the number of criteria. These averages provide an estimate of the relative weights of the criteria being compared. Furthermore, when combined with GIS, this technique offers quick access to satellite data and efficient management of large amounts of data to identify different indicators across various dimensions.[47]

3. Results and discussions

3.1 Surface water chemistry

The World Health Organization (WHO) standard recommendations values were contrasted with the chemical and physical characteristics of the analytical findings for surface water, in order to establish public health and drinking standards. A statistical summary of surface water quality parameters of Pre-monsoon season is presented and discussed below. The temperature recorded in this current study, varied as 25–30 °C. The recorded temperature demonstrated viability throughout the pre-monsoon season and thus, significantly increases downstream as a result of the hot springs. Based on WHO specification of pH, up to 8.5 is the greatest ideal and maximum amount that is allowed. Every year in the research region, the pH value ranges from a minimum of 7.51 to a maximum of 8.07, meaning that all seven stations are inside the upper limit that is allowed.

Based on WHO,[40] the benchmark for high-quality water is 6 mg/L, ensuring healthy aquatic life in a water body. The water samples' in-situ measured DO concentrations ranged from 6.7 to 7.6 mg/L. The alkalinity of the river running water contains about a reading, that vary between 51.6–72.89 mg/L, but they remain within the permissible limit of 200 mg/L. Additionally, the hydrolysis of these acidic compounds lowers the pH of the water, which is a key component of organic wastes that are dumped into rivers before the sites at P-(7), which raises the nitrate and sulphate levels in the river water. The obtained EC score spanned from 151 to 311 μS/cm, during the studied period. At P-(1), the EC value was found to be highest close to the riverbank and to decrease with increasing distance. The recorded TDS values range from 86 to 182 mg/L, for all the water samples collected throughout the study period. The desirable limit is taken as 500 mg/L. The study demonstrates that all 100 % of the sample is below the TDS level and is safe to consume.

From the overall results in the study, the order of anionic abundance, expressed in mg/L, is noted as: $SO_4^{2-} > Cl^- > NO_3^- > F^- > PO_4^{3-}$. The major anionic trend displays that sulphate (SO_4^{2-}) is the most dominant anion. The sulphate in surface water is mostly made available by the breakdown of rocks that include substances like gypsum, iron sulfides, and other things that include sulfur. The values of SO_4^{2-} concentration ranged from 7.78 to 40.12 mg/L, with a mean of concentration of 16.50 mg/L. The study's findings showed that none of the water samples that were gathered went above the allowable limit of SO_4^{2-} (200 mg/L) for drinking water. The Cl^- concentration varies between 8.2 and 23.5 mg/L, with an average of 13.57 mg/L. Still, the Cl^- ion content in the study area's surface water at every location was within the 250 mg/L upper limit that was permitted. From the above results, the high Cl^- values at the p-(1–3 and 7) show that organic materials are mostly not biodegradable. Moreover, moving downstream causes the NO_3^- concentrations to rise, particularly in water samples taken at point (7), which has the highest NO_3^- values, observed at p-(1), respectively. The river water's concentrations varied from 0.65 to 3.53 mg/L. The reported readings were deemed appropriate since they fell below the 45 mg/L cutoff amount. However, the study area's surface water has varying F^- concentrations between 0.3 and 1.16 mg/L. The desirable value of F^- for drinking, is taken as 1 mg/L, based on WHO (2004). Sometimes, the F^- value is recorded higher than 1 mg/L at one location (P-1). Evidently, phosphate (PO_4^{3-}) is termed as another crucial parameter for water quality that can significantly affect surface water quality overall. According to WHO (2004), PO_4^{3-} concentration in 5 water samples exceeds the contaminated levels (0.1 mg/L), for an aquatic organism. In this study, the computed PO_4^{3-} value recorded as 0.07–0.29 mg/L, whereas the phosphate mean is calculated to be 0.15 mg/L. In addition, the findings showed that water samples from two locations were determined to be within the advised limit and to be fit for human consumption without the need for additional treatment.

While analyzing the cationic chemistry, the cationic abundance hierarchy, represented in mg/L, is denoted as: $Ca^{2+} > Mg^{2+} > Na^{+} > K^{+}$. The primary cation trend observed in the surface water within the examined area is exhibited as Ca^{2+}. Thus, it acts as a dominant cation and the levels of Ca^{2+} ranged from 18.5 to 29.20 mg/L, where the uppermost allowable limit of Ca^{2+} is taken as 75 mg/L, for comparison purpose. The concentration of Ca^{2+} at site P-(1), (2), and (7), is greater in comparison to other places. This could be as a result of surface runoff and stream inflows into the reservoirs. Excess sodium often, cause the soil's permeability and infiltration to decline, creating issues for crop production. The Na^{+} content of the river water contents was measured as ranging from 5.3 to 15.3 mg/L. The observation was in good agreement with the WHO standards (200 mg/L). The concentration of Mg^{2+} varied as 18.7–29.80 mg/L. The reported readings were complying according to WHO limit (30 mg/L). Utilizing the WHO classification, all 100 % of the river water samples in this investigation were deemed safe for rural residents to drink. The K^{+} fluctuation takes place most likely as a result of house development, the usage of manure in cultivated regions, and the dumping of domestic and agricultural sewage. The calculated K^{+} values, varied between 1.75–7.24 mg/L. All of the stations' K^{+} levels fall below the range, as reported by WHO of 12 mg/L. The assessment of physicochemical analysis results at P-(1), which scored higher, determined that manmade and natural processes are the combined causes of this contamination. An IDW interpolation map was used to show the variability of all water quality indicator values at all sampling points, as displayed in Fig. 3A–O. Proceeding further, the purpose of this benchmark study was to use the CPI, GRA, and OWA to assess the river water quality. Multivariate statistical tools were employed to determine the origins of pollutants, which resulted in the suggestion of the treatment methods, that need to make the water suitable for human use.

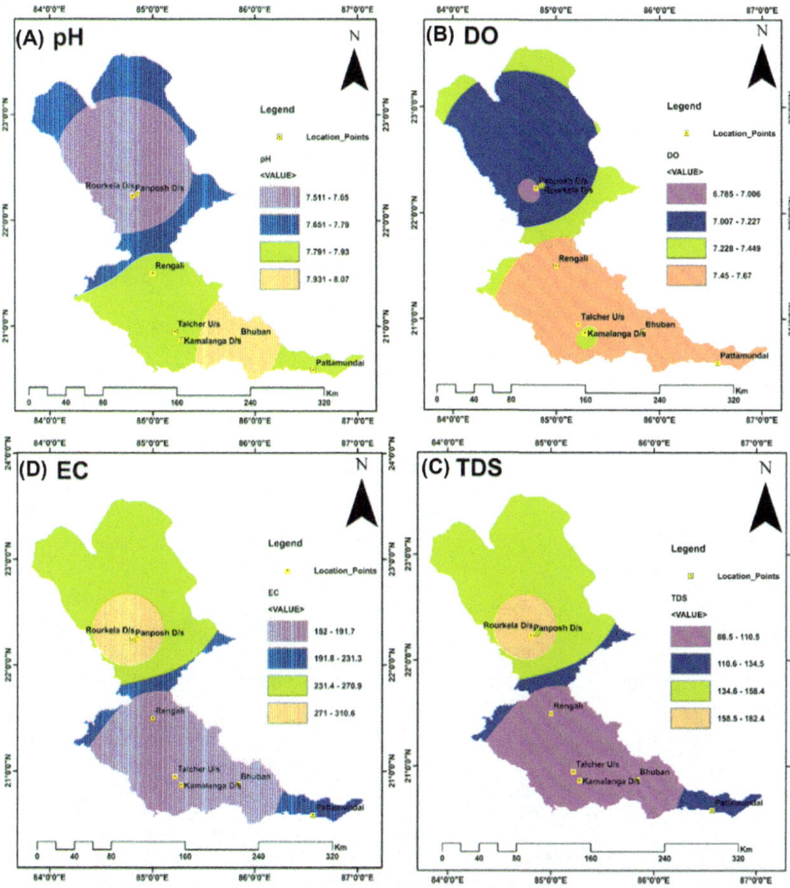

Fig. 3 Water quality interpolated maps drawn using Arc GIS software: (A) pH, (B) DO, (C) TDS, (D) EC, (E) Alkalinity, (F) Na^+, (G) K^+, (H) Ca^{2+}, (i) Mg^{2+}, (J) F^-, (K) Cl^-, (L) NO_3^-, (M) SO_4^{2-}, (N) PO_4^{3-}, and (O) Temperature. Map lines delineate study areas and do not necessarily depict accepted national boundaries.

3.2 Application of water quality (WQ) methods using CPI

In this ongoing work, the interpolated results of the CPI method, are shown in Fig. 4. The results are calculated as follows starting from Site 1–7: 1.40, 1.60, 1.20, 0.49, 0.51, 0.46 and 1.90. The average WQI value in the river water is 1.08. According to these results, the river has qualified to polluted water quality, in the present study. At two sites on the upstream face, namely, P-(1) and (2), are mostly influenced by climate, agricultural activities, and domestic wastes. Also, salinity parameter such as EC and

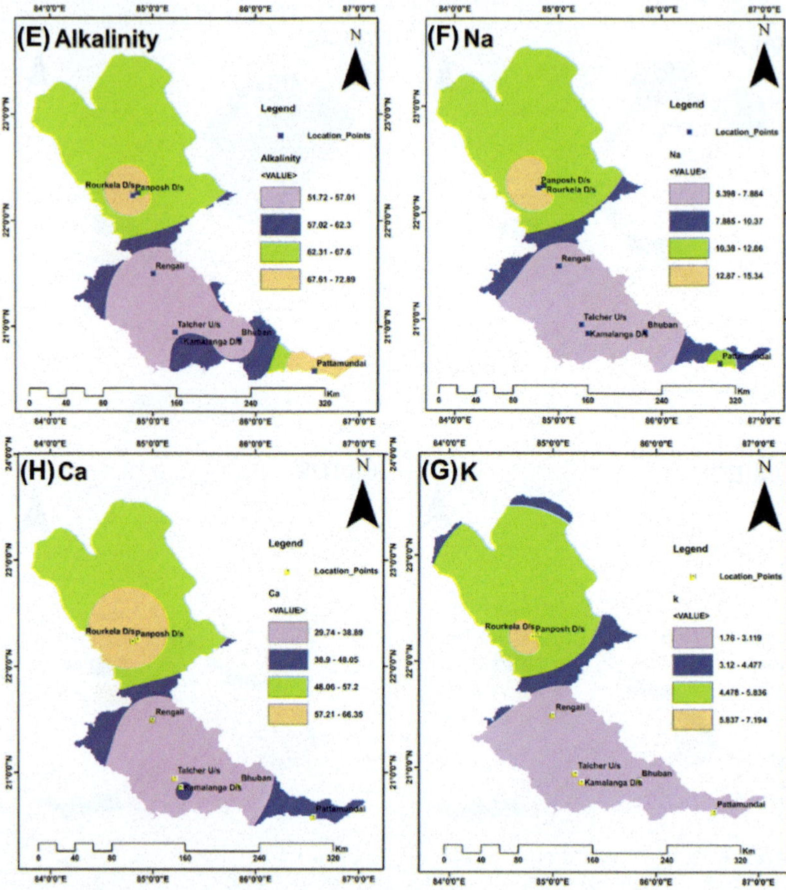

Fig. 3 (*continued*)

TDS, are also exhibited higher values at P-(3) and (7), which represent non-point inorganic pollution. This approach shared water points, that were studied in the study area, which contributes around 42.86 % as qualified class of water, and 57.14 % of locations are classified in the polluted category. Notably, the upstream and downstream stations are signified as deteriorated water quality, while the mid-stream stations projects towards safe drinking water. Therefore, this stream promotes suitability for industrial, agricultural, and drinking uses, at three places such as P-(4), (5) and P-(6), respectively.

The WQ methods such as CPI has some limitations and uncertainty. This model may be impacted by suppressing or overemphasizing a particular problematic parameter value. Consequently, it should be revised once

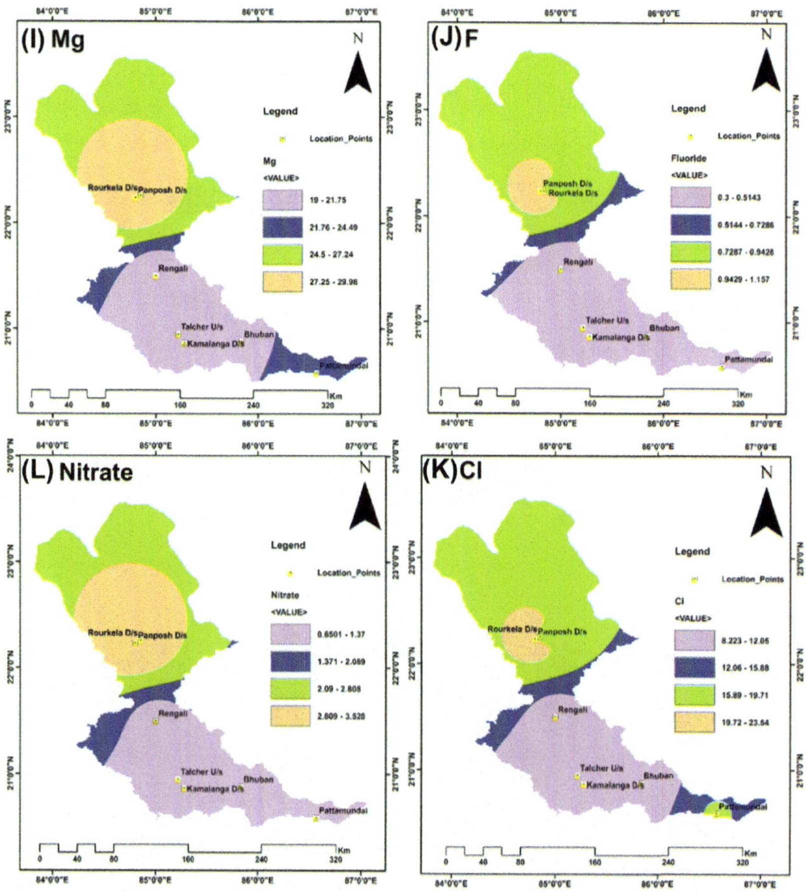

Fig. 3 (*continued*)

more using the OWA and GRA models in consideration of the aforementioned constraints and recommendations in order to achieve better results.

3.3 Water quality assessment using GRA and OWA

The 15 water quality parameters from the seven sites that were gathered during pre-monsoon season, and thereby, examined in the form of a matrix. Both aquatic ecosystems and living individuals are negatively impacted when contaminated water is consumed. It is noteworthy that Multi-Criteria Decision-Making Tools namely GRA and OWA, on the other hand, is employed to determine and calculate the matrix-based correlations between water quality parameters and other parameters, by

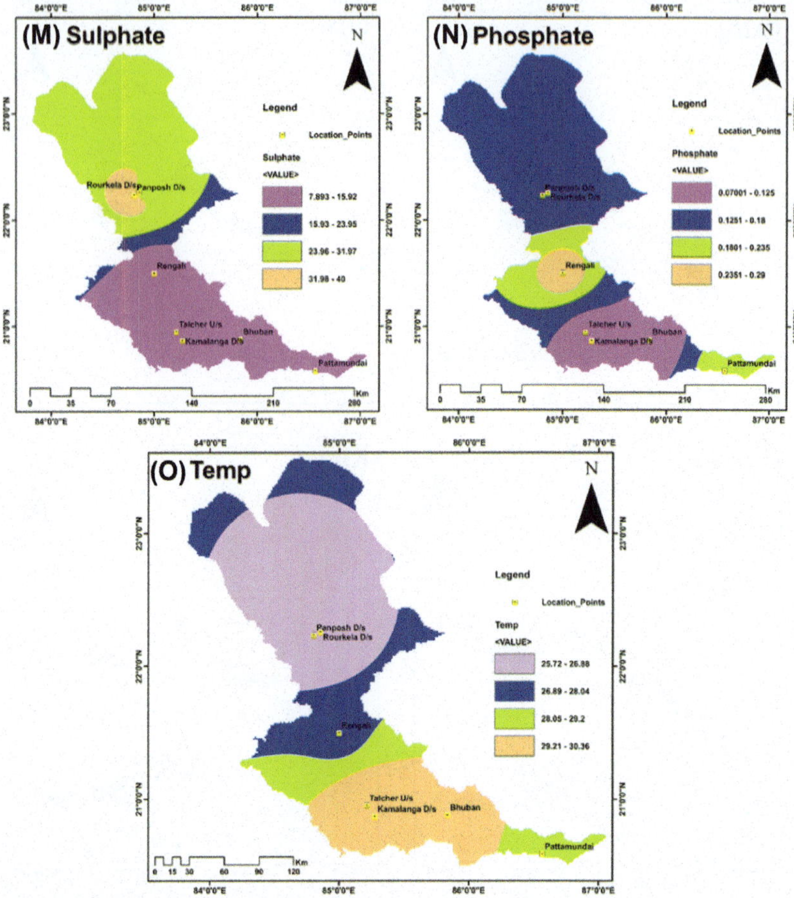

Fig. 3 (*continued*)

applying a rank, depending upon a performance degree score (R_i). Consequently, the identified GRA and OWA index of the river, was assessed for a period of six years for the Brahmani River in Eastern Odisha. To improve the results' interpretability, the data were interpolated using ArcGIS and IDW geo-statistics. The distribution of values along the sampling stations is shown in Fig. 5A and B. The computed value of GRA ranged from 0.03 to 0.059. Again, the evaluation of water quality using OWA method, for a period of 6 years, using 7 monitoring sites is carried out and the mean value of OWA is found as 0.62, ranging from 0.47 to 0.93. From the outcomes, good water zones at P-(4), (5) and (6) is being observed, containing a value of 0.469, 0.53, and 0.466 by OWA and 0.030,

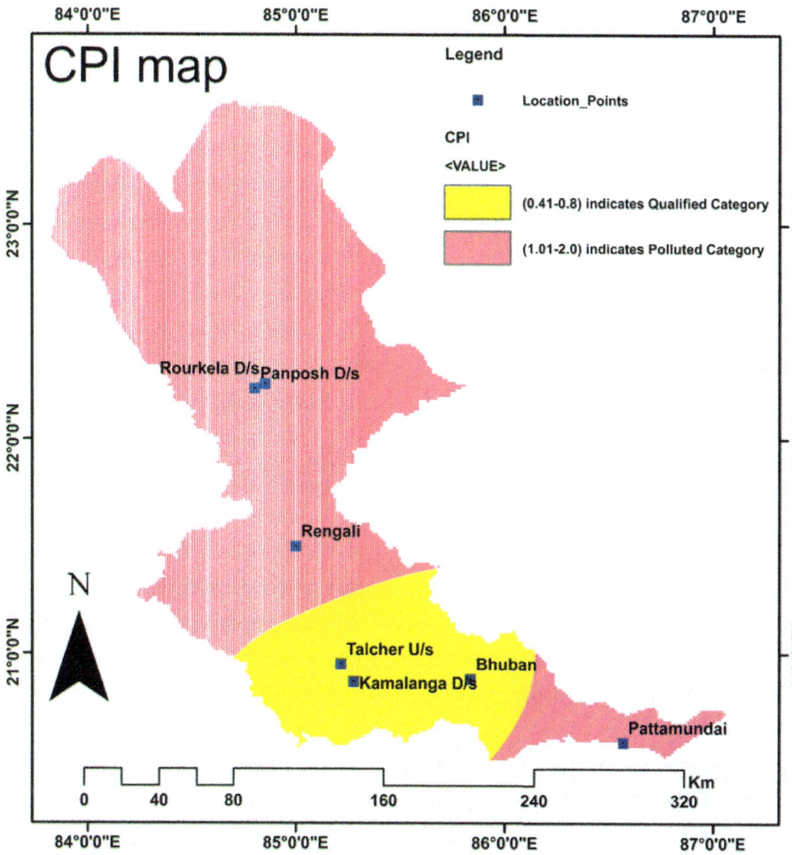

Fig. 4 Spatial plot of CPI map. Map lines delineate study areas and do not necessarily depict accepted national boundaries.

0.032 and 0.032 using GRA technique, indicating the use of first treatment, followed by disinfection, for drinking water. The river water quality had significantly improved as a result of the government's recent attention to these environmental issues and implementation of sensible management methods. Using two methods, it is determined that because of the infiltration of seawater from the Bay of Bengal during high tides, sampling site P-(1) had the lowest water quality. This location, evidently, contains a high CPI value of 1.40, followed by P-(2) as 1.6 and 1.90 correspond to P-(7). It is seen that at these locations, higher values were attributed to TDS, EC, PO_4^{3-}, Cl^-, and SO_4^{2-}. Since, the river continues to flow downstream, passing through a number of cities and towns until entering the state of

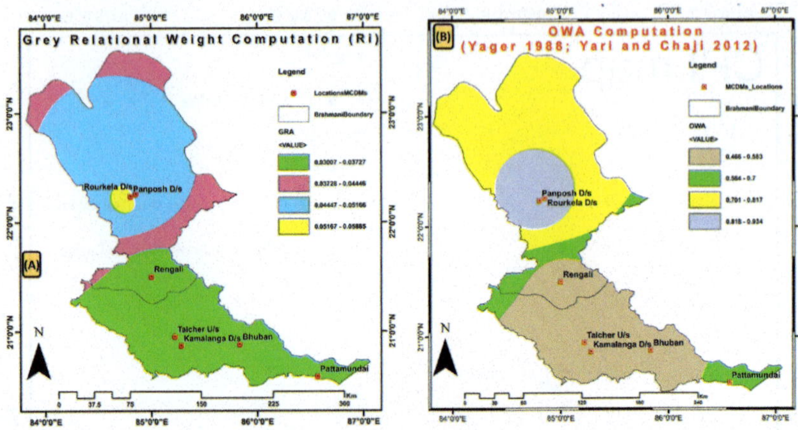

Fig. 5 Spatial distribution of (A) GRA, and (B) OWA map. Map lines delineate study areas and do not necessarily depict accepted national boundaries.

Odisha, before pouring into the sea, it appears to pick up more pollution. Finally, the significant finding from the MCDM analysis, clearly signify a distinct differentiation, that provides once again, another helpful use of water quality in environmental data analysis.

3.4 Climate change implications on surface water quality in the Brahmani River basin

Water is integrally tied to the pillars of sustainable development because of its close ties to the climate, which intersect with the United Nations' (UN) Sustainable Development Goals (SDGs). Recent global water crisis has demonstrated that climate change is more dangerous than previously believed, having a significant impact on life and sustainable development as well as the water cycle. Climate change opponents reply that climate change is a natural occurrence that is beyond human control and does not pose a serious threat, which is equally responsible to those who embrace it. In light of the ongoing threat posed by climate change, increased water contamination in recent years highlights the need for instruments with the special capacity to characterize future trends and predict possible variations in water quality. Therefore, this ground-breaking project is to examine the potential effects of climate change on the water resources of the selected watershed, which is situated in the Indian state of Odisha.

Water scarcity is a serious threat to the current global environment and a crucial issue for sustainable development, but because of human activity and conflicts of interest among pertinent stakeholders, issues related to water pollution, waste water treatment, greenhouse gas reduction, and water consumption reduction are still complex and intricately linked. According to earlier research, climate change will increase the discharge of suspended particles and nutrient loading to water bodies by causing high local temperatures and frequent or severe precipitation. Therefore, the methods and strategies used in this study also include modeling the many hydrologic processes that take place in the river basin and coming up with workable and relevant solutions. For this purpose, water quality index (WQI) and pollution indices such as CPI, combines physical, chemical, and biological elements to provide a single number between 0 and 2 that can be used to characterize the general quality of the water. By condensing the majority of the data into a single value, it is a useful method of characterizing the water's quality. 15 water quality indicators served as the foundation for the CPI's design.

Water stress is expected to have an effect on the population of Odisha by 2025, including millions of people who live in low-lying areas, such as small islands and large coastal cities, who would need to relocate in large numbers and become climate refugees if global warming is neglected. This massive population shift will probably put further strain on the already overburdened water supplies and put the population's most vulnerable segments at risk of exploiting water sources without the necessary protections to guarantee water security. Such data assesses water quality and reports on degradation that has been implemented just somewhat. This can be done by considering the effects of climate change, the movement of sediments and nutrients, the temperature-dependent variation of the hydrological cycle, population growth, and fecal contamination. Thus, the study concentrated on multi-criteria decision-making models for predicting future climate change scenario parameters and evaluating water quality in the river basin based on historical hydrometeorological circumstances. GRA and OWA feature analysis methods were used to assess the hydrometeorological parameters' appropriateness. They collaborated on the hydro-chemical variables, which showed that these chemical species in the surface water had a good association since they were created by the same geochemical processes and existed in a similar geochemical environment. As a result, the models enable the optimization of indexing and decision-making methods by giving decision-makers the essential information about the availability of surface water resources in various seasons.

Meanwhile, these strategies will aid in the creation of policies capable of overseeing activities in the event of climate change.

The surface water in the study area is often composed of low amounts of dissolved inorganic constituents, most of which are considerably below the WHO maximum allowable levels. The river basin's varied features also make it necessary to separate the entire basin's area into distinct zones for in-depth study. In order to make the drinking water levels sustainable, the effects of drinking and irrigation using surface water were also examined. Overall, the models applied in the current investigation, showed excellent accuracy in determining appropriateness in a range of seasonal circumstances, offering water managers strong decision-making assistance when developing different plans. A thorough seasonal analysis offers a scientific foundation for the wise use and preservation of water resources by enabling a greater comprehension of the complexities of surface water systems.

4. Conclusion, limitations, and its solutions/ recommendations

In this research, we offer a blend of Water quality (WQ) and Pollution indices (PIs), and multicriteria decision-making tools of water quality data, in addition with hydro-meteorological data, and also, chemical analyses of water samples, to evaluate the Brahmani reservoir's water quality and locate the pollution sources. The pH of the study area reflects alkaline nature of the samples. The main ions' abundances are arranged in the following order: $Ca^{2+} > Mg^{2+} > Na^{+} > K^{+}$ and $SO_4^{2-} > Cl^{-} > NO_3^{-} > F^{-} > PO_4^{3-}$. The calculated CPI ranged from 0.46 to 1.9. According to the CPI classification, the range of % give rise to 57.14 %, that suggests polluted water quality, throughout the whole period. The degree of value, that GRA reported, which vary between 0.03 and 0.059, indicating mid-stream sites, that promotes water suitability. However, the analytical results of OWA, containing physical and chemical indicators of surface water, validate a class that depicts P-(1) is the most affected place in this study area. This is mostly associated with soluble salts that fall into the natural category and the load of organic pollutants that is present in the anthropogenic group. Ultimately, the spatial distribution of the interpolated maps, showed that the study region's central high raised sections have the safest surface water quality for drinking. Hence, these results provide important information for understanding and managing water quality deterioration in the selected river basin. In this perspective, the suggested predictive method could be further upon by utilizing additional optimization strategies and used in massively parallel processing environments such as Hadoop and Spark. Thus, the future scope of automation is described by merging smart phone integration, weather remote sensing power, and big data execution environment into the proposed framework. Therefore, surface water level extraction and monitoring would be made easier by self-regulating surface water management systems that use environmental and surface parameter sensing devices obtained from the Internet of Things and artificial intelligence (AI). This would enable decisions linked to water use.

5. Limitations

The following limitations are discussed below:

- Obtaining relevant information, such as maps, specialized software, and groundwater samples, was difficult and hindered the investigation's thoroughness.
- The study's failure to account for seasonal variations in surface water samples may limit our knowledge of temporal changes in water quality.
- The requirement for high-quality training data and the variation in model performance across datasets are two enduring problems.

6. Recommendations

The study's recommendations for the river water quality assessment are described as follows:

- To boost the accuracy and dependability of the assessment, surface water monitoring should be improved, and the density and frequency of the site should be increased for more thorough studies on irrigation suitability, water quality, and replenishment.
- Since chloride has a negative impact on plants when it is present in water, it is highly recommended that surface water be treated and its levels reduced using techniques like reverse osmosis or ion exchange before being used for irrigation.

- Integrating machine learning models into surface water investigations will be essential and beneficial since they offer excellent prediction performance levels, saving both time and money.

Acknowledgements

The author wishes to express their sincere thanks to the C.V. Raman Global University (CGU), Bhubaneswar, Odisha, India, for providing necessary data and information that made this study possible. Special thanks go to the Miss Ranu-Prativa, Ruchika Square, Bhubaneswar, for providing necessary software and hardware support. **I am grateful to my family members: Ashok Kumar Das (Father), Urmila Mohakud (Mother), and Sarita Das (Sister), for their valuable suggestions and would like to express our sincere thanks to the editors and reviewers for their very helpful comments on the paper.**

Data availability

The data used to support the findings of this study are available from the corresponding author upon request.

Ethical approval

This article does not contain any studies with human participants or animals by any of the authors.

Conflict of interest

The authors declare that there are no conflicts of interest regarding the publication of this paper.

Declaration of generative AI and AI-assisted technologies in the writing process

No involvement of any AI tool in writing the whole manuscript.

References

1. Anh DT, Pandey M, Mishra VN, Singh KK, Ahmadi K, Janizadeh S, et al. Assessment of groundwater potential modeling using sup- port vector machine optimization based on Bayesian multi-objective hyperparameter algorithm. *Appl Soft Comput* 2023;**132**:109848. https://doi.org/10.1016/J.ASOC.2022. 109848.
2. Sud A, Kanga R, Singh SK, Meraj G, Kanga S, Kumar P. Simulating groundwater potential zones in mountainous Indian Himalayas—a case study of Himachal Pradesh. *Hydrology* 2023;**10**.
3. Li L, Knapp JL, Lintern A, Ng GHC, Perdrial J, Sullivan PL, et al. River water quality shaped by land–river connectivity in a changing climate. *Nat Clim Change* 2024;**14**(3):225–237.
4. Baccour S, Goelema G, Kahil T, Albiac J, van Vliet MT, Zhu X, et al. Water quality management could halve future water scarcity cost-effectively in the Pearl River Basin. *Nat Commun* 2024;**15**(1):5669.
5. Picos Corrales LA, Gundogdu S. Pollution and environmental water quality of river and streams. *Front Water* 2024;**6**:1441071.

6. Huang Y, Cai Y, He Y, Dai C, Wan H, Guo H. A water quality prediction approach for the downstream and delta of Dongjiang River basin under the joint effects of water intakes, pollution sources, and climate change. *J Hydrol* 2024;**640**:131686.
7. Guo X, Gui X, Xiong H, Hu X, Li Y, Cui H, et al. Critical role of climate factors for groundwater potential mapping in arid regions: insights from random forest, XG Boost, and LightGBM algorithms. *J Hydrol* 2023;**621**:129599. https://doi.org/10.1016/J.JHYDROL.2023.129599.
8. Ghachoui E, Taouil H, Naser RS, Said IA. Spatial assessment of groundwater quality for drinking In Souk El Arbaa, Morocco. *Water Supply* 2024;**24**(4):1267–1281.
9. Benyoussef S, Arabi M, El Yousfi Y, Makkaoui M, Gueddari H, El Ouarghi H, et al. Assessment of groundwater quality using hydrochemical process, GIS and multivariate statistical analysis at central Rif, North Morocco. *Environ Earth Sci* 2024;**83**(17):515.
10. Moayedi H, Salari M, Dehrashid AA, Le BN. Groundwater quality evaluation using hybrid model of the multi-layer perceptron combined with neural-evolutionary regression techniques: case study of Shiraz plain. *Stoch Environ Res Risk Assess* 2023;**37**(8):2961–2976.
11. Alattar MH. Mapping groundwater dynamics in Iraq: integrating multi-data sources for comprehensive analysis. *Modeling Earth Syst Environ* 2024:1–11.
12. Chen C, Guo Z, Judge J. A QGIS plugin for GIS-based multicriteria decision analysis: an application of developing alternative future land-use scenarios in Ghana. *Appl Spat Anal Policy* 2024:1–19.
13. Fei-Baffoe B, Badu E, Miezah K, Sackey LNA, Sulemana A, Amuah EEY. Contamination of groundwater by petroleum hydrocarbons: Impact of fuel stations in residential areas. *Heliyon* 2024;**10**(4).
14. Patel SS, Sahoo S. Investigation of water quality of Damanganga estuary using water quality index, south Gujarat, India. *Sustainability, Agri, Food Environ Res* 2021;**9**(2):148–167.
15. Anuja SA, Kavitha P, Jothi CH, Rose RJ. Seasonal assessment of water quality for irrigation using multiple indices in Thamirabarani river, Kanyakumari district, India. *Desalination Water Treat* 2024;**317**:100206.
16. Wibowo MA, Andarini P, Kurniatama DVP, Humaira NG. Methods for determining the water quality index in developing Asian countries: a review. *Ecol Eng & Environ Technol* 2024;**25**.
17. Saha S, Bera B, Shit PK, Bhattacharjee S, Sengupta N. Prediction of forest fire susceptibility applying machine and deep learning algorithms for conservation priorities of forest resources. *Remote Sens Applications: Soc Environ* 2023;**29**:100917. https://doi.org/10.1016/j.rsase.2022.100917.
18. Singh VK, Pandey HK, Singh SK. Groundwater storage change estimation using GRACE data and Google earth engine: a basin scale study. *Phys Chem Earth* 2023;**129**:103297. https://doi.org/10.1016/j.pce.2022.103297.
19. Jayaraman P, Nagarajan KK, Partheeban P, Krishnamurthy V. Critical review on water quality analysis using IoT and machine learning models. *Int J Inf Manag data insights* 2024;**4**(1):100210.
20. Sajib AM, Diganta MTM, Rahman A, Dabrowski T, Olbert AI, Uddin MG. Developing a novel tool for assessing the groundwater incorporating water quality index and machine learning approach. *Groundw Sustain Dev* 2023;**23**:101049.
21. Kazapoe RW, Addai MO, Amuah EEY, Sagoe SD. Fluid boundaries: a cross-country exploration of groundwater quality amid threats from climate change. *Environ Chall* 2024:100953.
22. Sharifian S, Mortazavi MS, Nozar SLM. Health risk assessment of commercial fish and shrimp from the North Persian gulf. *J Trace Elem Med Biol* 2022;**72**:127000.

23. Genjula W, Jothimani M, Gunalan J, Abebe A. *Applications of statistical and AHP models in groundwater potential mapping in the Mensa River catchment, Omo River valley, Ethiopia. Model. Earth Syst. Environ;* 2023pp. 1–19. https://doi.org/10.1007/S40808-023-01765-8/METRICS
24. Al-Humairi BAJ, Rahal NS. Variability of time duration analysis for rainfall water using precipitation indexes in Hai town. *Water Pract & Technol* 2024;**19**(2):519–531.
25. Zwain HH, Abed BS. Groundwater flow modeling and hydraulic assessment of Al-Ruhbah region, Iraq. *J Mech Behav Mater* 2023;**32**(1):20220214.
26. Workie MD, Hailu BT, Birhanu B, Suryabhagavan KV. Geospatial analysis of volcano-tectonic interactions: implications for spatiotemporal variations in water quality and quantity at Lake Beseka, East Africa, Ethiopia. *Int J River Basin Manag* 2024:1–18.
27. Masoudian E, Pahlavani P. Gis-based groundwater potential mapping using ma- chine learning models, a case study: Qom Province, Iran. *ISPRS Ann Photogramm Re- mote Sens Spat Inf Sci* 2023;**10**:489–494. https://doi.org/10.5194/isprs-annals-X-4-W1-2022-489-2023.
28. Morgan H, Madani A, Hussien HM, Nassar T. Using an ensemble machine learn- ing model to delineate groundwater potential zones in desert fringes of East Esnaldfu area, Nile valley, Upper Egypt. *Geosci Lett* 2023;**10**. https://doi.org/10.1186/s40562-023-00261-2.
29. Derakhshani R, Lankof L, GhasemiNejad A, Zaresefat M. Artificial intelligence-driven assessment of salt caverns for underground hydrogen storage in Poland. *Sci Rep* 2024;**14**(1):14246.
30. Mahammad S, Islam A, Shit PK. Geospatial assessment of groundwater quality using entropy-based irrigation water quality index and heavy metal pollution indices. *Environ Sci Pollut Res* 2023;**30**(55):116498–116521.
31. Miraj P, Berawi MA, Aninditya A, Sari M. Evaluating the impact of knowledge management and database management on decision-making process: A case study of subsea project services. *J Open Innovation: Technology, Market, Complex* 2024;**10**(3):100340.
32. Haggerty R, Sun J, Yu H, Li Y. Application of machine learning in groundwater quality modeling-A comprehensive review. *Water Res* 2023;**233**:119745.
33. Mallick S, Maity B, Das P, Rudra S. GIS-based groundwater recharge potentiality analysis using frequency ratio and weights of evidence models. *Case Stud Geospatial Appl Groundw Resour* 2023;**91–108**:00016-X. https://doi.org/10.1016/B978-0-323-99963-2.
34. Salamat A, Ardestani M, Malekmohamadi B. Modeling groundwater potential using machine learning models. *Watershed Manag Res J* 2023. https://doi.org/10.22092/WMRJ.2023.360675.1506.
35. Giao NT, Nhien HTH, Anh PK, Thuptimdang P. Groundwater quality assessment for drinking purposes: a case study in the Mekong Delta, Vietnam. *Sci Rep* 2023;**13**(1):4380.
36. Kouassi KJM, Lachassagne P, Mangoua OMJ, Sombo AP, Dibi B. Identifying the origin of springs in weathered-fractured crystalline aquifers using a hydrogeophysical approach. *Sci Rep* 2024;**14**(1):12977.
37. Das A. Water quality assessment using water quality index (WQI) under GIS framework in Brahmani Basin, Odisha. *International conference innovation in smart and sustainable infrastructure;* Springer Nature Singapore: Singapore, 2022, pp. 131–150.
38. Gupta G, Khan J, Singh NK, Bhave VN, Bharadwaj V, Singh RP. Evaluation of wastewater quality using multivariate analysis and water quality index for its sustainable management in Urban Area. *Pollution* 2024;**10**(3):973–984.

39. APHA. *Standard methods for the examination of water and wastewater;* 20th ed.; American Public Health Association, American Water Works Association, Water Environment Federation, 1999, https://doi.org/10.2105/AJPH.51.6.940-a
40. WHO. *Guidelines for drinking water quality: training pack;* WHO: Geneva, Switzerland, 2004, .
41. Das A. Multivariate statistical approach for the assessment of water quality of Mahanadi basin, Odisha. *Mater Today: Proc* 2022;**65**:A1–A11.
42. Laskar N. Statistical analysis and water quality modeling towards holistic health assessment of Himalayan and peninsular rivers of India. *Environ Chall* 2024;**15**:100917.
43. Khyalia P, Duhan SS, Laura JS, Nandal M. A comprehensive analysis of fluoride contamination in groundwater of rural area with special focus on India. *Water resources management for rural development;* Elsevier, 2024, pp. 201–212.
44. Sharma S, Banyal HS. Insight into aquatic potential of Kunah stream, Hamirpur (HP) by analysis of water quality parameters using weighted arithmetic water quality index. *Environ Ecol* 2024;**42**(2B):754–761.
45. Yager RR. On ordered weighted averaging aggregation operators in multi- criteria decision making. *IEEE Trans Syst Man Cybern* 1988;**18**(1):183e190.
46. Yari G, Chaji A. Determination of ordered weighted averaging operator weights based on the m-entropy measures. *Int J Intell Syst* 2012;**27**(12):1020e1033. https://doi.org/10.1002/int.21559.
47. Husain MS, Ahmad I, Umar R. Hydrochemistry, controlling mechanisms and water quality index in hot springs of the Kullu-Manali region, India. *Groundw Sustain Dev* 2024;**24**:101070.

CHAPTER TEN

Harnessing the potential of Sentinel-1 SAR data for flood inundation mapping

Arjita Maurya[a], Gaurav Tripathi[b,*], Achala Shakya[c,*], Ritambhara K. Upadhyay[d], Yateesh Ketholia[e], and Vipin Chander Pant[f]

[a]RMSI Pvt. Ltd., Dehradun, Uttarakhand, India
[b]Department of Civil Engineering, Indian Institute of Technology Bombay, Powai, Mumbai, Maharashtra, India
[c]Department of Computer Science and Engineering, School of Engineering and Sciences, SRM University-AP, Guntur, Andhra Pradesh, India
[d]Department of Geology, Central University of Himachal Pradesh, Dharamshala, Himachal Pradesh, India
[e]Geosciences & Geohazards Department, Indian Institute of Remote Sensing, Dehradun, Uttarakhand, India
[f]Council of India, New Delhi, India
*Corresponding authors. e-mail address: 20004109@iitb.ac.in; achala.shakya@ieee.org

Contents

Abstract

Floods are the most common and destructive category of natural calamities across the globe. Hence, satellite-based images are one of the most reliable sources of information related to disasters and natural calamities. Satellite imageries can cover large geographic regions even in adverse climate conditions, making them crucial tools for disaster management. The northeastern state of the country gets unsettled

Advances in Chemical Pollution, Environmental Management and Protection, Volume 15
ISSN 2468-9289, https://doi.org/10.1016/bs.apmp.2025.11.001

every year due to the substantial amount of water from the upstream catchment areas. In 2022, over 200 people died from the flooding, affecting 5.4 million people in 32 districts. The Silchar was the worst-hit town, where 90 % of the land was underwater. This study characterizes flood inundation over the Silchar district in Assam during June–July 2022 using Sentinel-1 (S-1) Synthetic Aperture Radar (SAR). SAR data is adequate in providing accurate and real-time responses for flood inundation mapping. For the data processing, the SNAP tool was used to process S-1 data to derive flood inundation extent and Landsat 9 data for validation. Climate Hazards Group InfraRed Precipitation Satellite data (CHIRPS) was used to assess the spatio-temporal rainfall variability over the study region. This study investigated continuous rainfall over the week in upper catchment areas of the Barak River, a major tributary of the Brahmaputra River, draining Silchar heavily and causing significant flooding. CWC-based gauge station at Badurpur ghat in Karimganj and Annapurna ghat in Cachar showed that the river flow was beyond the danger level during the flooding period. The total area inundated was 136.64 sq.km on 7th June, which increased to 517.85 sq.km (approximately 73.61 % of the total study area) by the 28th June. However, floodwater started receding after 12th July (flood extent was estimated as 256.61 sq.km on 12th July). The decision-makers can follow such outputs to employ mitigation and rehabilitation strategies in the affected regions.

Keywords: Flood inundation, Silchar, Assam, Brahmaputra river, Barak river, Sentinel-1, SAR, Landsat-9, Climate Hazards Group InfraRed Precipitation with Station data (CHIRPS).

1. Introduction

Flooding is the term used when water overtops the boundary and enters onto the normally dry terrain. It often caused by heavy rainfall, melting of snow, or the deterioration of water containment structures like dams or levees, and sometimes due to cyclone nearby to coastal areas etc. Floods typically occur in both inland, along the shore, and they can range from small-scale catastrophic disasters to localized flash floods. The most common devastating, frequent climatic and hydrological risks are floods, which can occur due to human-induced and natural reasons. Most importantly, it causes economic and human suffering significantly. The destruction due to natural calamities in the South Asian region was heavily inhabited.[1] Emergency Events Database (EM-DAT) reported that 399 natural disasters happened in 2023 that was more than if compared with the annual average between 2003–2022, i.e. 369. Specifically, 164 flood events have been recorded as occurred in the year 2023, while 170 was the annual average during 2003–2022.[2] Notably, natural catastrophes affected Asia, Africa, Europe, America, and Oceania by 46.7 %, 16.9 %, 8.2 %, 24.3 %, and

3.9 %, respectively.[3–5] Unplanned roads, barrage, bund construction, and public utilities built inside floodplains have negatively impacted the river's or drainage system's regular flow causing flood inundation. Therefore, overflowing flood waters are typically caused by rapid riverbed siltation.[3,6]

The measurement of the strength and frequency of floods resulted from the fact that flooding has always been a severe and persistent worry for humanity. Therefore, flood inundation models and the creation of flood maps using satellite data have become essential for monitoring and evaluating the effect of floods over the past three decades.[7–9] Since it is not that much feasible to monitor flood progression and regression with in-situ data, remote sensing-based satellite data might be a key option to provide information on flood progression across more extensive areas. Over the past decade, substantial technological developments in remote sensing-based flood inundation mapping have occurred.[10–13] From the literature, it is evident that use of remote sensing satellite data has enhanced prediction accuracy and, hence, leading to stakeholder awareness of flood dynamics and flood forecasting. In the literature, optical and SAR (Synthetic Aperture Radar) have been extensively applied to monitor surface water dynamics.[8,14,15]

The lower Brahmaputra plain faced severe flooding during 1954, 1962, 1972, 1977, 1984, 1988, 1998, 2002, 2004, 2012, 2013, 2015, 2016, 2017, 2018, 2019, 2020, and 2022 in the last half-a-century.[3,16,17] It is observed that in the last twenty years, the frequency of the flood has picked up considerably when compared to the second half of the twentieth century. Recent catastrophic floods in India have encompassed the 2008 Bihar flood in the river Kosi, the 2005 Mumbai flood in the river Mithi, the 2013 Uttarakhand cloudburst, the 2015 Chennai flood, and the 2022 Assam disaster.[18] Remote sensing-based flood maps are being used for hydrodynamic model assessment and calibration, as well as to assimilate and enhance predictions. The key reason for using SAR data is that it can distinguish between the land and water boundary as well as it can also assist in defining flood extent and control. Several studies have used SAR data efficiently to map and monitor the temporal flood extent due to its longer wavelength which can penetrate clouds and take images into adverse climate condition and visualize the extent of floods across vast geographic areas. Due to all-weather capability, number of studies are exploiting SAR data for undertaking observations in flooding conditions which makes it a vital resource. The quantity and type of satellite data that become available to various actors or organizations during the time of a flood have grown in the last few decades, further increasing flood mapping and monitoring.[1,8,16,19]

Traditional flood mapping methods are based on sophisticated hydrodynamic models or oversimplified static approaches.[14,20,21] Moreover, empirical techniques employ ground observations and optical remote-sensing images to map historical flood inundations.[22,23] The Copernicus Emergency Mapping Service (EMS) employs Sentinel-1 and other remote sensing information for swift flood disaster mapping. Meanwhile, the Finnish Flood Center has created the SAR-based Flood Detection Algorithm (FC-FloDA) to identify flood extent and depth in northern Finland's forested regions.[5,20,24–26] Automatic mapping techniques, such as the RAPID algorithm by Shen et al. (2019a), have provided extensive flood inundation datasets in the U.S. using multi-source remote sensing data.[9,25,27,28] Flood detection using SAR images remains a popular research focus, with methods categorized into supervised detection, using a threshold, digital image segmentation, change detection of pre- and post-disaster images, manual digitizing, delineation of flooded vegetation, and urban flood water detection. However, challenges persist, as water-like surfaces (e.g., smooth roads) can reduce accuracy, and threshold-based methods may struggle with non-ideal histogram distributions. Techniques like deep convolutional networks improve accuracy by extracting deeper semantic features. Moderate Resolution Imaging Spectroradiometer (MODIS) and Landsat images have contributed significantly to the historical as well as global scale flood inundation mapping and environmental impact assessments too. Machine learning (ML) techniques have recently emerged as fast alternatives for flood inundation modelling, demonstrating their efficacy in hydrological applications. Machine learning (ML) techniques have shown significant potential in hydrology, primarily for flood inundation mapping and modelling, providing faster alternatives to traditional methods. These data-driven approaches capture flood dynamics based on conditioning factors, effectively replacing simplified conceptual models. While studies demonstrate the efficacy of ML models in simulating flood maps, they face challenges in extrapolating beyond their training data range.

Furthermore, data availability issues often limit these models to small spaces (related to the spatial resolution of the satellite data used) or individual sites (having satellite pass on the event date). However, these limitations nowadays can be avoided by using unmanned air vehicles or drones. In contrast to traditional models, where observed flow or water levels are used, ML methods utilize observed flood inundation maps and conditioning factors, which are usually sparse. Nevertheless, advances in 2D hydrodynamic models (2D), along with enhanced computational capabilities, have improved simulated flood inundation data, providing the capability to train ML models for fast and accurate generation of flood maps. Once trained, these models can support real-time river operations and flood management applications. Hydrodynamic flood simulation models are one-dimensional (1D), two-dimensional (2D), and coupled 1D-2D models.[9,25,26,28] The 1D models simulate flow along a river's center line, 2D models account for lateral flow in the floodplain, and coupled models combine both. Popular software options include LISFLOOD-FP, MIKE-SHE Flood, TELEMAC-2D, TUFLOW, and HEC-RAS.[26,29] Although these models accurately simulate floods by incorporating physical parameters, they require significant computation time.[29] Techniques such as high-performance computing and mathematical conversions can minimize computation time but require specialized software, knowledge, and expense.[13,15,26,27,30] Therefore, there is a need to create quicker and easier substitutes for flood inundation mapping.[11]

The flooding event worstly affected the Cachar plain near the Barak Riverbank, started from mid of June 2022 onwards due to torrential rainfall followed by the dyke breach near to the study region impacted nearly 500,000 people. Hence, flooding water followed the path and inundated downward region and created havoc there. The main objectives of the present study are to determine flood inundation extent using Sentinel-1A SAR data and validate flood inundation extent with Landsat-9 optical data during 28th June 2022 flood events over Cachar, Silchar, Assam. Therefore, by examining the scope and severity of the flood using SAR data, our ultimate goal is to provide helpful information for disaster management and response. To map floods, the study emphasizes the value of using SAR data in flood-prone areas and shows how it may supplement to other remote sensing data sources.

2. Study area

This study focused on the flood-inundated mapping in the Cachar district in Assam, India. The Cachar District geographically extends between 24°22′ to 25°8′ N and 92°25′ to 93°16′ E (Fig. 1). Some of the extensively engaged locations in the Cachar district were highly affected during this event are Katigorah, Lakhipur, Silchar, Sonai, and Udharbond. Among others, Silchar is one of the significant locations that was highly influenced by the flooding.[2]

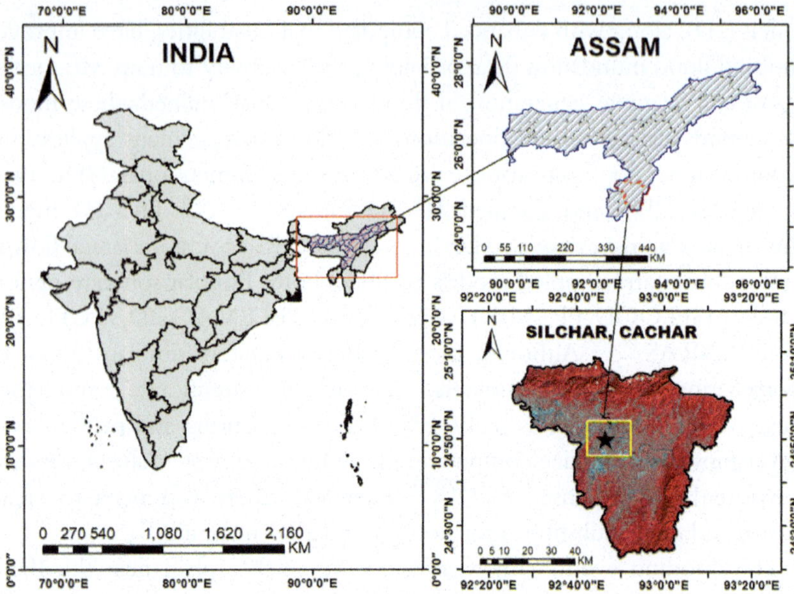

Fig. 1 Study area map of Silchar, Cachar, Assam. Map lines delineate study areas and do not necessarily depict accepted national boundaries.

The study area is located on the Barak River, which covers a total distance of about 564 kilometers long river. It is the second biggest river basin in North-East India with 2400–4100 mm of annual precipitation, ~85 % occurring during April to October, is located between latitudes 24°8′ and 25°8′ N and longitudes 92,015′ and 93,015′ E.[18] Its source is in Manipur's hill area, forming the region's largest and most important river. After Manipur, it passes into Assam and Mizoram and later, it crosses into Bangladesh, splitting into the Surma and Kushiyara rivers. One of the principal rivers in northeastern India is the Barak River, which passes through the states of Manipur, Mizoram, Assam, and Nagaland. It is 564 kilometers long and a tributary of the Brahmaputra River.[18] According to the Ministry of Water Resource's Report, at a height of 2331 m, the Manipur hills south of Mao in the Senapati district of Manipur give origin to the Barak River. Manipur controls 20.42 % of the total area, followed by Mizoram, Assam, and Tripura, with shares of 20.06 %, 16.70 %, and 18.31 %, respectively. Nagaland is the least of the basin's total area, at around 1.77 %.[1] It splits into the Surma and Kushiyara rivers on entering Bangladesh. Local runoffs of rainfall from the valley and surrounding hilly regions flow into the Barak river and its tributaries before being expelled into Bangladesh. The Katakhal, Jiri, Chiri, Modhura, Longai, Sonai, Rukni, and Singla are the major tributaries of the valley. The tributaries are mostly fed by rain and result in flooding problems during rainy seasons.

According to the seismic zone category, this basin is highly active. It lies in seismic zone 'V', a highly sensitive zone that has previously been the site of numerous catastrophic earthquakes. In contrast to the rest of India, there is barely a change in air temperature across the region, which has a humid climate with a 90 % humidity ratio. The area around the basin watershed receives the most rain globally. The mean annual rainfall in the basin watershed was approximately 2083.50 mm, and the temperature ranges from 9.25 °C to 30.66 °C.[1,20] India, Bangladesh, and Burma all receive water from the Barak sub-basin. The drainage area comprising India is 41,723 sq. km or around 1.38 % of the country's land area. It is bounded to the north by Bangladesh, to the east by the Na Lushai hills, and to the south and west by the Barail range, which separates it from the Brahmaputra sub-basin. The sub-basin is located in the Meghalayan States. Nagaland, Manipur, Mizoram, Assam, and Tripura.[1,3]

The valley and hill local rainfall runoff in the locality flows through the Barak River and its numerous tributaries before it flows into Bangladesh. The drainage basin that covers India is 41,723 sq. km or approximately 1.38 % of India's geographical area.[1] The Barak River and its tributaries form the second-largest river system of India's northeast. The river system in the basin is about 5381.19 km long in total. It spreads across Meghalaya, Manipur, Mizoram, Assam, Tripura, and Nagaland. It spans an area of 45,622.12 sq. km (area calculated in ArcGIS), or probably 1.38 % of the country's total geographic area.[18] The river's regime in India is divided into two distinct regions: the highland catchment and the alluvial plain, generally referred to as the "Barak Valley."

Considering the fact that Barak River causes inundation every year for its lower catchment areas, this study is conducted to evaluate flood inundation extent by exploiting SAR-based remote sensing technology, and further, flood maps were validated with remote-sensing-based imageries and ground-based datasets.

3. Data used and methodology

3.1 Data used

Sentinel-1C-band SAR satellite data, a dual-satellite constellation mission consisting of Sentinel-1A and Sentinel-1B, was launched on 4th March

2014 and 25th April 2016, respectively. The receptivity period is 11 days, but considering ascending and descending passes, it is 5–6 days. The mission employs single (HH and VV) and dual polarization (VV + VH and HH + HV) bands with incidence angles of 29° to 46°, allowing for continuous monitoring of dynamic processes. Sentinel-1A C-band GRD product is employed in this research to map the progression and regression of the flood during the event (Table 1).

The various spatial resolutions of 25 × 40 m, 5 × 5 m, and 5 × 20 m allow for detailed analysis of numerous land and marine activities. Its all-weather, day-and-night monitoring capability provides an advantage over conventional optical satellites. The spectral resolution, ranging from 1 to 180 nm, facilitates the analysis of various events across the electromagnetic spectrum.[1,23,31]

Landsat-9 optical satellite data is used here to validate the extent of flood inundation. This mission, launched on 27th September 2021, significantly improves the ability to observe the Earth with 30-meter spatial and 8-day temporal resolution, enabling more frequent and consistent surveillance. Further upgrading spectral bands improves its ability to recognize and characterize numerous natural and man-made objects, allowing it to detect a broader range of intensity levels.[21]

Climate Hazards Group Infrared Precipitation with Station Data (CHIRPS) produces the quasi-global rainfall dataset known as CHIRPS. The satellite-based precipitation product spans roughly three decades. At integrated station data, this gridded rainfall time series information exists from 1981 to nearly the current year at a spatial resolution of 0.05°. Various time scales (daily, pentad,10-day window, and monthly) are adopted in calculations of CHIRPS rainfall.[17] It is employed herein to reflect spatio-temporal rainfall distribution over study area to arrest flood forward and backward movement.

State Disaster Management Authority produces daily/weekly/bimonthly/monthly (as per the flood progression and regression) flood impact reports considered in this study to validate satellite-based flood maps. The flood impact report illustrates flood inundation extent with affected villages/towns/cities, flood depth, discharge rate, etc., making it a vital source to validate remote sensing-based outputs.

3.2 Methodology

The procedure involved in deriving flood inundation pixels is illustrated in Fig. 2. Sentinel-1 GRD products were utilized to generate flood

Table 1 Data acquisition used in this study.

Acquisition date	Temporal resolution	Spatial resolution	Acquisition dates	Source
Sentinel-1A (C-band SAR)	12 days	20*22m (range* azimuth)	7th, 16th, 28th June and 12th, 22nd July 2022	ASF
Landsat 9	8 days	30 m	28th June 2022	USGS
CHIRPS	Daily	0.05°	Pentadal Data (June and July) 2022	USGS
Flood impact reports	07/15 days	—	June-July 2022	ASDMA

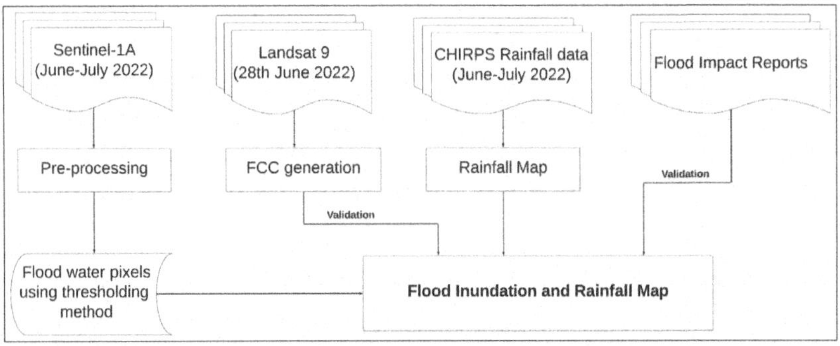

Fig. 2 Shows the methodology flowchart for flood inundation mapping using S-1 satellite data over Cachar, Assam, during June-July 2022.

inundation based on Otsu's algorithm after following the necessary pre-processing steps. The pre-processing steps included applying the orbit file (if unavailable with the data file, not necessary with all the data), speckle filtering, calibration, and terrain correction. Some steps, such as applying the orbit file and terrain correction, required downloading help files from the server, while the rest were algorithm-based corrections performed through standalone application software.

Image binarization was carried out on the backscatter image generated after pre-processing the satellite data using Otsu's thresholding method to extract flood inundation pixels.[6] The Sentinel Application Platform (SNAP), an open-access toolkit from the European Space Agency, was used for pre-processing and visualization procedures.

The following steps were performed during the pre-processing and extraction of flood pixels using Sentinel-1 SAR data:

a. The SNAP Desktop application imported Level-1 GRD images during the Sentinel-1 pre-processing. In processing multiple images, we automated the process in Google Earth Engine for producing flood inundation pixels.
b. To get technically significant radar backscatter pixel values, Sentinel-1 satellite images need to be calibrated and radiometrically adjusted.
c. Speckle filtering was used to minimize the grainy noise typical of SAR data.
d. Multi-look processing was applied to reduce speckles and further enhance image interpretation.
e. Image Doppler terrain correction has been implied to minimize geometric distortions in the SAR images and assign a coordinate system.
f. Image binarization was conducted using Otsu's thresholding algorithm. The threshold value was determined based on available resources such as flood impact reports, optical satellite data, and a permanent water body layer to differentiate between permanent and seasonal (flood patch) water bodies.

The resulting raster files were converted to TIFF/Geo TIFF format (flood: 1, non-flood: 0) for flood mapping and estimating areal flood extent. Rainfall data in raster format were integrated into the analysis to improve the understanding of flood characteristics during the event. Landsat 9 optical satellite data from 28th June 2022 was also used to validate the flood inundation extent derived from the SAR satellite data.[3,6,23]

4. Results

4.1 Spatio-temporal rainfall dynamics over Cachar, Assam

CHIRPS-based rainfall data of pentad temporal resolution have been used to illustrate spatio-temporal rainfall during the flooding period (Fig. 3). The rainfall patterns showed substantial variability across June 2022 (Fig. 3 & Fig. A1). Fig. 3A shows extensive rainfall (200 mm to >300 mm), whereas

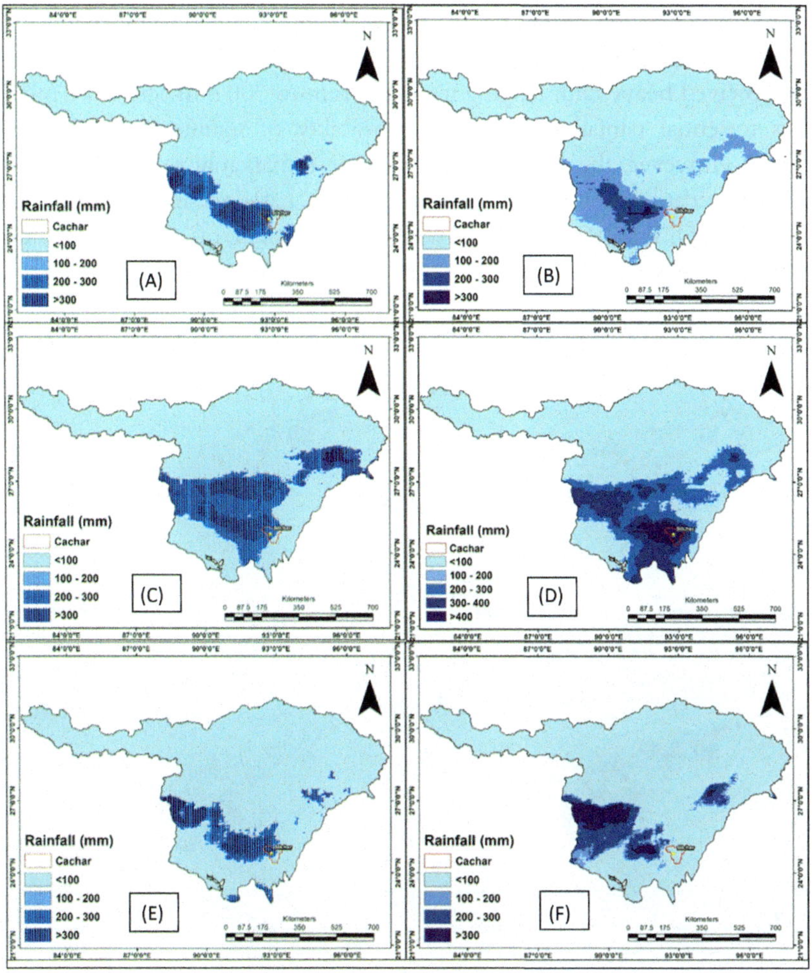

Fig. 3 CHIRPS based rainfall data showing rainfall pattern over upper and lower catchment areas including Silchar, Cachar (A) 1 June 2022–5 June 2022 (B) 6 June 2022–10 June 2022 (C) 11 June 2022–15 June 2022 (D) 16 June 2022–20 June 2022 (E) 21 July 2022–25 July 2022 (F) 26 July 2022–31 July 2022. Map lines delineate study areas and do not necessarily depict accepted national boundaries.

Fig. 3B shows contrasting weather (100 mm). Fig. 3D shows significant rainfall (>400 mm), which has caused severe flooding along the Barak River, while Fig. 3C shows moderate rainfall (200–300 mm). The significant flooding is likely to be caused by this intense rainfall event. Fig. A1B shows lighter rainfall (100 mm), while Fig. A1A denotes significant precipitation (200 mm to >300 mm).

According to the pentad CHIRPS rainfall data, substantial rainfall occurred in July 2022 too. Figs. A1C, D, E and F and 3E and F images showed areas that received heavy rain, ranging from 200 mm to 300 mm across the month. This torrential rainfall indicated the possibility of significant floods in the region. However, the Fig. 4D and E image remarkably reflected rapid and heavy inundation near the Barak River, which has seen significant floods.

Fig. A1D, E and F and 3F showing lower rainfall, often below 100 mm, reflected the preceding period of flood in the region also can be seen

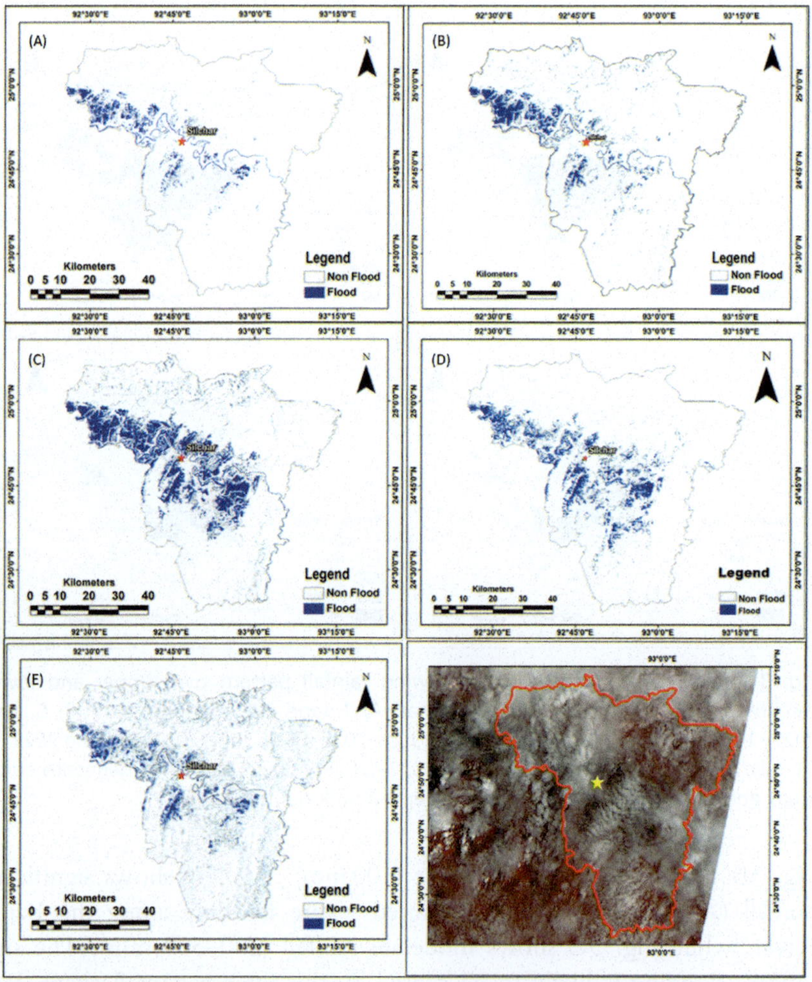

Fig. 4 Flood Inundation extent derived using Sentinel-1 SAR data during June and July 2022 (A) 7 June 2022, (B) 16 June 2022, (C) 28 June 2022, (D) 12 July 2022, (E) 22 July 2022, (F) 28 June 2022. Map lines delineate study areas and do not necessarily depict accepted national boundaries.

through flood maps (Fig. 4). The comparison of the satellite images during periods of heavy and light rain brings out the variation in precipitation over the pentad period. Fig. A1D and 3E change from moderate to heavy rainfall, illustrating the connection between heavy rain and floods.

Rainfall was concentrated over the central region, consistently resulting in flood progression. The heavy downpours that started between 1st-5th June 2022 triggered flooding in the central part, carried out with more than 300 mm of torrential rainfall between 11th–20th June, making it an even worse condition in downstream areas of the catchment. During this period, the central part and the eastern, western, and southern regions also received rainfall between 200–300 mm, creating havoc.

4.2 Flood Inundation Map over Cachar, Assam

S-1 satellite data derived flood-inundation shows severe impact of flooding in downstream areas of the basin (Fig. 4). Flooding started from the first week of June 2022 onwards after receiving heavy downpours more than 200 mm (Fig. 3). The inundation map in Fig. 4A shows that flood inundated region was estimated as 138.64 sq.km on June 7 whereas by June 16 the extent has been expanded to 149.68 sq.km, indicating a gradual increase in inundation. However, the flood inundation area immensely expanded to 517.85 sq.km on June 28 as shown in Fig. 3C. It was also the peak flooding period of the event. The noticeable growth in the flooded area indicates the severity and duration of the flood event throughout the three weeks. However, as of July 12, the area underwater has decreased to 256.61 sq.km, indicating a significant decrease in floods. Heavy rainfall, terrain, river releases, and local water management systems are only a few of the variables that may be responsible for the observed changes in the flood-affected area. The intensity of the flooding catastrophe during that period is indicated by the quick increase in the waterlogged region between June 16 and June 28, 2022 (Fig. 5). In Fig. 4F, flood inundation extent has been validated using optical satellite data for better visual interpretation.

On the other hand, a smaller area is inundated by Fig. 3D because of efficient government relief operations and receding floodwaters. The flooded area showed a slight rebound in Fig. 3E, reaching 322.07 sq. km, but it was still below the high recorded on June 28. During the event, the central region was impacted severely, followed by the western and northern regions.

For river systems like the Barak River, the study underscores the significance of ongoing monitoring and flow forecasting models for effectively predicting and managing possible flooding occurrences. During the research period, we observed both an extension and contraction of the flooded region, underscoring

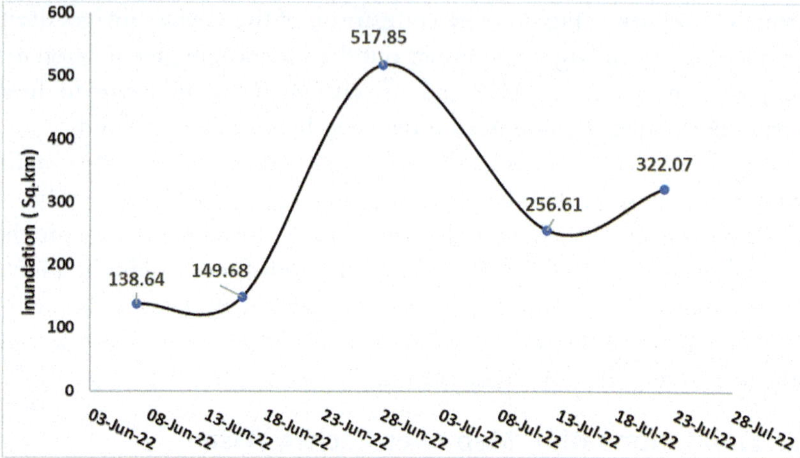

Fig. 5 Graph Showing Inundated area during flooding event between June and July 2022.

the dynamic character of flood occurrences and the demand for flexible disaster management plans. On average, a 524.87 sq.km area was inundated during the event. With 14.30 lakh affected residents, Cachar was the worst-hit district, followed by Barpeta (5.49 lakh) and Nagaon (5.19 lakh), where a total of 10,630 ha of cropland was negatively impacted by the flood, with Silchar being the town most severely hit, with 1922 ha of cropland submerged.

5. Discussion

This study shows the importance of SAR satellite data for flood inundation mapping over optical because of its all-weather capability, making it a unique assessment and recovery tool. It also can separate flooded areas from high-moisture zones. The study emphasizes the need for improved infrastructure development and flood prevention measures to lessen the detriment of such catastrophic weather occurrences in this region.

The study showed that flooding triggered by torrential rainfall started in the first week of June 2022, and the trend followed till mid-July, resulting in flood inundation across the region. By the first week of June 2022, the northern part of the Silchar received >300 mm rainfall, which continued till 25th June 2022, extensively triggered flooding, and the maximum or peak inundation was recorded as ~518 sq.km on 28th June 2022. Flood water stayed till 15th July after then after it started receding following the rainfall. The optical satellite data was used to validate flood extent and found that

Otsu's thresholding method is reliable and robust for flood water extraction. The methodology adopted here is simple and robust, providing an easy-to-use framework for both community and non-community users. It is beneficial in rapid flood monitoring and assessment. The current framework may assist in data-scarce regions or areas where real-time data is unavailable. The future scope of this research is to incorporate UAV data with a better spatial and temporal resolution for better understanding and analysis purposes.

5.1 Integration of satellite data for real-time flood monitoring

This study perfectly fulfills the objective of doing "Climate-Smart Water Resource Management: A Paradigm Shift from Ancient to Modern Practices with Integrated Technologies" and highlighting how water management strategies have evolved with modern tools such as flood inundation mapping employing SAR (Synthetic Aperture Radar) data strengthening our response towards the disaster. SAR technology gives essential information for mapping inundation areas and enables real-time monitoring of flood events regardless of the weather or time of day. In contrast, combining satellite observations with hydrological models provides improved real-time flood monitoring in terms of comprehensive spatial and temporal information. Satellite imagery provides near real-time information on precipitation, soil moisture, and water levels over broad areas. By combining this with hydrological models such as LISFLOOD-FP, precise simulation of flood behavior is made possible, with enhanced predictions and response measures. Satellite remote sensing data from missions like Sentinel-1, MODIS, and SMOS can validate and calibrate models for more accurate assessments of flood extent and impact. This integration enhances early warning systems, allowing authorities to make informed decisions, allocate resources efficiently, and mitigate disaster risks in vulnerable regions.

This development uses state-of-the-art modeling and forecasting techniques to support climate-smart strategies, improving flood prediction, response, and resilience. Integrated approaches to water management offer adaptable solutions to meet the growing challenges of climate change by combining cutting-edge technology with conventional wisdom.

6. Conclusion

The study aimed to evaluate flood extent by considering Sentinel-1 SAR data and validated it with optical data and flood impact reports. SAR data have been chosen over optical data because of its longer wavelength and accurate and timely information for disaster management and response. To create flow forecasting models, rainfall data from the CHIRPS satellites were used. Continuous rain caused significant flooding in the Barak River, a tributary of the Brahmaputra, which resulted in extensive flooding in the Cachar areas. The study uses SAR data because of its inherent features rather than optical data for mapping flood inundation. Moreover, the capacity of SAR to collect data during cloud cover and rainfall events was crucial in providing fast and accurate information for disaster response.

This study has demonstrated the effectiveness of Sentinel-1 data for the Cachar district of Assam's potential flood events in India during the June-July 2022 flooding period. The analysis of SAR data has shown that peak flood was recorded on 28th June 2022, where the area projected to be inundated was 517.85 sq.km, indicating that around 73.61 % of the entire study area was submerged. The flood report from Assam on 28th June 2022 provides information on the substantial effects of flooding in the Cachar district, especially in the Silchar area. The data presents a disturbing image of how the flood's aftermath caused severe problems for the public and the agricultural industry. The inundation's human toll was apparent, with a total population of 862,166 impacted in Cachar and 437,145 in Silchar. This study also highlighted that flood water stayed around a month over the affected region, causing severe crop damage. In Cachar, 10,630 ha of cropland were submerged by flooding. However, in Silchar, 1922 ha of cropland were submerged. These numbers highlighted the disastrous effects on local agriculture and food security. The tragic loss of three lives in Silchar highlights the necessity of rapid and efficient catastrophe planning and response procedures. It highlights the importance of utilizing SAR data for flood mapping in areas prone to frequent floods and provides valuable disaster management and response information. The findings of this research paper can contribute to the development of effective flood management strategies in the region, which can ultimately help minimize the losses and damage caused by floods.

CRediT authorship contribution statement

Arjita Maurya: Writing – review & editing, Writing – original draft, Validation, Software, Resources, Methodology, Investigation, Formal analysis, Data curation.

Gaurav Tripathi: Writing – original draft, Writing – review & editing, Supervision, Conceptualization, Resources, Methodology, Investigation, Data curation.

Achala Shakya: Writing – review & editing, Resources, Methodology, Investigation, Data curation.

Ritambhara K Upadhyay: Writing – review & editing, Methodology, Investigation, Data curation.

Yateesh Ketholia: Writing – review & editing.
Vipin Chander Pant: Writing – review & editing

Conflict of interest

Authors declare no conflict of interest.

Appendix

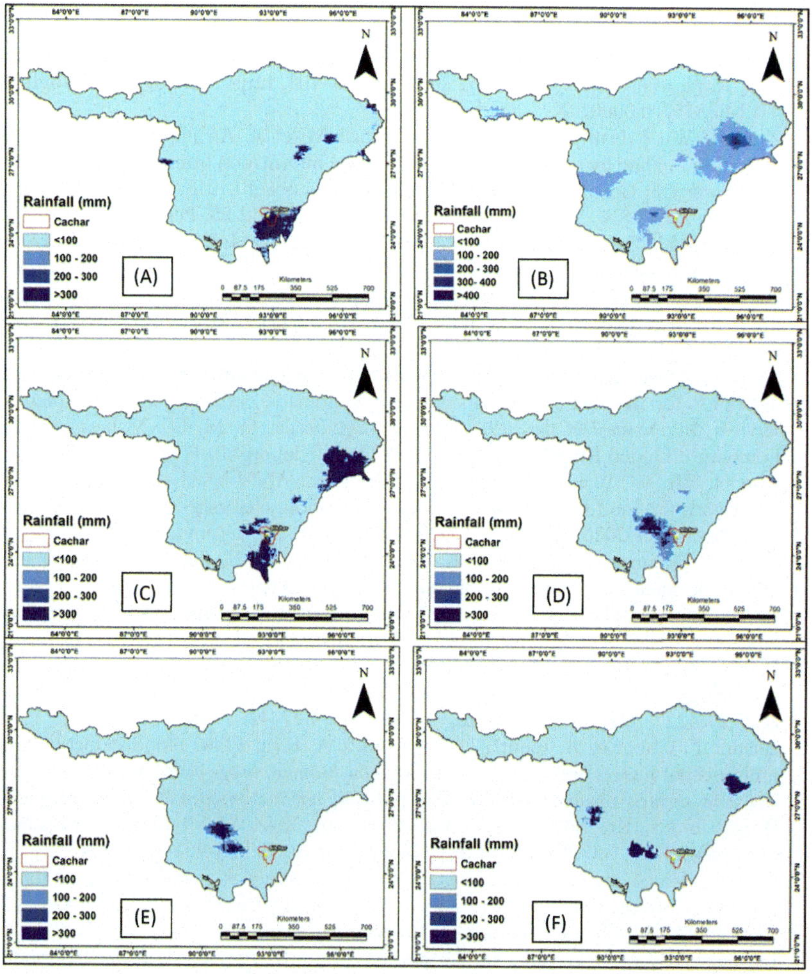

Fig. A1 CHIRPS based rainfall data showing rainfall pattern over upper and lower catchment areas including Silchar, Cachar (A) 21 June 2022–25 June 2022 (B) 26 June 2022–30 June 2022 (C) 1 July 2022–5 July 2022 (D) 6 July 2022–10 July 2022 (E) 11 July 2022–15 July 2022 (F) 16 July 2022–20 July 2022. Map lines delineate study areas and do not necessarily depict accepted national boundaries.

References

1. Borah SB, Sivasankar T, Ramya MNS, Raju PLN. Flood inundation mapping and monitoring in Kaziranga National Park, Assam using Sentinel-1 SAR data. *Env Monit Assess* 2018;**190**:520. https://doi.org/10.1007/s10661-018-6893-y.
2. Donatti CI, Nicholas K, Fedele G, Delforge D, Speybroeck N, Moraga P, et al. Global hotspots of climate-related disasters. *Int J Disaster Risk Reduct* 2024;**108**:104488. https://doi.org/10.1016/j.ijdrr.2024.104488.
3. Halder K, Ghosh A, Srivastava AK, Pal SC, Chatterjee U, Bisai D, et al. SAR-driven flood inventory and multi-factor ensemble susceptibility modelling using machine learning frameworks. *Geomat Nat Hazards Risk* 2024;**15**:2409202. https://doi.org/10.1080/19475705.2024.2409202.
4. Parida BR, Pandey AC, Kumar S, Tripathi G. Comparative flood area analysis based on change detection and binarization methods using Sentinel-1 synthetic aperture radar data. *Radar remote sensing;* Elsevier, 2022; pp. 93–108. https://doi.org/10.1016/B978-0-12-823457-0.00001-X
5. Zhao B, Wu J, Han X, Tian F, Liu M, Chen M, et al. An improved surface water extraction method by integrating multi-type priori information from remote sensing. *Int J Appl Earth Obs Geoinf* 2023;**124**:103529. https://doi.org/10.1016/j.jag.2023.103529.
6. Goh TY, Basah SN, Yazid H, Aziz Safar MJ, Ahmad Saad FS. Performance analysis of image thresholding: Otsu technique. *Measurement* 2018;**114**:298–307. https://doi.org/10.1016/j.measurement.2017.09.052.
7. Chen S, Huang W, Chen Y, Feng M. An adaptive thresholding approach toward rapid flood coverage extraction from Sentinel-1 SAR imagery. *Remote Sens* 2021;**13**:4899. https://doi.org/10.3390/rs13234899.
8. Chow C, Twele A, Martinis S. An assessment of the height above nearest drainage terrain descriptor for the thematic enhancement of automatic SAR-based flood monitoring services. In *Presented at the SPIE remote sensing;* Neale, C. M. U., Maltese, A., Eds.; Edinburgh: United Kingdom, 2016999808. https://doi.org/10.1117/12.2240766
9. Chu H, Wu W, Wang QJ, Nathan R, Wei J. An ANN-based emulation modelling framework for flood inundation modelling: application, challenges and future directions. *Env Model Softw* 2020;**124**:104587. https://doi.org/10.1016/j.envsoft.2019.104587.
10. Bellos V, Costanzo C, Costabile P, Kalogiros J. Calibrating 2D flood models in the era of high performance computing. In *Advances in hydroinformatics—SimHydro 2023 Volume 1, Springer water;* Gourbesville, P., Caignaert, G., Eds.; Springer Nature Singapore: Singapore, 2024; pp. 355–365. https://doi.org/10.1007/978-981-97-4072-7_23
11. Bentivoglio R, Isufi E, Jonkman SN, Taormina R. Rapid spatio-temporal flood modelling via hydraulics-based graph neural networks. *Hydrol Earth Syst Sci* 2023;**27**:4227–4246. https://doi.org/10.5194/hess-27-4227-2023.
12. Bhadra T, Chouhan A, Chutia D, Bhowmick A, Raju PLN. Flood detection using multispectral images and SAR data. In *Machine learning, image processing, network security and data sciences, communications in computer and information science;* Bhattacharjee, A., Borgohain, S. Kr, Soni, B., Verma, G., Gao, X.-Z., Eds.; Springer Singapore: Singapore, 2020; pp. 294–303. https://doi.org/10.1007/978-981-15-6315-7_24
13. Caviedes-Voullième D, Shepherd TG. Climate storylines as a way of bridging the gap between information and decision-making in hydrological risk. *PLOS Clim* 2023;**2**:e0000270. https://doi.org/10.1371/journal.pclm.0000270.
14. Enriquez AR, Wahl T, Talke SA, Orton PM, Booth JF, Agulles M, et al. MatFlood: an efficient algorithm for mapping flood extent and depth. *Env Model Softw* 2023;**169**:105829. https://doi.org/10.1016/j.envsoft.2023.105829.
15. Fraehr N, Wang QJ, Wu W, Nathan R. Supercharging hydrodynamic inundation models for instant flood insight. *Nat Water* 2023;**1**:835–843. https://doi.org/10.1038/s44221-023-00132-2.

16. Krishna Vanama VS, Rao YS. Change detection based flood mapping of 2015 flood event of Chennai City using Sentinel-1 SAR Images, in: IGARSS 2019 - 2019 IEEE International Geoscience and Remote Sensing Symposium. Presented at the IGARSS 2019 - 2019 IEEE International Geoscience and Remote Sensing Symposium, IEEE, Yokohama, Japan; 2019, pp. 9729–9732. https://doi.org/10.1109/IGARSS.2019.8899282.
17. Sulugodu B, Deka PC. Evaluating the performance of CHIRPS satellite rainfall data for streamflow forecasting. *Water Resour Manag* 2019;**33**:3913–3927. https://doi.org/10.1007/s11269-019-02340-6.
18. Samantaray D, Chatterjee C, Singh R, Gupta PK, Panigrahy S. Flood risk modeling for optimal rice planning for delta region of Mahanadi river basin in India. *Nat Hazards* 2015;**76**:347–372. https://doi.org/10.1007/s11069-014-1493-9.
19. Tripathi G, Parida BR, Pandey AC. Spatio-temporal rainfall variability and flood prognosis analysis using satellite data over North Bihar during the August 2017 flood event. *Hydrology* 2019;**6**:38. https://doi.org/10.3390/hydrology6020038.
20. Mangukiya NK, Kushwaha S, Sharma A. A novel multi-model ensemble framework for fluvial flood inundation mapping. *Env Model Softw* 2024;**180**:106163. https://doi.org/10.1016/j.envsoft.2024.106163.
21. Niroumand-Jadidi M, Legleiter CJ, Bovolo F. Bathymetry retrieval from CubeSat image sequences with short time lags. *Int J Appl Earth Obs Geoinf* 2022;**112**:102958. https://doi.org/10.1016/j.jag.2022.102958.
22. Teng J, Jakeman AJ, Vaze J, Croke BFW, Dutta D, Kim S. Flood inundation modelling: a review of methods, recent advances and uncertainty analysis. *Env Model Softw* 2017;**90**:201–216. https://doi.org/10.1016/j.envsoft.2017.01.006.
23. Tripathi G, Pandey AC, Parida BR, Kumar A. Flood inundation mapping and impact assessment using multi-temporal optical and SAR satellite data: a case study of 2017 flood in Darbhanga District, Bihar, India. *Water Resour Manag* 2020;**34**:1871–1892. https://doi.org/10.1007/s11269-020-02534-3.
24. Gangani P, Mangukiya NK, Mehta DJ, Muttil N, Rathnayake U. Evaluating the efficacy of different DEMs for application in flood frequency and risk mapping of the Indian Coastal River basin. *Climate* 2023;**11**:114. https://doi.org/10.3390/cli11050114.
25. Gomes MN, Giacomoni MH, Navarro FAR, Mendiondo EM. Global optimization-based calibration algorithm for a 2D distributed hydrologic-hydrodynamic and water quality model. *Env Model Softw* 2024;**179**:106128. https://doi.org/10.1016/j.envsoft.2024.106128.
26. Sharifian MK, Kesserwani G, Chowdhury AA, Neal J, Bates P. LISFLOOD-FP 8.1: new GPU-accelerated solvers for faster fluvial/pluvial flood simulations. *Geosci Model Dev* 2023;**16**:2391–2413. https://doi.org/10.5194/gmd-16-2391-2023.
27. Sanders BF, Schubert JE. PRIMo: Parallel raster inundation model. *Adv Water Resour* 2019;**126**:79–95. https://doi.org/10.1016/j.advwatres.2019.02.007.
28. Shen C, Yan L. Recent development of hydrodynamic modeling in heavy-ion collisions. *Nucl Sci Tech* 2020;**31**:122. https://doi.org/10.1007/s41365-020-00829-z.
29. Horritt MS, Bates PD. Evaluation of 1D and 2D numerical models for predicting river flood inundation. *J Hydrol* 2002;**268**:87–99. https://doi.org/10.1016/S0022-1694(02)00121-X.
30. Mehta A, Rawat K. Monitoring flash flood using optical and microwave sensor dataset:- A case study of Cachar district, Assam (India). In: 2023 International Conference on Device Intelligence, Computing and Communication Technologies, (DICCT). Presented at the 2023 International Conference on Device Intelligence, Computing and Communication Technologies, (DICCT), IEEE, Dehradun, India; 2023. pp. 90–94. https://doi.org/10.1109/DICCT56244.2023.10110185.
31. Twele A, Cao W, Plank S, Martinis S. Sentinel-1-based flood mapping: a fully automated processing chain. *Int J Remote Sens* 2016;**37**(13):2990–3004. https://doi.org/10.1080/01431161.2016.1192304.

CHAPTER ELEVEN

Prioritizing sub-watersheds for soil conservation in the Dikhu watershed, North-East India using morphometric and LULC analysis

Liza G. Kiba[a], Grace Nengzouzam[b,*], Y. Nyamnyei K[b], Sheihnon R. Yeanganghu[b], and Jubikali[b]

[a]Department of Agricultural Engineering and Technology, School of Agricultural Sciences, Nagaland University, Medziphema, Nagaland, India

[b]Department of Agricultural Engineering and Technology, School of Engineering and Technology, Nagaland University, Kohima, Nagaland, India

*Corresponding author. e-mail address: grace.zamie@gmail.com

Contents

Abstract

Prioritization of watersheds has become increasingly important in natural resources management, especially in the context of watershed management. The present work prioritizes the sub-watershed (SW) of the Dikhu river basin using Morphometric analysis and land use/land cover (LULC) analysis. The execution of these methods can be facilitated through remote sensing and geographic information system (GIS) approaches. Using ArcSWAT, 14 sub-watersheds have been established in the Dikhu river basin (SW1 to SW14), and these SWs are prioritized for soil and water

Advances in Chemical Pollution, Environmental Management and Protection, Volume 15
ISSN 2468-9289, https://doi.org/10.1016/bs.apmp.2025.11.002

conservation based on morphometric analysis and LULC information collected from the state department. After assigning the ranks using the two methods, the compound value was calculated by taking the aggregate value of each ranking, based on which the final prioritization was done. The SWs were further categorized into low, medium, and high priority based on morphometric and LULC analysis. Among the SWs identified under high priority, immediate priority was assigned to SW12 as it was categorized as high priority under both methods of analysis. The majority of the SWs were categorized under low and medium-priority classes, indicating that a higher percentage of the area is stable and under low erosion risk. This could be due to the large forest land area coverage (> 80 %), which protects the raindrops from causing the detachment of soil particles. Though the work suggested that most of the SWs were stable, it is advisable to implement some soil and water conservation measures in SWs categorized as high-priority. The findings from this study will be beneficial to the research community and stakeholders in decision-making and undertaking necessary steps for conservation measures of soil and water.

Keywords: Prioritization, Watersheds, Morphometric analysis, LULC, RS, GIS.

1. Introduction

A watershed is a fundamental component of all hydrological studies and the design of structures on the Earth's surface. The response of various hydrological processes is dependent on different geomorphologic, hydrological, and physiographic parameters of a watershed. Out of the three, the parameter that closely represents the hydrologic behavior is the geomorphological parameter. Under morphometric analysis, various geomorphometric variables are quantified, and these are found to be crucial in studies related to flood frequency analysis, prioritization of watersheds, and hydrological modeling. In many regions, especially regions with undulating topography like the North-Eastern part of India, basins are mostly ungauged and therefore, geomorphologic features studies of these basins are needed. In recent times, there has been a marked increase in the level of interest towards watershed management, for which a deeper insight into the watershed's geomorphometry is needed.

Management of natural resources is important as plans are made to expand water and land areas and to prevent the deterioration of soil to keep the environment and ecosystems in balance.[1] The application of remote sensing (RS) and geographic information systems (GIS) has been successful in tackling specific challenges related to managing natural resources.[2–4] To develop effective plans for managing land and water resources, it is important to first understand the geological features and the river systems in the area. Drainage basins are key areas for managing water and land resources, offering a framework for efficient management of these resources.[5] For instance, watersheds are seen as essential areas for erosion, where water and land interact. Managing watersheds is vital for safeguarding water and soil resources.[6] To comprehend the drainage patterns of watersheds, information on land use and land cover (LULC), hydrology, and geomorphology is found to be very useful.[7]

Morphometry involves using mathematical formulas to describe the shape and elevation changes.[8,9] The shape and size of a watershed's features are crucial for understanding the water flow within it. These features show how water behaves in the basin and help in evaluating how the basin responds. Areas with more features per unit area tend to experience quicker water runoff and more erosion of river channels.[10,11] For preventing soil erosion and prioritizing the watershed, the findings from morphometric studies are valuable. Studies on the prioritization of watersheds have been conducted since the mid-20th century, initially using a method based on traditional topographic maps.[12–14] Mapping contours on a topographic map or conducting an extensive field examination, creating digital representations of river networks, dividing rivers into their respective orders, and manually determining the length of the rivers are the tasks involved in the traditional method of ranking watersheds. Consequently, this conventional approach is time-consuming, costly, and requires a significant amount of manual labor.

Remote sensing can help in obtaining a comprehensive view of a large area and is found to be useful in analyzing morphometric parameters of a watershed. Various researchers carried out conventional morphometric analysis by employing quantitative measurements of drainage network, basin geometry, and relief parameters to understand watershed characteristics and processes, including techniques like bifurcation ratio, drainage density, and stream frequency, which is tedious and time-consuming. However, with the development of geospatial technologies, it has become easier to make more accurate and detailed analyses. Many studies have focused on the shape and size of river basin features using GIS and RS.[15–17] Many researchers have carried out geomorphological analysis using RS and GIS techniques, as these techniques are efficient and expeditious.[18–20] Many studies have been carried out for prioritizing sub-watersheds (SWs) to evaluate erosion-prone regions based on morphometry, land use/land cover (LULC), soil type, and other climatic factors.[21–24] To fulfil the growing demand for natural resources,[25,26] a major part of the Earth's surface has been transformed, and therefore, the LULC of a watershed is an essential consideration when prioritizing SWs.[27–29]

A well-designed strategy for managing watersheds can help mitigate natural disasters, as well as address issues within the watershed.[30] To achieve this, decision-makers need to have a good understanding of the watershed's parameters and the hydrological processes that occur within it. For identifying the high priority area in a watershed, information obtained from geomorphometric analysis can be very helpful.[31] Keeping this in mind, in this study, an effort has been put forth to identify SWs with common priorities using both geomorphometric analysis and LULC of Dikhu watershed. By analyzing the morphometric characteristics and LULC classes, the study prioritized the SWs of the Dikhu watershed, and the final ranking of 14 SWs was done in light of the analysis results.

2. Materials and methods

2.1 Description of the study area

Nagaland, a state in the North-East India, bounded within 25°6′N to 27°4′N latitude and 93°20′E to 95°15′E longitude, has a geographical area of 16,579 km^2, (Fig. 1). The state enjoys a sub-tropical to warm temperate monsoon type and receives approximately 2600 mm of annual mean

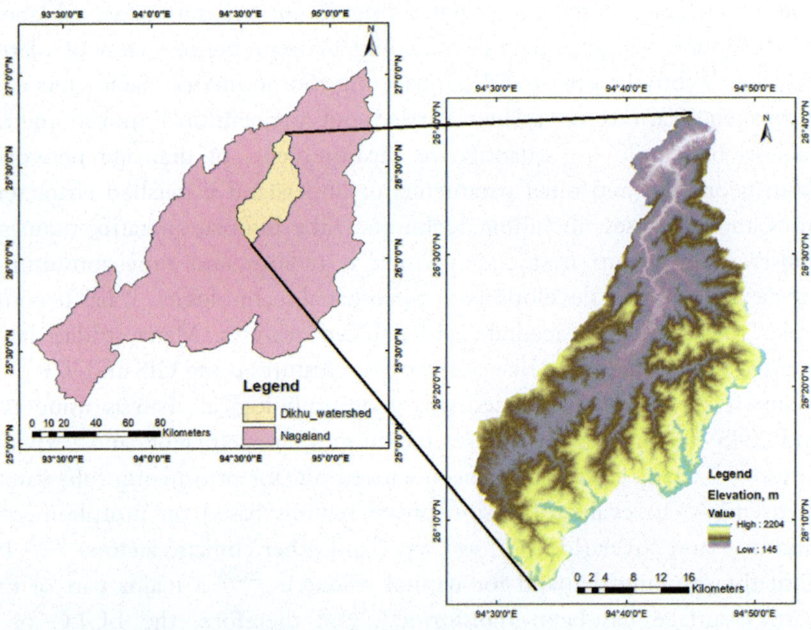

Fig. 1 Location and elevation of the study area. Map lines delineate study areas and do not necessarily depict accepted national boundaries.

rainfall. The Dikhu River, which is a tributary of the mighty Brahmaputra, flows close to the field located in the Mokokchung district and is one of the most prominent rivers of Nagaland. It passes through five districts, viz., Longleng, Mokokchung, Mon, Tuensang, and Zunheboto. The river originates from Nuroto Hill area under Zunheboto district, flows further north-ward and leaves the hill near Naginimora, and finally joins with the Brahmaputra River in Assam. The total length is about 160 km, and the total catchment area of 1080.62 km^2. The elevation of the study area extends from as low as 145 m at the outlet to as high as 2204 m.

2.2 Pre-processing of data

The present study created the Dikhu watershed base map using the CARTOSAT Digital Elevation Model (DEM). CARTOSAT produces the DEM with 30 m × 30 m spatial resolution, and comparatively, it has an adequate resolution (downloaded from https://earthexplorer.usgs.gov). The Dikhu watershed was delineated using ArcSWAT. For the present study, the watershed was sub-divided into 14 sub-watersheds (SWs) viz., SW1–SW14 (Fig. 2). The land use/land cover (LULC) for the study area, i.e., Dikhu watershed, Nagaland, was also acquired from the RS and GIS department, Nagaland. The LULC of the watershed was classified under 5 main classes, viz., Agricultural land, Forest land, Built-up, Wasteland, and Water bodies.

2.3 Methodology

2.3.1 Morphometric analysis

This analysis is done considering the Linear, Areal, and Relief aspects. After reviewing the literature and considering the data, a subset of 18 parameters was selected. Under the linear aspect: bifurcation ratio (R_b), drainage density (D_d), stream frequency (F_s), texture ratio (R_t), and overland flow (L_{om}); underareal aspects: form factor (F_f), elongation ratio (R_e), circulatory ratio (R_c), and compactness coefficient (C_c); under relief aspects: relief ratio (R_r), relative relief (R_{hp}), and ruggedness number (R_n), were considered. The formula for computing these parameters is presented in Table 1. To determine the parameters considered under these three aspects mentioned above, a few basic morphometric parameters, such as the total number of streams, the length of streams, the drainage area, the basin length, and the basin perimeter, were computed using ArcGIS software. The stream order generated from ArcGIS for all the sub-watersheds (SWs) is illustrated in Fig. 3.

Table 2 illustrates the calculated parameters under the three aspects. After estimating all the parameters, the ranking of each parameter for all

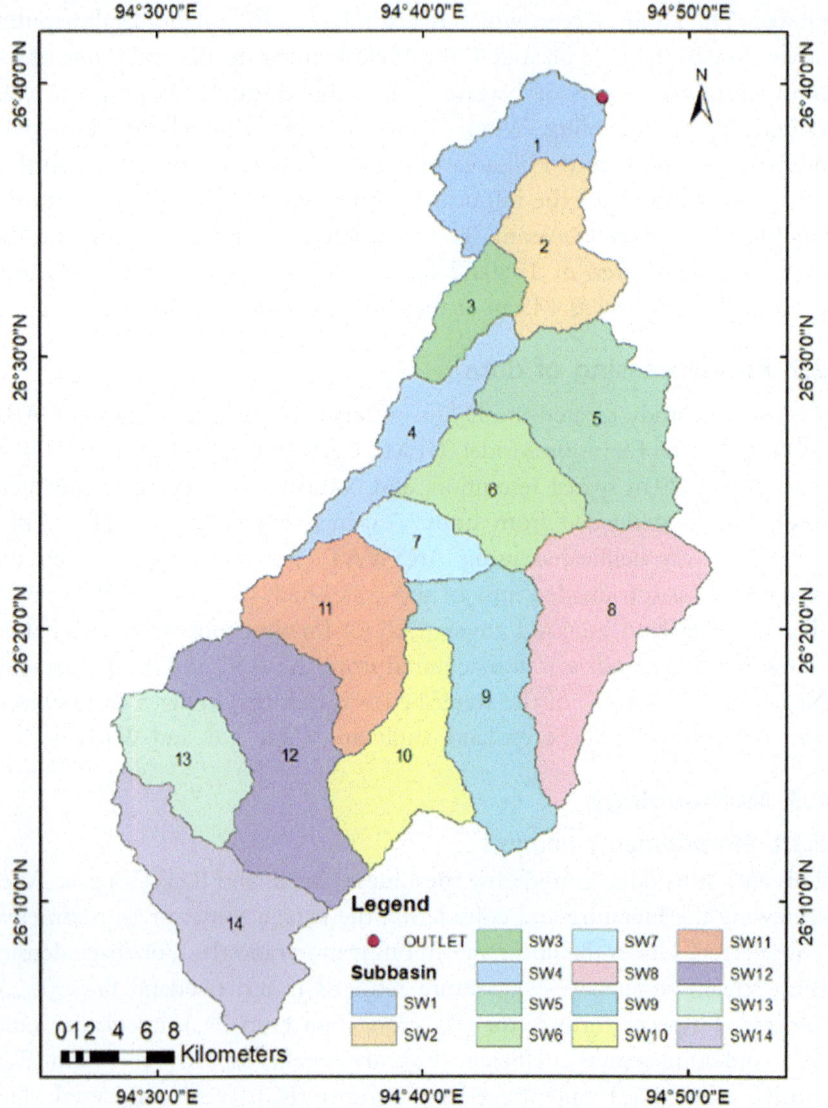

Fig. 2 Sub-watersheds of Dikhu generated from ArcSWAT. Map lines delineate study areas and do not necessarily depict accepted national boundaries.

SWs is determined. Since the parameters that fall under linear and relief aspects are directly related to soil erosion, the SW with the highest value of these parameters is given the highest priority (1st rank), and so on. Parameters under the shape aspect are indirectly related to soil erosion, and therefore, the SW with the lowest value is assigned the highest priority

Table 1 Various formulae of morphometric parameters.

Morphologic parameters	**Formula**	**References**
Stream order(U)	Hierarchical rank	Strahler[14]
Stream number (N_u)	$N_u = N_{u1} + N_{u2} + N_{u3} + \ldots + N_{um}$	Horton[12]
Basin length (L_b)	$L_b = 1.312 \times A^{0.568}$ Where, A = area of the basin (km^2)	Gregory and Walling[32]
Stream length (L_u)	$L_u = L_{u1} + L_{u2} + \ldots + L_{um}$	Horton[12]
Mean stream length (L_{sm})	$L_{sm} = \frac{L_u}{N_u}$ Where, L_u = total stream length of order u N_u = total number of stream segment of order u	Strahler[14]
Bifurcation Ratio (R_b)	$R_b = \frac{N_u}{N_{u+1}}$ Where, N_{u+1} = no. of stream segment of the order (u+1)	Schumm[13]
Drainage density (D_d)	$D_d = \sum \frac{L_u}{A}$	Horton[12]
Stream frequency (F_s)	$F_s = \sum \frac{N_u}{A}$	Horton[12]
Texture ratio (R_t)	$R_t = \sum \frac{N_u}{P}$	Horton[12]
Mean length of overland flow (L_{om})	$L_{om} = \frac{1}{2D_d}$	Horton[12]
Form factor (F_f)	$F_f = \frac{A}{L_b^2}$	Schumm[13]
Elongation ratio (R_e)	$R_e = \frac{\sqrt{\frac{4A}{\pi}}}{L_b}$	Schumm[13]
Circulatory ratio (R_c)	$R_c = \frac{4\pi A}{p^2}$ Where, P = perimeter of the basin, km	Miller[33]
Compactness Coefficient (C_c)	$C_c = 0.2821 \times \frac{P}{A^{0.5}}$	Horton[12]
Relief basin (B_h)	$B_h = H - h$ Where, H & h = maximum and minimum elevation (km)	Strahler[34]

(continued)

Table 1 Various formulae of morphometric parameters. (*cont'd*)

Morphologic parameters	**Formula**	**References**
Relief ratio (R_r)	$R_r = \frac{B_h}{L_b}$	Schumm[13]
Relative relief (R_h)	$R_h = H \times \frac{100}{P}$	Melton[35]
Ruggedness number (R_n)	$R_n = B_h \times D_d$	Strahler[34]

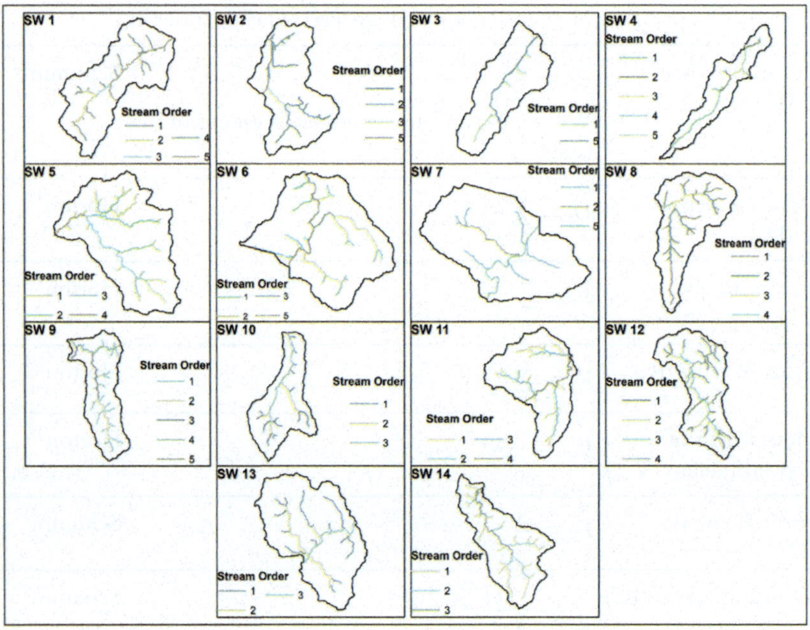

Fig. 3 Stream order generated for the sub-watersheds. Map lines delineate study areas and do not necessarily depict accepted national boundaries.

(1st rank), and so on. Once the rank has been assigned for all the parameters, the compound parameter (Cp) value for each SW was calculated by using the formula

$$C_p = \frac{M}{N} \tag{1}$$

Table 2 Morphometric parameters of the sub-watersheds.

SWs	Linear aspect									Areal aspect				Relief aspect			
	N_u	L_u	L_b	L_{sm}	R_b	D_d	F_s	R_t	L_{om}	F_f	R_e	R_c	C_c	B_h	R_r	R_h	R_n
SW1	101	35.99	19.24	0.36	2.95	0.57	1.61	1.77	0.29	3.26	0.46	0.34	1.72	986	0.513	1.732	732.19
SW2	93	32.70	15.28	0.35	3.83	0.55	1.57	1.81	0.28	3.87	0.57	0.23	2.08	1113	0.728	2.161	865.07
SW3	81	14.00	12.61	0.17	1.00	0.46	2.64	2.25	0.23	2.43	0.50	0.49	1.43	983	0.780	2.734	669.35
SW4	54	26.04	22.70	0.48	3.44	0.49	1.02	0.75	0.25	2.33	0.36	0.57	1.33	1115	0.491	1.547	853.22
SW5	123	60.13	21.77	0.49	1.53	0.56	1.14	1.88	0.28	4.96	0.54	0.14	2.66	1566	0.719	2.391	1215.12
SW6	105	43.61	17.46	0.42	2.33	0.60	1.44	2.07	0.30	4.19	0.55	0.21	2.16	1272	0.728	2.509	1126.70
SW7	45	19.87	9.55	0.44	7.00	0.64	1.46	1.43	0.32	3.23	0.66	0.39	1.61	981	1.027	3.111	631.68
SW8	172	59.68	21.01	0.35	3.55	0.52	1.49	2.28	0.26	5.48	0.58	0.11	3.04	1540	0.733	2.040	799.66
SW9	142	58.14	26.82	0.41	9.98	0.57	1.40	1.91	0.29	3.78	0.42	0.25	1.99	1826	0.681	2.461	1046.89
SW10	94	37.72	20.51	0.40	1.58	0.50	1.26	1.54	0.25	3.64	0.48	0.24	2.05	1504	0.733	2.465	762.19
SW11	130	48.40	19.06	0.37	1.32	0.55	1.48	2.11	0.28	4.60	0.55	0.16	2.47	1268	0.665	2.059	700.87
SW12	220	74.26	26.17	0.34	1.96	0.61	1.81	2.78	0.31	4.64	0.48	0.18	2.37	1622	0.620	2.048	991.19
SW13	99	36.90	16.40	0.37	1.88	0.64	1.73	2.13	0.32	3.49	0.52	0.33	1.73	953	0.581	2.047	614.62
SW14	144	58.94	25.74	0.41	1.92	0.56	1.37	1.91	0.28	4.09	0.45	0.21	2.18	1440	0.559	1.907	805.93

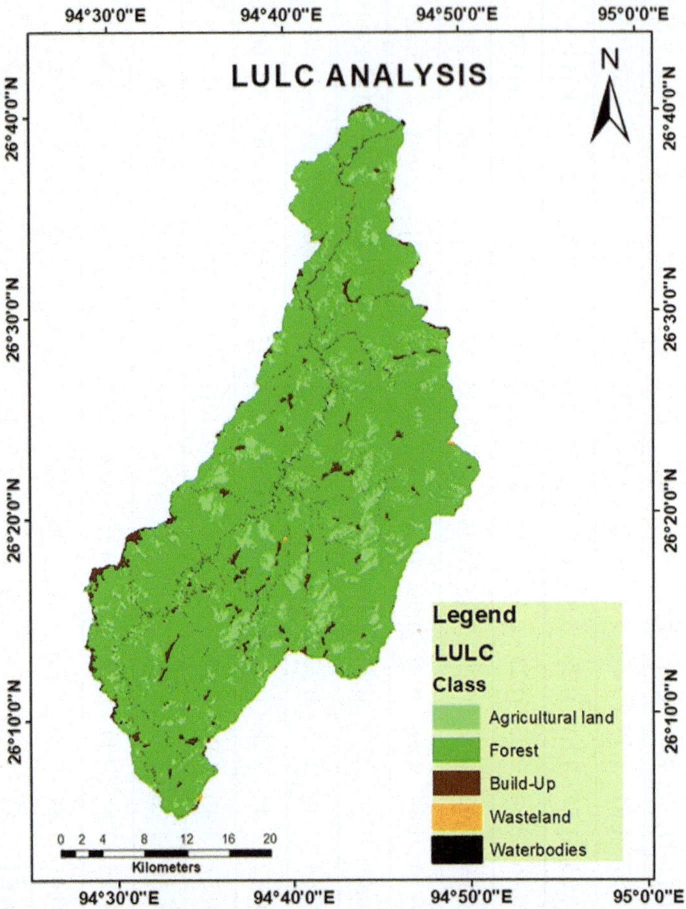

Fig. 4 LULC map of Dikhu watershed. Map lines delineate study areas and do not necessarily depict accepted national boundaries.

where M is the sum of ranks of all parameters for a particular SW and N is the number of parameters considered.

The SWs are then categorized as either high, medium, or low priority based on the value of C_p.

2.3.2 Land-use/land cover (LULC) analysis

LULC refers to how people are utilizing the land and the physical characteristics of the Earth. The LULC map obtained from the state department classified the entire study area into five main classes, viz., agricultural land, forestland, built-up, wasteland, and waterbodies, as shown in Fig. 4. Area under forest land was found as the dominant land cover class, covering an area of

901.21 km^2 (83.40 %), followed by agriculture land of 152.55 km^2 (14.12 %), built-up area of 22.01 km^2 (2.04 %), waterbodies of 4.53 km^2 (0.42 %), and wasteland of 0.32 km^2 (0.03 %) (Table 2). For this study, agricultural land, forestland, and built-up areas were considered for prioritization. The SWs with the lowest area under agricultural land and forest land were assigned rank one (highest priority). In the case of built-up, the SWs with the maximum area under this class were assigned rank one (highest priority), and so on.

3. Results and discussion

3.1 Morphometric analysis

The parameters under the three aspects were estimated with equations shown in Table 1 and the values obtained for each parameter are shown in a tabular format in Table 3.

3.1.1 Linear parameters

Stream order (u): For determining the stream order, the method proposed by Strahler (1964)[14] was used. Outermost streams are designated as first-order streams (1). In this method, subsequent stream orders combine, creating the next higher order of the stream. There are 3, 4, and 5 order streams in Dikhu watershed (Fig. 3).

Stream number (N_u): This parameter computes the total number of streams present in each SW. In this study, the highest stream number is found in SW12 (220) whereas the lowest value is in SW7 (45) (Table 2).

Stream length (L_u): It characterizes runoff and basin features. Usually, streams of shorter lengths are located on upper, steeper slopes, and streams of longer lengths are in the areas with lesser slopes.[34] In this current study, SW12 (74.26 km) has the highest value whereas SW3(14.00 km) has the lowest value.

Basin length (L_b): It is the total distance from the basin outlet to the divide streams.[13] The longest basin length among the 14 SWs is found as 26.82 km (SW9), and the shortest length is 9.55 km (SW7).

Mean stream length (L_{sm}): It is the total length of the stream of any order divided by the total segments in that order.[14] The highest value was calculated as 0.49 (SW5) and the lowest as 0.17 (SW3) for the study area.

Bifurcation ratio (R_b): It is the total number of stream segments of a given order divided by the next higher order.[13] Lower values indicate greater branching.[36] It is also an indication of the basin shape. In the current study, SW9 has the maximum bifurcation ratio (9.98).

Table 3 Ranking and prioritization of Dikhu sub-watersheds based on morphometric analysis.

Sub-watershed (SW)	Linear aspects					Areal aspects				Relief aspects						
	Bifurcation ratio (R_b)	Drainage density (D_d)	Stream frequency (F_s)	Texture ratio (R_t)	Mean length of overland flow (L_{om})	Form factor (F_f)	Circularity ratio (R_c)	Compactness coefficient (C_c)	Elongation ratio (R_e)	Relief (B_h)	Relief ratio (R_r)	Relative ration (R_h)	Ruggedness number (R_n)	C_p value	Ranking	Priority
SW1	6	5	4	11	5	4	11	4	4	11	13	13	10	7.85	8	medium
SW2	3	9	5	10	9	8	7	8	12	10	5	7	5	7.46	6	medium
SW3	14	14	1	3	14	2	13	2	7	12	2	2	12	7.54	7	medium
SW4	5	13	14	14	13	1	14	1	1	9	14	14	6	9.15	14	low
SW5	12	8	13	9	8	13	2	13	9	3	7	6	1	8.08	10	low
SW6	7	4	9	6	4	10	6	9	10	7	6	3	2	6.39	3	high
SW7	2	2	8	13	2	3	12	3	14	13	1	1	13	6.69	4	medium
SW8	4	11	6	2	11	14	1	14	13	4	4	11	8	7.92	9	medium
SW9	1	6	10	7	6	7	9	6	2	1	8	5	3	5.39	1	high
SW10	11	12	12	12	12	6	8	7	6	5	3	4	9	8.23	12	low
SW11	13	10	7	5	10	11	3	12	11	8	9	8	11	9.08	13	low
SW12	8	3	2	1	3	12	4	11	5	2	10	9	4	5.69	2	high
SW13	10	1	3	4	1	5	10	5	8	14	11	10	14	7.39	5	medium
SW14	9	7	11	8	7	9	5	10	3	6	12	12	7	8.15	11	low

Drainage density (D_d): It shows the total length of stream orders per area.[12] Low drainage density indicates low relief and abundant vegetation, while high drainage density indicates areas with rugged terrain and limited vegetation.[37] The D_d of the study area falls within 0.456–0.64 km/km^2, which indicates a coarser drainage pattern. The highest drainage density was found at SW13 (0.64) and the lowest at SW3 (0.46).

Stream frequency (F_s): The ratio between the total number of stream segments of all orders and the basin area.[12] The value of Fs for the study area ranges from 1.02 (SW4) to 2.64 (SW3).

Texture Ratio (R_t): It is the number of stream segments of all orders present per perimeter of that area.[12] The SW12 (2.78) had the highest value, whereas SW4 (0.75) was found to have the lowest.

Length of overland flow (L_{om}): It indicates the total length of the flow of water over the ground before concentration. Three sub-watersheds, viz., SW7, SW12, and SW7 showed the highest L_{om} value of 0.322 while SW3 (0.228) showed the lowest L_{om} which indicates a comparatively high relief in the study area.

3.1.2 Areal aspects

Form Factor (F_f): The watershed area divided by the square of the watershed length gives the Form Factor.[32] Table 2 indicates that the F_f values range from 2.33 to 5.48 signifying mainly elongated sub-basins. The highest F_f value was obtained in SW8 (5.48) and SW4 (2.33) was found to have the lowest.

Elongation Ratio (R_e): It is the measurement of the circular shape of the watershed about its greatest length.[13] The highest R_e value was observed in SW7 (0.66), whereas SW4 (0.36) was found to have the lowest.

Circularity ratio (R_c): It is the ratio of the basin area to the area of a circle having the same perimeter as the basin.[33] In this study, R_c values varied from 0.11 (SW8) to 0.57 (SW4).

Compactness Coefficient (C_c): It is the ratio between the watershed's perimeter and a sphere's circumference having the same area as the watershed.[12] The SW8 (3.04) reported the highest C_c value, whereas SW4 (1.33) showed the lowest C_c value for the study area.

3.1.3 Relief aspects

Four morphometric parameters, viz., Relief basin, Relief ratio, Relative relief, and Ruggedness number, were considered under relief aspects and are discussed below.

Relief basin (B_h): It defines the grade of the river. The relief of the entire Dikhu watershed was found to be 2060 m. Considering the SWs, the maximum relief was found in SW9 (1826 m) and the minimum relief in SW13 (953 m).

Relief Ratio (R_r): is the ratio of the basin relief to the basin length. The values of R_r for the present study varied from 0.491 to 1.027 with the SW4 and SW7 having the lowest and the highest values of relief ratio, respectively.

Relative relief ratio (R_h): It is the relief divided by the perimeter of the watershed.[35] It is not dependent on the basin length. For this study, the estimated relative relief values varied from 1.547 (SW4) to 3.111 (SW7).

Ruggedness number (R_n): The basin relief multiplied by the drainage density gives the R_n.[34] The highest R_n values indicate sharp and longer slopes. In the present study, SW5 (1215.12) obtained the highest value and SW13 (614.62) showed the lowest value of ruggedness number.

3.2 Prioritization based on morphometric parameters

There are 13 morphometric parameters under different aspects considered in this study. After estimating all the values and calculating the compound parameter (Cp), it was found that the SW9 obtained the lowest Cp value of 5.39 and, therefore was given rank 1 (highest priority) while the SW4 obtained the highest Cp value of 9.15, hence, given rank 14 (least priority) (Table 3). The SWs were further categorized under three priority classes based on Cp value. Those SWs with Cp value <6.5 were considered under the high priority zone, Cp value of 6.5–8.0 under medium, and Cp > 8.00 as the low priority zone (Table 3). Hence, based on these criteria, three SWs (i.e., SW6, SW9, and SW12) are classified under the high priority zone, six SWs (i.e., SW1, SW2, SW3, SW7, SW8, and SW13) under the medium priority zone, and five SWs (i.e., SW4, SW5, SW10, SW11, and SW14) under the low priority zone. The SWs categorized as high priority are vulnerable, and they face the greatest risk of soil erosion and overland flow. Therefore, it is necessary to prioritize soil conservation measures in particular regions, for the development and management of natural resources.[38] The final prioritized map of Dikhu SWs based on the morphometric analysis is shown in Fig. 5.

3.3 LULC analysis

The area coverage (%) under various land use/land classes (LULC) for all the SWs of the Dikhu watershed was calculated from the LULC map obtained from the Soil and Water Conservation Department, Nagaland, and is shown in Table 4. *Forests*.

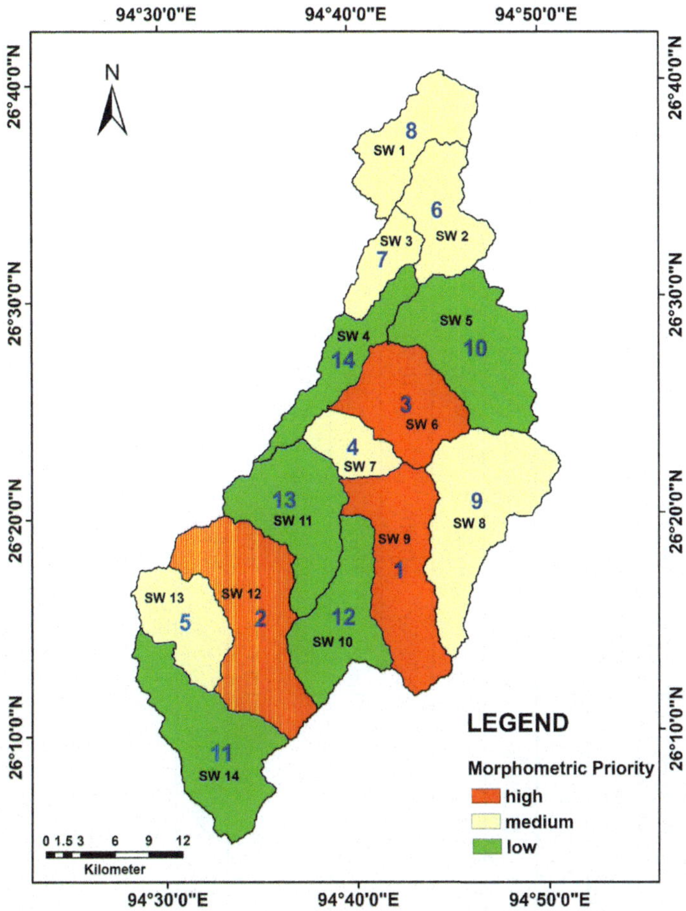

Fig. 5 Prioritization and ranking of Dikhu SWs based on morphometric analysis. Map lines delineate study areas and do not necessarily depict accepted national boundaries.

The study area has the maximum area under forest (83.40 %) as shown in Fig. 4, which is a clear indication of ecosystem stability. From Table 4, it is clear that the highest and lowest areas under forests fall within SW12 with a land area of 108.37 km^2 (89.18 %) and SW7 with an area of 23.58 km^2 (76.43 %), respectively. The SWs with the lowest percentage of forest cover were given high priority, i.e., SW10 (75.80 %), and vice-versa.

3.3.1 Agricultural land

The SWs with the lowest area under Agricultural land were assigned rank one (high priority) and vice versa. Dikhu watershed has a total area of 152.55 km^2 (14.12 %) under Agricultural land. The area occupied by

Table 4 Area coverage under different LULC classes with percentage for Dikhu sub-watersheds.

SWs	Agriculture land	Forest	Build-up	Wasteland	Waterbodies	Total area cover (km^2)
SW1	5.21	52.39	1.02	0.02	0.51	59.15
	(8.81 %)	(88.57 %)	(1.73 %)	(0.03 %)	(0.85 %)	
SW2	7.32	54.10	0.92	0.08	0.30	62.72
	(11.67 %)	(86.26 %)	(1.46 %)	(0.1 %3)	(0.49 %)	
SW3	4.33	25.44	0.52	0.00	0.34	30.63
	(14.13 %)	(83.06 %)	(1.71 %)	(0.00)	(1.10 %)	
SW4	5.77	45.90	1.02	0.01	0.22	52.97
	(10.91 %)	(86.73 %)	(1.93 %)	(0.01 %)	(0.42 %)	
SW5	12.94	93.37	1.46	0.00	0.30	108.07
	(11.97 %)	(86.40 %)	(1.35 %)	(0.00)	(0.28 %)	
SW6	12.30	59.33	0.90	0.03	0.59	73.16
	(16.81 %)	(81.10 %)	(1.23 %)	(0.04 %)	(0.81 %)	
SW7	6.70	23.58	0.30	0.00	0.28	30.86
	(21.72 %)	(76.43 %)	(0.96 %)	(0.00)	(0.89 %)	
SW8	20.77	93.68	0.53	0.07	0.11	115.17

	(18.03 %)	(81.34 %)	(0.46 %)	(0.06 %)	(0.10 %)	
SW9	18.17	81.27	1.69	0.00	0.33	101.46
	(17.90 %)	(80.10 %)	(1.66 %)	(0.00)	(0.33 %)	
SW10	16.05	56.65	1.83	0.03	0.18	74.73
	(21.48 %)	(75.80 %)	(2.45 %)	(0.04 %)	(0.24 %)	
SW11	14.58	71.02	1.75	0.00	0.36	87.70
	(16.62 %)	(80.98 %)	(1.99 %)	(0.00)	(0.41 %)	
SW12	8.79	108.37	3.91	0.00	0.45	121.52
	(7.23 %)	(89.18 %)	(3.22 %)	(0.00)	(0.37 %)	
SW13	8.08	46.28	2.67	0.00	0.20	57.22
	(14.12 %)	(80.87 %)	(4.66 %)	(0.00)	(0.35 %)	
SW14	11.55	89.82	3.50	0.08	0.35	105.31
	(10.97 %)	(85.29 %)	(3.32 %)	(0.08 %)	(0.34 %)	
Overall Total Area covered	152.55 (14.12 %)	901.21 (83.40 %)	22.01 (2.04 %)	0.32 (0.03 %)	4.53 (0.42 %)	1080.62

Note: 1. Bold values are the total area cover in km^2.
2. The values under the brackets are the total percentage (%) cover

agricultural land for all 14 SWs, along with their percentage, was calculated and presented in Table 4. The largest area under agricultural land use was reported in SW8 (20.77 km^2, which is 18.03 %), while SW3 reported the least area under agricultural land use (4.33 km^2, which is 14.13 %).

3.3.2 *Build-Up*

Build-up covers a total area of 22.01 km^2 (2.04 % of the entire watershed area) in the Dikhu watershed (Table 4). The SW with the highest percentage of build-up was reported in SW13 (4.66 %), followed by SW12 (3.22 %), and is found to be the lowest in SW8 (0.46 %), followed by SW7(0.96 %). Those SWs with highest % of the build-up area were assigned rank one (highest priority) and so on.

3.3.3 *Water-bodies*

In Dikhu watershed, the water bodies comprise only 0.42 % (4.54 km) of the total area. The SW3 (1.1 %) has the highest water bodies area, while SW8 (0.10 %) has the least area and was therefore given rank one (highest priority).

3.4 Prioritization based on LULC

Prioritization of all 14 SWs of the Dikhu watershed was also carried out based on the LULC layer collected from the state department. As mentioned earlier, the classes of land use considered for prioritization are Agricultural land, Forest land, Built-up, and Water bodies. By considering the percentage area under each land-use category, ranks were assigned as per area (in %) under each land-use category and the final ranking and prioritization of the SWs is shown in Table 5. Prioritization was done by categorizing the highest and lowest range of Cp values into three classes: high priority ranking ($Cp < 6.5$), medium priority zone (Cp between 6.5–8), and low priority ranking ($Cp > 8$). In doing so, as many as six SWs (viz., SW1, SW2, SW3, SW6, SW7, and SW8) were classified under low-priority area; four SWs (viz., SW4, SW5, SW9, and SW11) were classified under medium priority, and four SWs (viz., SW10, SW12, SW13, and SW14) under high priority (Table 5). The result indicates that most of the area of Dikhu watershed is under low to medium soil erosion risks. This may be because more than 80 % of the land area is under thick forest cover even though most of the watershed comprises of moderate to steep sloping areas. However, few of the SWs are under high soil erosion risks, which implies that soil conservation measures are needed to reduce the land degradation occurring in these SWs. Fig. 6 shows the prioritization based on LULC.

Table 5 Ranking and prioritization of Dikhu SWs based on percent area under different LULC classes.

SWs	Agricultural land	Forest land	Built-up	Water bodies	Cp value	Ranking	Priority
SW1	2	13	7	12	8.50	9	Low
SW2	5	10	10	10	8.75	11	Low
SW3	8	8	8	14	9.50	12	Low
SW4	3	12	6	9	7.50	7	Medium
SW5	6	11	11	3	7.75	8	Medium
SW6	10	6	12	11	9.75	13	Low
SW7	14	2	13	13	10.50	14	Low
SW8	12	7	14	1	8.50	10	Low
SW9	11	3	9	4	6.75	6	Medium
SW10	13	1	4	2	5.00	3	High
SW11	9	5	5	8	6.75	5	Medium
SW12	1	14	3	7	6.25	4	High
SW13	7	4	1	6	4.50	1	High
SW14	4	9	2	5	5.00	2	High

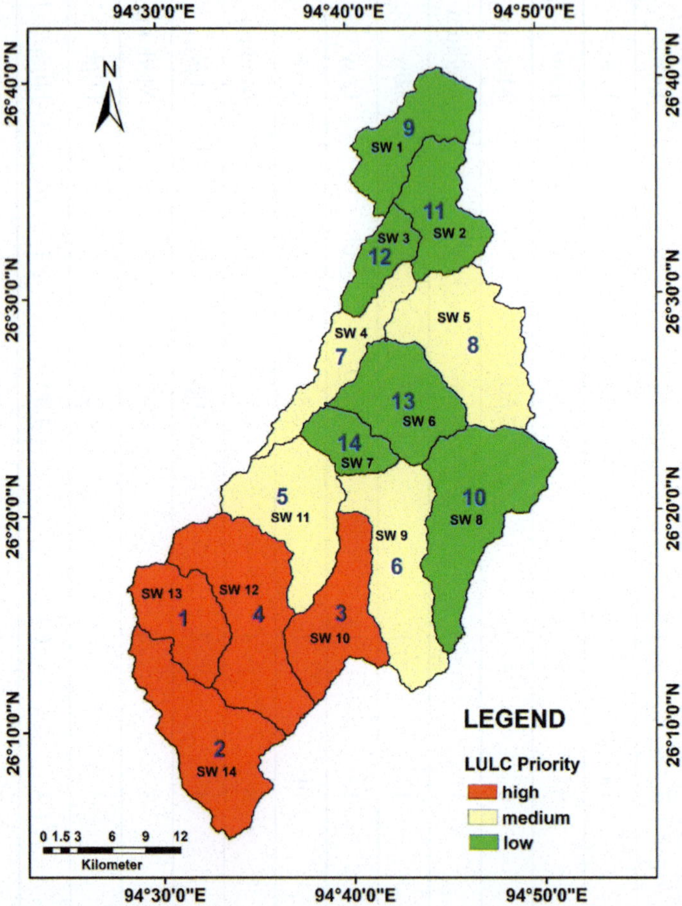

Fig. 6 Prioritization and ranking of Dikhu sub-watershed based on LULC analysis. Map lines delineate study areas and do not necessarily depict accepted national boundaries.

3.5 Comparison of the two methods of analysis

In this study, two methods for prioritization were used since significance of the parameters may vary depending on characteristics of different SWs. It was found that both methods yield varying priority levels. The comparison of the final priority class of SWs based on the two methods is shown in Table 6. Prioritization based on morphometric analysis classified only three SWs under high-priority areas while LULC analysis classified four SWs under high priority areas. This shows that the majority of the area of the Dikhu watershed is under a low erosion risk zone. Even though the study area is highly undulated, 83.40 % of the land area is under dense forest land,

Table 6 Comparison of morphometric and LULC prioritization.

Sub-watershed	**Prioritization**	
	Morphometric	**LULC**
SW1	Medium	Low
SW2	Medium	Low
SW3	Medium	Low
SW4	Low	Medium
SW5	Low	Medium
SW6	High	Low
SW7	Medium	Low
SW8	Medium	Low
SW9	High	Medium
SW10	Low	High
SW11	Low	Medium
SW12	High	High
SW13	Medium	High
SW14	Low	High

which could be the main reason why many portions of Dikhu showed low soil erosion risk zones. Considering the final prioritization from both morphometric and LULC analysis, SW12 was prioritized as high priority area. This shows that importance should be given to SW12. Different measures for soil conservation and effective land management practices should be put into place to control soil erosion and ensuing flooding in high priority SWs.

3.6 Implications of climate-smart watershed management

The combination of climate change adaptation and mitigation measures to ensure sustainable water management, ecosystem vitality, and combat long-term climate change in a watershed is what climate-smart watershed management is all about. Prioritizing watersheds by identifying and ranking them based on their vulnerability and potential for sustainable intervention can ensure climate-smart watershed management. This helps in the efficient allocation of resources and targeted interventions for sustainable water management and climate change mitigation measures.

In this study, prioritization of the sub-watersheds was done based on Morphometric analysis and Land Use Land Cover (LULC) analysis. Morphometric analysis aids in understanding runoff, infiltration patterns, and hydrological patterns, and is therefore crucial for predicting runoff, identifying flood-prone areas, and developing appropriate water resources management strategies. LULC analysis, further, evaluates changes in LULC patterns, as these changes can impact water quality and quantity. Understanding them is crucial for informed planning and adaptive management. Morphometric parameters like drainage density, stream frequency, and relief ratio indicate erosion susceptibility, while the reduction in Forest area land cover can exacerbate soil erosion. All these considered, the application of both Morphometric analysis and LULC analysis offers a potent framework for understanding the watershed dynamics and creating a climate-smart watershed planning and management.

4. Conclusion

Watershed prioritization involves the identification and ranking of environmentally degraded SWs, which are then treated to conserve soil and degraded land at a priority level. Morphometric parameters can easily identify areas prone to erosion and deserving prioritization. Comparison of erosion severity and sediment yields allows for prioritization of watersheds, with emphasis on the LULC of the area. This study aims to prioritize watersheds to conserve soil and water, as well as efficiently managing water resources. By classifying degraded watersheds or sub-watersheds based on risk level, this method helps target conservation efforts and allocate budgets effectively. This approach ensures that investments and interventions are optimized, resulting in the most impact for sustainable water management.

For the present study, Dikhu River basin was delineated using ArcSWAT and was further divided into 14 sub-watersheds, numbered from SW1 to SW14. GIS and RS approaches were used for both Morphometric and LULC analysis. The morphometric analysis was carried out under three aspects, i.e., Linear, Areal, and Relief aspects. Based on the computation and analysis of geomorphological parameters for different SWs, priority ranking is done for 14 SWs. To prioritize the SWs based on LULC, the LULC layer of the study area was analyzed. The study area was divided into five main classes, viz., agricultural land, forest, build-up, wasteland, and waterbodies. Prioritization based on LULC was carried out by calculating the land area and the percentage of land area covered by the agricultural land, forest land, built-up area, and water bodies, and ranking was given to each SW depending on the percentage area coverage.

Under morphometric analysis, higher priorities were assigned to SW6, SW9, and SW12, while higher priorities using LULC analysis were assigned to SW10, SW12, SW13, and SW14. Among the SWs identified under high priority, immediate priority is assigned to SW12 as it was categorized as high priority under both methods of analysis. As a result, the high-priority SW may be used by decision-makers for conservation, land management, planning, and to protect the land from erosion and flooding. Various strategies can be put into practice to reduce soil erosion and the resulting floods in highly prioritized sub-watersheds, including: Encouraging eco-friendly practices in land use and building structures designed to protect soil, such as: setting up check dams, wire baskets, and infiltration tanks. These structures are effective in capturing and slowing the flow of water, which in turn reduces erosion and the amount of sediment that ends up in water bodies. They also help to keep water within the natural environment, encouraging it to seep into the ground and decreasing the amount that flows over the surface. By adopting these approaches, it is feasible to lessen soil erosion and the flooding that follows, leading to better management of land and water resources, increased agricultural output, and a lower risk to human communities and infrastructure.

As an extension of the current study, the authors plan to broaden the data collection and employ a more modern technique, like Machine Learning (ML), for prioritizing the sub-watersheds to encompass more micro-level watersheds. This will provide opportunities for a more comprehensive analysis and insights.

Acknowledgments

The authors would like to thank the department of Remote Sensing and GIS, Govt. of Nagaland, for providing land use land cover (LULC) data for the study area. The author would also like to thank the USGS Earth Explorer for making available the satellite data.

Declaration of generative AI and AI-assisted technologies in the writing process

The author(s) declared that we have not used any generative artificial intelligence (AI) tools or technologies in the preparation of this article. This choice was made to ensure the originality of the work and to maintain the integrity of the writing process. We take full responsibility for the content of this article.

References

1. Nookaratnam K, Srivastava YK, Venkateswara Rao V, Amminedu E, Murthy KSR. Check dam positioning by prioritization micro-watersheds using SYI model and morphometric analysis – remote sensing and GIS perspective. *J Indian Soc Remote Sens* 2005;**33**:25–38.
2. Redvan G, Mustafa U. Flood prioritization of basin based on geomorphometric properties using principal component analysis, morphometric analysis and Redvan's priority methods: a case study of Harshit river basin. *J Hydrol* 2021;**603**.
3. Srivastava A, Jain S. Remote sensing for environment assessment: multispectral, hyperspectral, and thermal imaging applications. In *Remote Sensing for Environmental Monitoring;* Saritha, V, Pande, C B, Singh, R, Shahid, M, Eds.; Springer: Singapore, 2025https://doi.org/10.1007/978-981-96-5546-5_1
4. Srivastava A, Chinnasamy P. Investigating impact of land-use and land cover changes on hydro-ecological balance using GIS: insights from IIT Bombay, India. *SN Appl. Sci.* 2021;**3**:343. https://doi.org/10.1007/s42452-021-04328-7.
5. Moore ID, Grayson RB, Ladson AR. Digital terrain modelling. In: A Review of hydrological, geomorphological and biological applications, Vol. 7, No. 31. Chichester: Wiley; 1994, p. 249.
6. Poongodi R, Venkateswaran S. Prioritization of the micro-watersheds through morphometric analysis in the Vasishta Sub Basin of the Vellar River, Tamil Nadu using ASTER digital elevation model (DEM) data. *Data in Brief* 2018;**20**:1353–1359.
7. Bhattacharya RK, Chatterjee DN, Das K. Sub-basin prioritization for assessment of soil erosion susceptibility in Kangsabati, a plateau basin: a comparison between MCDM and SWAT models. *Sci Total Environ* 1966;**734**:139474.
8. Clarke JI. *Morphometry from Maps. Essays in geomorphology;* Elsevier Publ, Co: New York, 1966.
9. Obi Reddy GP, Sambasiva Rao M, Maji AK. Delineation and prioritization of macro watersheds in semi-arid Anantapur district Andhra Pradesh. *Geogr Rev India* 2004;**66**(1):72–83.
10. Withanage NS, Dayawansa NDK, De-Silva RP. Morphometric analysis of the Gal Oya river basin using spatial data derived from GIS. *Trop Agric Res* 2014;**26**(1):175–188.
11. Samal DR, Gedam SS, Nagarajan R. GIS based drainage morphometry and its influence on hydrology in parts of Western Ghats region, Maharashtra, India. *Geocarto Int* 2015;**30**:755–778.
12. Horton R. Erosional development of streams and their drainage basins; hydro physical approach to quantitative morphology. *Geol Soc Am Bull* 1945;**56**:275–370.
13. Schumm SA. Evolution of drainage system and slope in Badlands at Perth Amboy, New Jersey. *Geol Soc Am Bull* 1956;**67**:597–646.
14. Strahler AN. Quantitative geomorphology of drainage basins and channel networks. In *Handbook of applied hydrology;* Chow V T, Ed.; McGraw Hill Book Company: New York Section, 1964; pp. 4–11.
15. Mishra C, Jaiswal R, Nema A, Chandola V, Chouksey A. Priority assessment of sub-watershed based on optimum number of parameters using fuzzy-AHP decision support system in the environment of RS and GIS. *J Indian Soc Remote Sens* 2018;**47**(4):603–617.
16. Kadam AK, Jaweed TH, Kale SS, Umrikar BN, Sankhua RN. Identification of erosion-prone areas using modified morphometric prioritization method and sediment production rate: a remote sensing and GIS approach. *Geomat Nat Haz Risk* 2019;**10**:986–1006.
17. Srivastava A, Chinnasamy P. Watershed development interventions for rural water safety, security, and sustainability in semi-arid region of Western-India. *Environ Dev Sustain* 2024;**26**:18231–18265. https://doi.org/10.1007/s10668-023-03387-7.
18. Biswas S, Sudhakar S, Desai VR. Prioritization of sub watersheds based on morphometric analysis of drainage basin. A remote sensing and GIS approach. *J Indian Soc Remote Sens* 1999;**27**:155–166.

19. Gajbhiye S, Mishra SK, Pandey A. Prioritizing erosion-prone area through morphometric analysis: an RS and GIS perspective. *Appl Water Sci* 2014;**4**:51–61.
20. Shekar PR, Mathew A, Arun PS, Gopi VP. Sub-watershed prioritization using morphometric analysis, principal component analysis, hypsometric analysis, land use/land cover analysis, and machine learning approaches in the Peddavagu River Basin, India. *J Water Clim Change* 2023;**14**(7).
21. Sreedhar Y, Nagaraju A, Krishna GM. An appraisal of land use/land cover change scenario of Tummalapalle, Cuddapah Region, India – a remote sensing and GIS perspective. *Advances in Remote Sensing* 2016;**5**:232–245.
22. Said S, Siddique R, Shakeel M. Morphometric analysis and sub-watersheds prioritization of Nagmati River watershed, Kutch District, Gujarat using GIS based approach. *J Water Land Dev* 2018;**39**:131–139.
23. Adhami M, Sadeghi SH, Duttmann R, Sheikhmohammady M. Changes in watershed hydrological behaviour due to land use management scenarios. *J Hydrol* 2019;**577**:124001.
24. Sharma S, Mahajan AK. GIS-based sub-watershed prioritization through morphometric analysis in the outer Himalayan region of India. *Appl Water Sci* 2020;**10**:163.
25. Song X, Hansen MC, Stehman SV, Potapov PV, Tyunkvina A, Vermote EF. Global land change from 1982 to 2016. *Nature* 2018;**560**:639–643.
26. Gupta N, Mathew A, Khandelwal S. Spatio-temporal impact assessment of land use/land cover (LU-LC) change on land surface temperatures over Jaipur city in India. *Int J Urban Sustain Dev* 2020;**12**(3):283–299.
27. Javed A, Khanday MY, Ahmed R. Prioritization of sub-watersheds based on morphometric and land use analysis using remote sensing and GIS techniques. *J Indian Soc Remote Sens* 2009;**37**:261–274.
28. Puno GR, Puno RCC. Watershed conservation prioritization using geomorphometric and land use-land cover parameters. *Glob J Environ Sci Manag* 2019;**5**(3):279–294.
29. Shekar PR, Mathew A. Morphometric analysis for prioritizing sub-watersheds of Murredu River basin, Telangana State, India, using a geographical information system. *J Eng Appl Sci* 2022;**69**:44.
30. Meshram SG, Sharma SK. Prioritization of watershed through morphometric parameters: a PCA-based approach. *Appl Water Sci* 2017;**7**:1505–1519.
31. Ahmed R, Sajjad H, Husain I. Morphometric parameters-based prioritization of sub-watersheds using fuzzy analytical hierarchy process: a case study of lower Barpani watershed, India. *Nat Resour Res* 2017;**27**(1):67–75.
32. Gregory KJ, Walling DE. *Drainage basin form and process: a geomorphological approach;* Edward Arnold: London, 1973.
33. Miller VC. A quantitative geomorphic study of drainage basin characteristics in the clinch mountain Area, Virginia and Tennessee. Project NR 389042, Tech. Rept.3, Columbia University, Department of Geology, ONR, Geography Branch, New York; 1953.
34. Strahler AN. Quantitative analysis of watershed geomorphology. *Trans Am Geophys Union* 1957;**38**:913–920.
35. Melton MA. An analysis of the relations among elements of climate, surface properties, and geomorphology. DTIC Document; 1957.
36. Suji VR, Sheeja RV, Karuppasamy S. Prioritization using morphometric analysis and land use/land cover parameters for Vazhichal watershed using remote sensing and GIS techniques. *Int J Innov Res in Sci & Tech* 2015;**2**(1):61–68.
37. Nag SK. Morphometric analysis using remote sensing techniques in the Chaka subbasin, Purulia district, West Bengal. *J Indian Soc Remote Sens* 1998;**1**:69–76.
38. Gosain AK, Rao S. GIS based technologies for watershed management. *Curr Sci* 2004;**87**:948–953.

CHAPTER TWELVE

Assessing the impact of long-term rainfall and temperature variability on land use and land cover

Nibedita Guru*, **Manimoy Das, and Dhrubajyoti Das**
Forest Ecology and Climate Change Division, Rain Forest Research Institute, Jorhat, Assam, India
*Corresponding author. e-mail address: nibeditaguru149@gmail.com

Contents

Abstract

Nowadays, the impact of climate change on terrestrial ecosystems has been a focus point. The Northeast region (NER) is one of the highest rainfall-receiving regions having huge water and hydropower potential. Hence, analysis of rainfall and temperature variability would be of interest to water and energy planners. This study analysed the variability and trends in rainfall and temperature time series between 2010–2022 at monthly and monsoon season scales using IMD gridded datasets, manually installed automatic weather station at Jeypore Forest IB of Dihing Patkai National Park (DPNP), and some other meteorological stations. A comparison of climatic variation has been examined with nearby available meteorological stations to capture the variation in both time series. Further, as the study area represents tropical rainforest, it is imperative to analyze the variation and inter-relationship between land use cover and climate pattern (temperature and rainfall) with all stations. The results show an increasing trend at Dihing Patkai landscape during monsoon due to dense

Advances in Chemical Pollution, Environmental Management and Protection, Volume 15
ISSN 2468-9289, https://doi.org/10.1016/bs.apmp.2025.11.008

forest cover as compared to other stations. It was also observed that, variation in rainfall (maximum in June-Aug all stations, 600–800 mm) and temperature follows the same pattern for all stations except Namsang and Dillighat on a monthly scale due to the different land use cover. Similarly, maximum temperature is showing an increasing trend in July-Aug (37.8°) months for most of the year whereas minimum temperature follows a decreasing trend in the study area.

Keywords: Climate change, Trend, Rainfall pattern, Temperature pattern, Land use cover, DPNP.

1. Introduction

Present day climate change is a major threat to the world's water resources since it is changing hydrological cycles and making water-related hazards like droughts, floods, and water scarcity. Hence, climate-smart water resource management techniques that improve resilience and assurance sustainable access to water are urgently needed. This strategy combines effective water governance, mitigation, and climate adaptation to protect water security for present and future generations. The amount and quality of freshwater resources are being impacted by changing precipitation patterns, rising temperatures, and an increase in the frequency of extreme weather events. The dynamics of surface and groundwater systems are shifting as a result of rising evapotranspiration rates, sea level rise, and glacial retreat. It is imperative to create solutions that manage both short-term and long-term climate threats since these impacts provide serious threats to ecosystems, energy production, agriculture, and urban water supplies. A comprehensive evaluation of current water resources is the first step towards effective climate-smart management. This entails assessing the availability of surface and groundwater, recognizing regional and seasonal variability, and comprehending the utilization trends that are currently observed across many sectors. Here rainfall and temperature are the two most important weather parameters which characterized the climatic condition. In last few decades, it has been reported that the earth average temperature increased by 0.74°C.[1] For the last 50 years Antarctica has been experienced in highest warming due to rapid deforestation, emission of greenhouse gases especially carbon dioxide in the atmosphere.[2] These changes of climatic activity have changes the growth rate, life style, food chain, etc. of the ecosystem which have been noticed in different landscapes especially in terrestrial ecosystem. Rainfall and temperature both parameters are effect the entire cycle of ecosystem. Impacts of changing pattern of rainfall and temperature in terrestrial ecosystem has got special attention worldwide in recent years. Several studies of rainfall and temperature variability and long term trend analysis have been done in India using through weather station and grid dataset.[3–14] In India, many regions are received considerable rainfall during monsoon season, while some region receive huge rainfall especially the North-eastern region and some region receive less rainfall due to the variation of land cover and temperature. The north-eastern region of India is well known for having an abundance of rare, unusual, and unique flora and fauna that are not found anywhere else in the nation. In every climatic change activity were started in small changes of land cover (for example-the difference between forest land and agricultural land) which have influence the weather activity. Various land uses, such as forestry, industry, urbanization, agriculture, livestock, and forestry, have an impact on water resources.[15–18] Climate change and land use and land cover are acknowledged as the primary drivers of environmental change.[19,20] The processes of runoff and percolation, groundwater replenishment, sediment transport, and water loss through evaporation are all impacted by changes in land use.[21-23]

In this study the major objective is to analyze the variation of local climatic fluctuations especially natural forest land and other nearest lands concerning rainfall and temperature (average, maximum, and minimum). For climate-smart land use planning, evaluating temperature and rainfall variability is essential because it offers the scientific foundation for making well-informed, climate change-adaptive decisions. Variability in these climate elements has a direct impact on soil health, crop productivity, water availability, and the likelihood of natural disasters like floods and droughts. For this, descriptive statistics of long-term variation of rainfall and temperature were analyzed in different land use cover using monthly and monsoon season data of IMD grid datasets, manually installed automatic weather station, and some other meteorological stations for the period 2010–2023. Furthermore, a non-parametric test (Man-Kendall test) and linear regression technique were used to analyze the trend of temperature and rainfall on different land use cover. Hence, the outcome of this study would help to understand the facts of the changes in weather patterns of the different landscapes and help in future projection studies. Further by understanding these patterns, planners can identify climate-vulnerable areas, improve water and soil resources management, design resilient infrastructure and settlement plans etc.

2. Materials and methods

2.1 Study area

Dihing Patkai National Park (DPNP), which covers 234.26 km^2 of rainforest, is situated between 95°22′0″ E and 27°21′0″ N in the districts Dibrugarh and Tinsukia of Assam (Fig. 1). It is located near the Namdapha Wildlife Sanctuary, on the banks of the Dihing River, and in the foothills of the Patkai Hill Ranges. This lowland rainforest, which is dominated by dipterocarps, is regarded as one of the few lowland tropical rainforests that still exist in mainland India. The actual forest is separated into three adjacent areas: Upper-Dihing, Dirok, and Jeypore, and it spans 500 square kilometres. The Tirap and Changlang districts of Arunachal Pradesh are further covered by the forest. With about 4000 mm of annual precipitation, the region has a primarily tropical climate. There is at least 60 mm of precipitation per month. July sees the most rainfall, while September marks the end of the monsoon season. The usual temperature ranges from 6 to 38° Celsius. The fact that this National Park is a rainforest contributes to its great diversity.

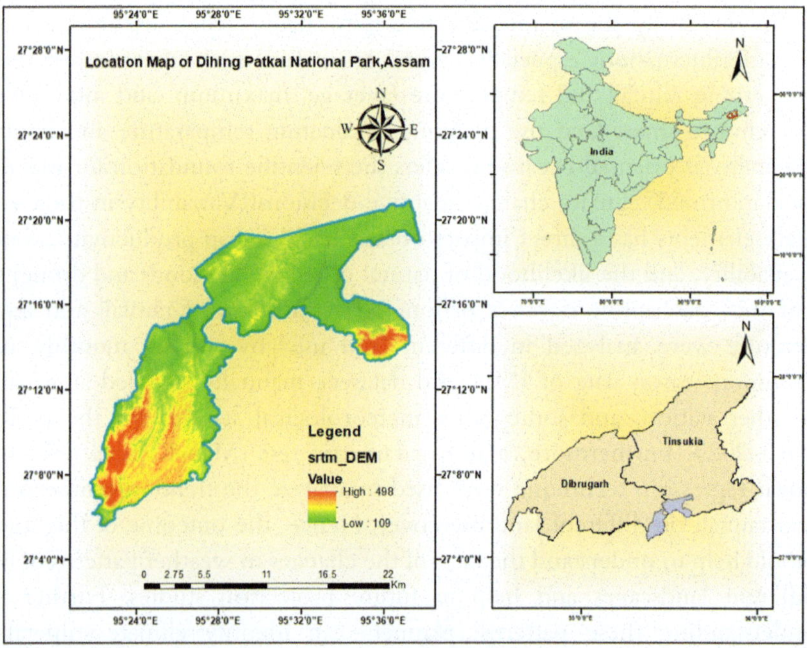

Fig. 1 Location map of study area (DPNP). Map lines delineate study areas and do not necessarily depict accepted national boundaries.

2.2 Source of data

In the present study, daily rainfall and temperature data of the gauged station from May 2022 to August 2023 was obtained from the Automatic weather station which is installed within the vicinity of the DPNP forested area at Jeypore IB (AWS). Further, the gridded historical daily rainfall (0.25° × 0.25°) at four locations i.e. S1(27.25, 95.50), S2 (27.50, 95.50), S3(27.25, 95.25), and S4 (27.50, 95.25) near the DPNP forested area, and temperature (1° × 1°) at one location (S2) datasets were downloaded from the India Meteorological Department (IMD) for a period of 13 years i.e. 2010–2022 to analyze the effect of climate change (Fig. 2). Additionally, temperature and rainfall records were gathered at the Dillighat and Namsang stations to compare and illustrate the impact of various land-use patterns on climate data.

2.3 Trend analysis

A statistically significant ± change in a random variable over time is called a trend. Both parametric and non-parametric statistical tests can be used to evaluate trends. A non-parametric test was employed in this study to

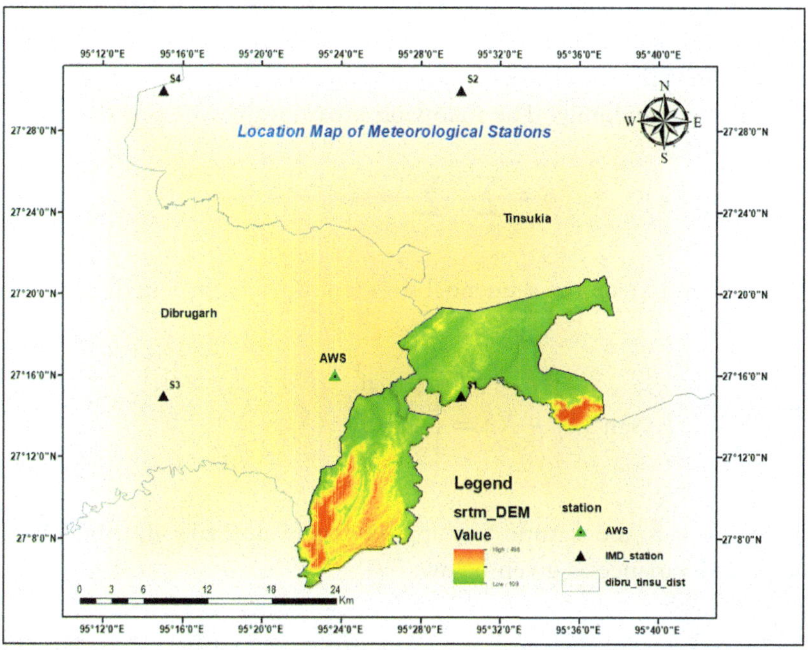

Fig. 2 Location map of IMD meteorological stations of study area (DPNP). Map lines delineate study areas and do not necessarily depict accepted national boundaries.

evaluate the climate variables. The Mann-Kendall test has been used to find trends in monthly and seasonal time series data of average rainfall and temperature.[24,25] Four climatic seasons, namely winter (January–February), pre-monsoon (March–May), monsoon (June–September), and post-monsoon (October–December) were famed for the purposes of seasonal study. Descriptive statistics (mean, standard deviation, coefficient of variation (CV), skewness, and kurtosis) were considered for the long-term data using standard statistical processes.[26] For example, the variation in rainfall can be assessed by calculating CV using the method specified in the equation.

$$CV = \frac{\sigma}{\mu} \times 100$$

Where, μ is the mean precipitation, σ is the standard deviation, and CV is the coefficient of variation. Greater variability is indicated by a higher CV value, while less variability is indicated by a lower CV value.

2.3.1 Mann-Kendall test

The Mann-Kendall test was employed for trend analysis. A popular non-parametric technique for identifying monotonic patterns in hydro-meteorological data is the Mann-Kendall (MK) test, which is based on rank.[25] The MK test makes the assumptions that the data are independent and arranged randomly. The following provides the MK test statistic:

$$S = \sum_{i=1}^{n-1} \sum_{j=i+1}^{n} \text{sig}\,(X_j - X_i),$$

Where xi and xj are the sequential data points, N is the length of the data series, and

$$\text{sgn}\,(X_j - X_i) = \begin{cases} +1 & \text{if} \quad (X_j - X_i) > 0 \\ 0 & \text{if} \quad (X_j - X_i) = 0 \\ -1 & \text{if} \quad (X_j - X_i) < 0, \end{cases}$$

For N > 8, the statistic S is approximately normally distributed with mean, and variance as given below:

$$\text{V}(S) = \frac{1}{18}\left[n\,(n-1)\,(2n+5) - \sum_{p=1}^{q} t_p\,(t_p - 1)\,(2t_p + 5)\right]$$

Where t_p is the number of data points belonging to the kth tied group and t is no of tied groups in the dataset. The Mann-Kendall Z statistic is computed as follows:

$$Z = \begin{cases} \frac{S-1}{\sqrt{\text{VAR}(S)}} & \text{if} \quad S > 0 \\ 0 & \text{if} \quad S = 0 \\ \frac{S+1}{\sqrt{\text{VAR}(S)}} & \text{if} \quad S < 0. \end{cases}$$

Z represents the Mann-Kendall test statistic, which is based on a conventional normal distribution with a variance of one and a mean of zero. If $Z_{jj} > Z\alpha/2$, the trend is significant and the null hypothesis is rejected in a two-sided trend test for the significance level of alpha (α). At the chosen significance level (α), a positive (negative) Z value indicates a positive (negative) trend. The normalized test statistics (Z) value was used to determine whether a monotonic increasing or decreasing trend was present using the non-parametric Mann-Kendall test. If Z is negative, the trend is said to be decreasing; if Z is positive, the trend is said to be increasing.

2.3.2 Linear regression

A linear regression technique was used to analyze the trends in the climatic time series data. The regression analysis's slope represents the average temporal change in the variable under consideration. A linear regression model can be stated simply as

$$Y = a + bX$$

The equation can be interpreted as; if b is positive, then X variables increase and vice versa. Where, the least squares method is used to find "a" and "b".

2.3.3 Effect of rainfall and temperature variability on land use cover

Temperature and rainfall are two important climatic variables that affect patterns of land use and land cover (LULC) in different geographical areas. As the effects of climate change intensify, changes in these variables such as unpredictable rainfall patterns, extended dry spells, and higher temperatures are changing agricultural practices, influencing forest dynamics, and modifying ecosystems. Hence the relationship between changes in land use and cover and climate variability was examined.

3. Results and discussion

A descriptive statistic of monthly and seasonal time series of rainfall from 2010 to 2022 for IMD stationsis shown in Table 1. Over a 13-year period, the average annual rainfall was 2135 mm, of which 60.91 % (1299.72 mm) fell during the monsoon season. This rainfall has been considered as the maximum seasonal rainfall due to the high precipitation in June (350 mm) and August (400 mm), followed by July and September (Fig. 3). On the other hand, the post-monsoon season had a 6.22 % contribution to yearly rainfall, while the minimum seasonal rainfall (50 mm) collected during the winter season only provided 2.34 %.

A coefficient of variation (CV) can be used to understand rainfall variability; a high CV indicates substantial rainfall variability[27] and if CV is varying between 20–30 then the variability is moderate followed by less variability having CV value less than 20. Based on this, S1 and S3 have moderate monsoon variability whereas S2 and S4 have less monsoon variability.

With the exception of July and September, the majority of the months showed slightly positive kurtosis and skewness in the monsoon, according to the probability distribution analysis of the time series data. In general, this indicated that the extreme variations in rainfall were smaller than usual.

Further, a descriptive statistic of monthly and seasonal time series of rainfall from May 2022 to September 2023 at AWS-DPNP station, 2010–2017 at Namsang, and 2014–2021 at Dillighat station is shown in Table 2. It has been observed that, AWS-DPNP and Namsang station have less monsoon variability whereas Dillighat station has high monsoon variability. Also, slightly positive kurtosis and skewness in monsoon and most of the month have been observed by concluding less changes in rainfall from a normal distribution.

It has been observed that the average monthly rainfall in all stations has a similar kind of pattern showing maximum distribution during the month of June to September, which also represents as monsoon season. Also, a maximum rainfall has occurred at AWS-DPNP station in all months as compared to others (Fig. 4). Moreover, monsoon rainfall data has also followed the same pattern over different periods having maximum value at AWS-DPNP station (Figs. 5 and 6).

Table 1 Descriptive statistics of average monthly and seasonal rainfall (mm) of IMD stations (S1 to S4) from 2010 to 2022.

Month/ Season	S1					S2				
	Average (mm)	**Standard deviation**	**CV (%)**	**Skewness**	**Kurtosis**	**Average (mm)**	**Standard deviation**	**CV (%)**	**Skewness**	**Kurtosis**
January	15.33	13.33	86.95	0.60	−0.75	16.02	13.37	83.45	0.48	−1.39
February	30.52	25.38	83.18	0.55	−1.84	39.54	33.74	85.33	0.91	−0.34
March	91.92	56.16	61.09	0.50	−0.58	108.27	64.48	59.55	0.05	−1.58
April	245.70	158.48	64.50	1.05	0.87	269.58	181.71	67.41	0.90	0.01
May	263.71	90.45	34.30	0.62	0.54	305.02	97.71	32.03	−0.09	−0.66
June	338.38	94.41	27.90	1.09	0.73	356.44	107.65	30.20	1.24	0.41
July	380.15	100.86	26.53	−0.30	−0.88	378.72	103.19	27.25	−0.96	−0.32
August	285.26	103.55	36.30	0.58	−0.44	304.49	107.76	35.39	0.46	−0.69
September	229.93	105.23	45.77	−0.87	0.25	241.17	106.43	44.13	−1.26	1.34
October	96.35	51.92	53.89	0.01	−0.70	107.33	67.42	62.82	0.81	1.22
November	13.16	14.44	109.72	1.47	2.48	15.75	18.44	117.08	1.60	2.88
December	9.49	12.26	129.19	1.03	−0.60	11.32	14.65	129.33	1.03	−0.57

(continued)

Table 1 Descriptive statistics of average monthly and seasonal rainfall (mm) of IMD stations (S1 to S4) from 2010 to 2022. (*cont'd*)

Monsoon	1159.13	252.76	21.81	−0.51	2.69	320.21	65.60	19.55	0.48	−0.16
	S3					**S4**				
Month/ season	**Average (mm)**	**Standard deviation**	**CV (%)**	**Skewness**	**Kurtosis**	**Average (mm)**	**Standard deviation**	**CV (%)**	**Skewness**	**Kurtosis**
January	15.24	13.15	86.28	0.64	−0.57	14.64	11.83	80.86	0.65	−0.89
February	30.78	28.40	92.26	0.76	−1.19	37.42	33.75	90.20	0.85	−0.57
March	97.18	61.71	63.50	0.46	−0.82	102.57	57.46	56.02	0.18	−1.38
April	258.48	164.54	63.66	0.75	−0.10	262.58	165.45	63.01	0.64	−0.69
May	287.51	85.95	29.89	0.09	0.18	315.02	97.44	30.93	−0.05	0.33
June	365.42	86.57	23.69	0.84	−0.10	360.36	103.77	28.80	1.63	2.31
July	419.69	117.63	28.03	−0.64	−0.75	400.17	112.60	28.14	−0.79	−0.12
August	309.86	114.50	36.95	0.41	−0.85	325.19	123.23	37.89	0.46	−0.17
September	252.03	112.10	44.48	−0.77	0.69	251.60	110.65	43.98	−1.09	1.09
October	112.64	76.77	68.16	1.05	1.58	114.31	72.21	63.17	0.52	0.20
November	14.80	15.24	102.95	0.70	−0.65	17.01	19.94	117.22	1.29	0.97
December	9.14	14.04	153.64	1.35	0.11	10.68	14.29	133.83	1.09	−0.47
Monsoon	333.97	68.25	20.44	1.13	1.20	334.33	64.58	19.32	0.58	0.30

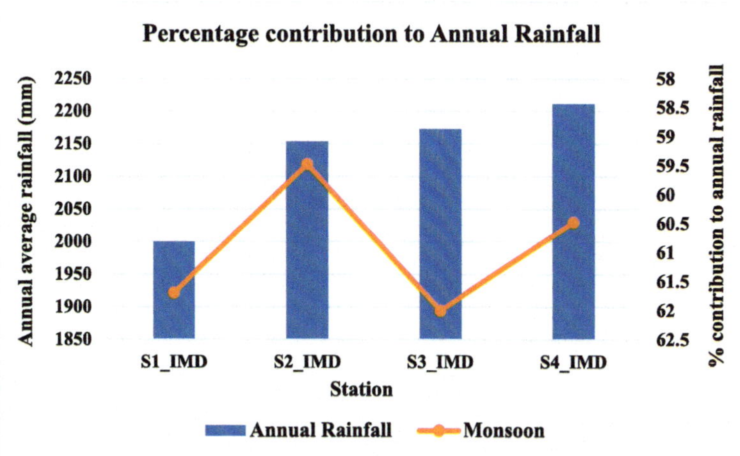

Fig. 3 Monsoon percentage contribution to annual rainfall of IMD station.

4. Mann-Kendall trend analysis for rainfall and temperature

The trend in the time series was found using Mann-Kendall trend analysis. The average rainfall (Table 3) and temperature (Table 4) trend analysis results revealed monthly and monsoon trend patterns from 2010 to 2023. During the period of (2010–2022) at IMD stations S1 of May and June and S3 of June, the average rainfall showed significantly an increasing trend and did not show any significant trend in the other months (Fig. 7A, B and C). Whereas, at AWS-DPNP the average rainfall of May and September showed a decreasing trend over a period from 2022 to 2023 (Fig. 7D and E) and at Namsang a decreasing trend (Fig. 7F) has been observed in December month over a period of 8 years (2010–2017), which has been plotted using linear regresion trend. Increased rainfall intensity can lead to more frequent and severe flooding, affecting water quality and availability. Increased runoff from heavy rains can lead to higher sedimentation and pollution levels in water bodies.

The average temperature trend analysis result (Table 4) showed monthly and monsoon trend patterns over a period from 2010 to 2023 at one station of IMD, AWS-DPNP and Namsang meteorological station. For AWS-DPNP station linear regression trend analysis for a period of May 2022 to Sep 2023 was carried out whereas for other two stations mann-kandall trend test was performed. Based on the table, it can be seen that the

Table 2 Descriptive statistics of average monthly and seasonal rainfall (mm) of gauge stations from different periods.

Month/ season	AWS-DPNP (May2022-Oct23) Namsang (2010-2017)										Dillighat (2014-2021)				
	Average (mm)	Standard deviation	CV (%)	Skewness	Kurtosis	Average (mm)	Standard deviation	CV (%)	Skewness	Kurtosis	Average (mm)	Standard deviation	CV (%)	Skewness	Kurtosis
May	334.46	95.57	28.57	—	—	361.33	202.17	55.95	2.53	7.02	460.00	656.13	142.64	2.61	7.09
June	725.17	178.64	24.63	—	—	226.75	151.13	66.65	−0.41	−1.06	226.75	151.13	66.65	−0.41	−1.06
July	578.52	221.50	38.29	—	—	431.52	182.15	42.18	0.47	−0.85	1262.13	2931.19	232.24	2.81	7.92
August	773.79	192.74	24.91	—	—	398.91	97.07	24.33	−0.76	0.15	329.91	323.80	98.15	2.05	4.82
September	470.56	4.60	0.98	—	—	311.25	158.12	50.80	1.50	2.57	404.24	410.37	101.52	2.12	4.90
October	318.51	—	—	—	—	89.93	47.29	52.58	0.09	−0.46	279.68	353.09	126.25	1.32	0.02
November	0.00	—	—	—	—	9.81	9.70	98.88	1.39	1.21	143.90	321.81	223.63	2.44	5.95
December	0.90	—	—	—	—	10.68	14.29	133.82	1.08	−0.47	4.64	6.37	137.24	0.63	−3.22
Monsoon	2370.41	337.07	14.22	—	—	379.86	57.81	15.22	−0.23	−2.05	236.51	106.81	45.16	−1.18	−0.12

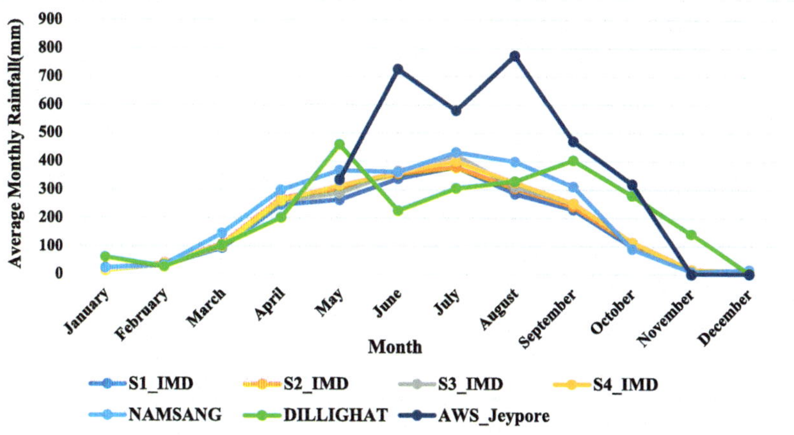

Fig. 4 Average monthly rainfall of all station.

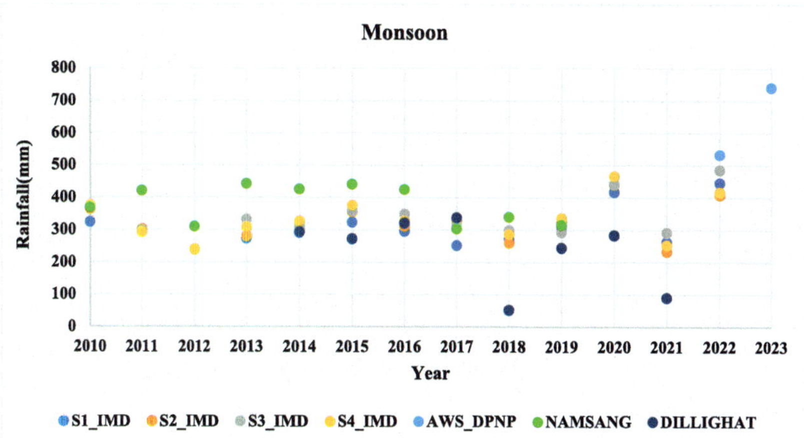

Fig. 5 Monsoon rainfall of all station over different periods.

average temperature at Namsang station decreased in May and did not significantly change in any other months; however, the average temperature at IMD station showed a nonsignificant trend in both monthly and monsoon season (Fig. 8). Similarly, maximum temperature is showing an increasing trend whereas minimum temperature follows a decreasing trend in the study area. Increased temperatures and altered precipitation patterns may stress forest ecosystems, affecting biodiversity and forest health (Fig. 9).

5. Effect of rainfall and temperature variability on land use cover

A land cover map for all the stations with a 2 km buffer was prepared using ArcGIS software for a better understanding of microclimate rainfall and temperature variation on different land use patterns as shown in (Fig. 10). Out of four stations of IMD one station (S1) grid lieswithin the DPNP landscape cover of dense forest and the other three are present near the AWS-DPNP atJeypore station, which generally coversthe agriculture area. The Namsang station has tea-dominatedagriculture plantation, whereas Dillighat also falls within the forest area inthe southern part of the DPNP landscape. The installed AWS-DPNP gauge station represents the Dihing forest landscape (Fig. 11).

The maximum area of evergreen forest was found at station S1 of IMD (7.8 sq km) followed by Namsang (3.9 sq km), and Dillighat(3.9 sq km) within the 2 km buffer zone. Based on the LULC map, the AWS-DPNP station has a maximum forest tree-clad area (1 sq km) besides the main Dihing River flowing in its vicinity (Fig. 12). However, it has been observed that in both monthly and monsoon seasons (Figs. 5 and 6), the average rainfall was maximum at the AWS-DPNP station as compared to S1 of IMD and Namsang which has having high forest cover area within the buffer zone. This can be due to the high resolution of capturing

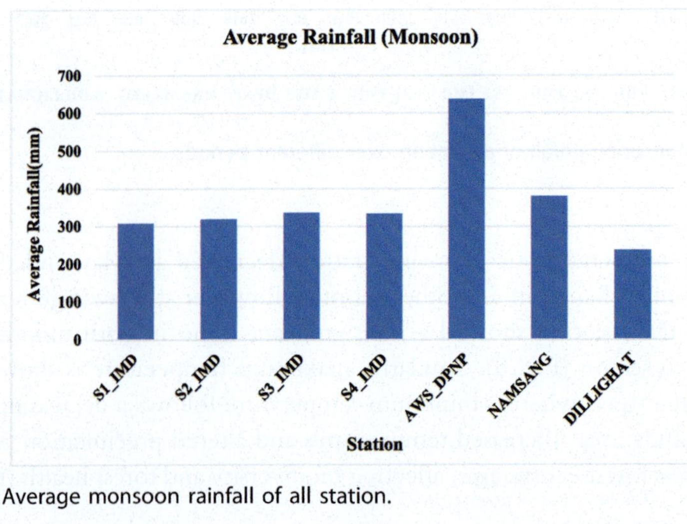

Fig. 6 Average monsoon rainfall of all station.

Table 3 Results of Mann-Kendall trend analysis for rainfall time series.

	S1_IMD (2010–2022)		S2_IMD (2010–2022)		S3_IMD (2010–2022)		S4_IMD (2010–2022)	
Month/ season	Z-Test statistics	Trend	Z-Test statistics	Trend	Z-Test statistics	Trend	Z-Test statistics	Trend
January	1.40	NS	1.65	NS	1.28	NS	1.65	NS
February	1.16	NS	1.28	NS	0.55	NS	1.28	NS
March	−0.79	NS	−1.53	NS	−1.28	NS	−1.16	NS
April	−1.04	NS	−1.28	NS	−1.28	NS	−1.40	NS
May	2.01	+	0.67	NS	1.65	NS	0.43	NS
June	2.14	+	1.28	NS	2.14	+	0.92	NS
July	0.92	NS	0.18	NS	0.31	NS	0.67	NS
August	0.55	NS	0.31	NS	0.67	NS	0.43	NS
September	−0.06	NS	−0.55	NS	0.18	NS	−0.18	NS
October	1.04	NS	1.40	NS	0.43	NS	0.55	NS
November	0.49	NS	0.61	NS	0.85	NS	0.49	NS
December	0.67	NS	0.43	NS	0.06	NS	0.18	NS
Monsoon	0.67	NS	0.67	NS	0.18	NS	1.16	NS

(continued)

Table 3 Results of Mann-Kendall trend analysis for rainfall time series. (*cont'd*)

	AWS_DPNP		Namsang (2010–2017)	Dillighat (2014–2021)		
Month/ Season	**Z-Test statistics**	**Trend**	**Z-Test statistics**	**Trend**	**Z-Test statistics**	**Trend**
January			−0.49	NS	0.56	NS
February			1.36	NS	−0.56	NS
March			−0.62	NS	−0.38	NS
April			−0.12	NS	−0.60	NS
May	−2.00	—	−1.11	NS	0.37	NS
June	0.00	NS	−0.87	NS	−1.11	NS
July	0.00	NS	−0.62	NS	−0.37	NS
August	0.00	NS	−0.37	NS	−0.87	NS
September	−2.00	—	−0.37	NS	0.37	NS
October			−0.87	NS	1.11	NS
November			−0.74	NS	−1.11	NS
December			−1.98	—	−0.49	NS
Monsoon			−0.12	NS	−0.37	NS

NS: Non Significant; − Negetive significant trend; + Positive significant trend Z-test statistics at 95 % level of significance= |1.96|.

Table 4 Results of Mann-Kendall trend analysis for average temperature.

	$S1_T$_IMD (2010-2022)		Namsang (2010-2021)	
Month/Season	**Z-Test statistics**	**Trend**	**Z-Test statistics**	**Trend**
January	−0.55	NS	−1.10	NS
February	−0.18	NS	−0.89	NS
March	−0.31	NS	0.07	NS
April	0.18	NS	1.03	NS
May	−0.79	NS	−2.33	—
June	0.43	NS	0.27	NS
July	1.65	NS	−0.75	NS
August	0.67	NS	0.48	NS
September	0.06	NS	−0.27	NS
October	1.89	NS	0.48	NS
November	1.53	NS	−1.03	NS
December	0.92	NS	−1.44	NS
Monsoon	0.18	NS	−0.27	NS

NS: Non Significant; − Negetive significant trend; + Positive significant trend Z-test statistics at 95 % level of significance= |1.96|.

surrounding land use at the micro level in AWS- DPNP. Further, the S1-IMD station has $0.25^0 \times 0.25^0$ grid information, and the existing AWS station at Namsang records tea plantation land use weather information. Forests contribute to local rainfall cycles through transpiration and moisture recycling, which can moderate extreme changes.

Similarly, the average monthly and monsoon temperature showed less value at AWS-DPNP followed by Namsang and IMD station. The forest land cover shows less variation due to the dense canopy cover which regulates surface temperature through shade and evapo-transpiration.

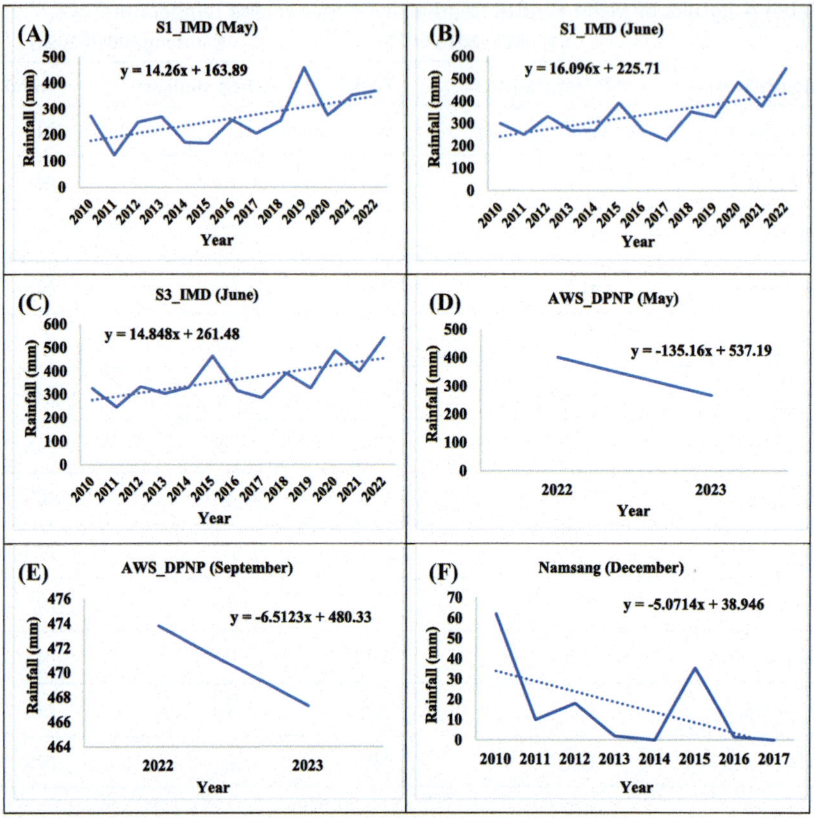

Fig. 7 Linear trend of rainfall time series at IMD(S1 &S3), AWS-DPNP and Namsang. (A) trend of rainfall timeseries at S1_IMD (May). (B) trend of rainfall timeseries at S1_IMD (June). (C) trend of rainfall timeseries at S3_IMD (June). (D) trend of rainfall timeseries at AWS_DPNP (May). (E) trend of rainfall timeseries at AWS_DPNP (september). (F) trend of rainfall timeseries at Namsang (December).

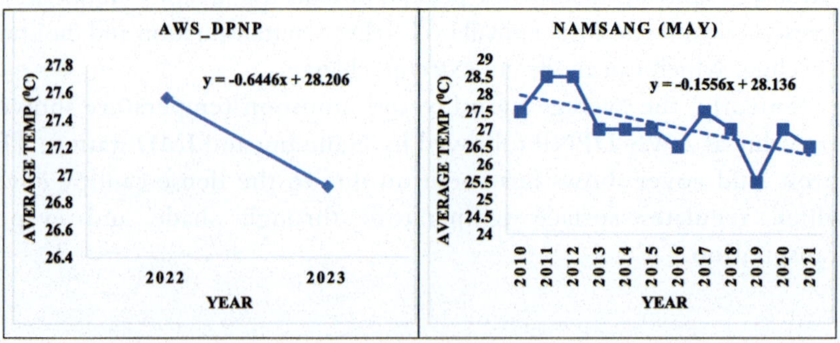

Fig. 8 Linear trend of average temperature time series at AWS_DPNP and Namsang.

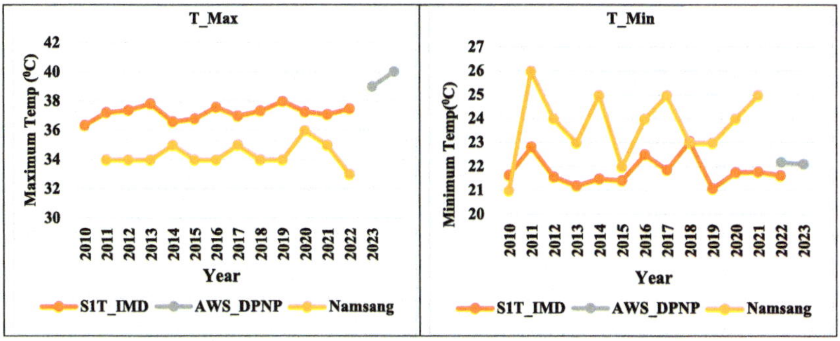

Fig. 9 Linear trend of maximum and minimum temperature.

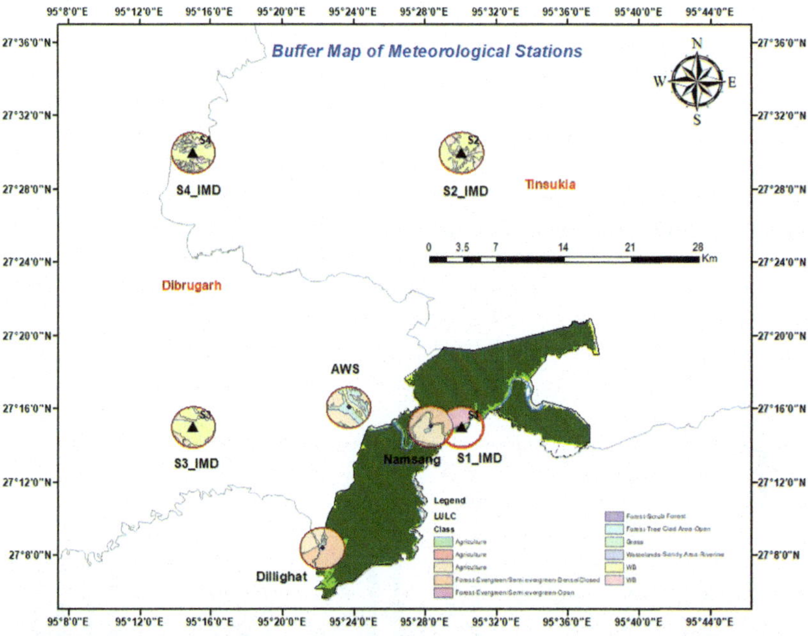

Fig. 10 Buffer map of all stations. Map lines delineate study areas and do not necessarily depict accepted national boundaries.

6. Implications of rainfall and temperature variability for climate-smart land use planning

Rainfall and temperature variability are key indicators of climate change that directly impact how land can be used sustainably. Understanding their implications is essential for climate-smart land use

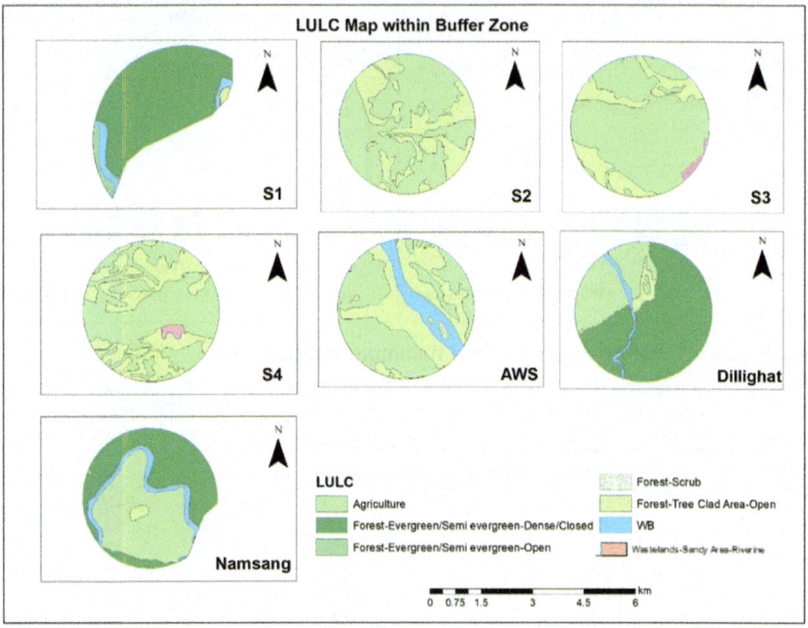

Fig. 11 Land use buffer map of all stations.

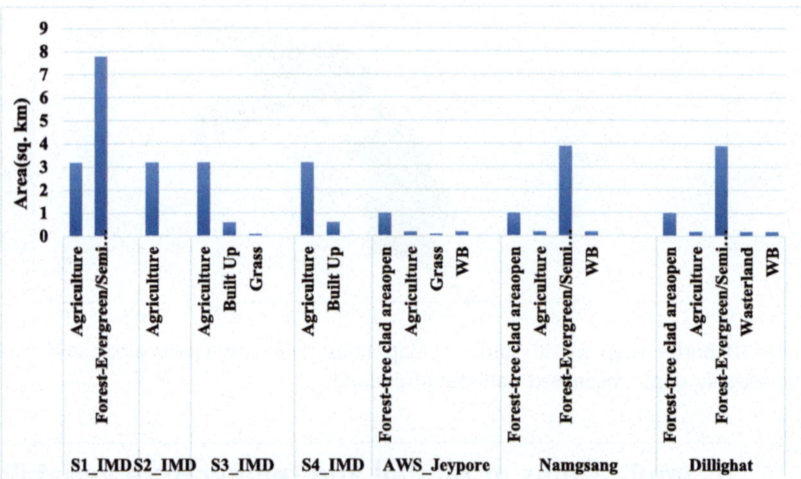

Fig. 12 Areal distribution of land use cover within the buffer at all stations.

planning, which aims to enhance resilience, productivity, and environmental sustainability. While rising temperatures can alter growing seasons, increase evapotranspiration, and raise the risk of wildfires, pests, and disease outbreaks, changes in rainfall patterns affect soil health, crop compatibility, and water availability. Adaptive measures include choosing crops that are adaptable to climate change, improving irrigation systems, preserving natural ecosystems, and addressing land degradation.[28,29] From observations, rainfall and temperature variations are generally lower in forested areas than in nearby non-forest land uses. This difference plays a significant role in shaping local microclimates and guiding climate-smart land use planning.

7. Conclusions

This study used the Mann-Kendall test and linear regression trend test methods to examine how temperature and rainfall affect land cover at monthly and monsoon season scales. According to the z statistics derived from the Mann-Kendall test, the rainfall distribution displayed a moderate degree of variance between monthly rainfalls in the majority of the years and no noticeable trend in the majority of the stations. Further, the variation and inter-relationship between land use cover and climate pattern (temperature and rainfall) with all stations showed a positive trend. The results show an increasing trend at DPNP during monsoon and most of the months due to dense forest cover as compared to other stations. It was also observed that, variation in rainfall and temperature follows the same pattern for all stations except Namsang and Dillighat in a monthly scale due to the different land use cover. Similarly, maximum temperature is showing an increasing trend whereas minimum temperature follows a decreasing trend in the study area. Climate-induced changes may lead to habitat fragmentation, threatening species survival. Therefore, in the event of anticipated climate change, trend analysis of temperature and rainfall patterns may provide information for developing strategic adaptation plans to climatic variability. Moreover, understanding and addressing the impacts of rainfall and temperature variability are crucial for sustainable development in the Dihing Patkai landscape.

Acknowledgment

The authors express their gratitude to the Ministry of Environment and Forests & Climate Change, Government of India, Compensatory Afforestation Fund Management and Planning Authority (CAMPA) for providing financial assistance for this research project. The Director, ICFRE- Rain Forest Research Institute, Jorhat, for his prompt advice, is greatly acknowledged by the authors. The authors would also like to express their gratitude to the Head of the Forest Ecology and Climate Change Division at RFRI Jorhat and Group Coordinator (Research) for their invaluable assistance.

References

1. UNFCCC. *Climate change: impacts, vulnerabilities and adaptation in developing countries;* United Nations Framework Convention on Climate Change: Bonn, Germany, 2007, .
2. Hughes GL, Rao SS, Rao TS. Statistical analysis and time-series models for minimum/maximum temperatures in the Antarctic Peninsula. *Proc R Soc A: Mathematical, Phys Eng Sci* 2006;**463**:241–259. https://doi.org/10.1098/rspa.2006.1766.
3. Jain SK, Kumar V. Trend analysis of rainfall and temperature data for India. *Curr Sci* 2012;**102**(1).
4. Jain SK, Kumar V, Saharia M. Analysis of rainfall and temperature trends in northeast India. *Int J Climatol* 2012. https://doi.org/10.1002/joc.3483.
5. Parthasarathy B, Dhar ON. Trend analysis of annual Indian rainfall. *Hydrol Sci Bull* 1975;**20**:257–260.
6. Prabhu A, Oh J, Kim IW, Kripalani RH, Mitra AK, Pandithurai G. Summer monsoon rainfall variability over North East regions of India and its association with Eurasian snow, Atlantic sea surface temperature and Arctic Oscillation. *Clim Dyn* 2017;**49**: 2545–2556. https://doi.org/10.1007/s00382-016-34445-4.
7. Bora SL, Bhuyan K, Hazarika PJ, Gogoi J, Goswami K. Analysis of rainfall trend using non-parametric methods and innovative trend analysis during 1901-2020 in seven states of North East India. *Curr Sci* 2022;**122**(7).
8. Saini A, Sahu N, Kumar P, Nayak S, Duan W, Avtar R, et al. Advanced rainfall trend analysis of 117 years over west coast plain and hill agro-climate region India. *Atmosphere* 2020;**11**:1225. https://doi.org/10.3390/atmos11111225.
9. Singh P, Kumar N. Analysis of trend and temporal variability in rainfall over Northeast India. *Mausam* 2022;**73**(2). https://doi.org/10.54302/mausam.v73i2.5479.
10. Choudhury BU, Das A, Ngachan SV, Slong A, Bordoloi LJ, Choudhury P. Trend analysis of long term weather variables in mid altitude Meghalaya, North-East India. *J Agric Phys* 2012;**12**(1):12–22.
11. Deb S, Sil BS. Trend analysis of climate variables for the subtropical region of Assam and its neighbouring states in NE India. *Indian J Geo Mar Sci* 2019;**49**(8):1359–1371.
12. Gharphalia BJ, Deka RL, Islam AN, Dutta P, Medhi K. Variability and trends of rainfall events in the Brahmaputra Valley of Assam, India. ISSN: 2319-7706 *Int J Curr Microbiology Appl Sci* 2012;**7**(11). https://doi.org/10.20546/ijcmas.711.215.
13. Gogo K. Analysis of Assam, North East India's rainfall trends. *J EnvirSci* 2022.
14. Chougule PS, Patil SV, Shinde PV, Pratiksha PS. A study trends of rainfall pattern of Maharashtra state based on statistical tolls. *EPRA Int J Agriculture Rural Economic Res (ARER)-Peer-Reviewed J* 2021;**9**(1). https://doi.org/10.36713/epra0813.
15. Abbas A, Waseem M, Ullah W, Zhao C, Zhu J. Spatiotemporal analysis of meteorological and hydrological droughts and their propagations. *Water* 2021;**13**:2237. https://doi.org/10.3390/w13162237.

16. Li S, Liu Y, Elahi E, Meng X, Deng W. A new type of urbanization policy and transition of low-carbon society: A 'local- neighborhood' perspective. *Land Use Policy* 2023;**131**:106709. https://doi.org/10.1016/j.landusepol.2023.106709.
17. Waseem M, Khurshid T, Abbas A, Ahmad I, Javed Z. Impact of meteorological drought on agriculture production at different scales in Punjab, Pakistan. *J Water Clim Change* 2022;**13**:113–124. https://doi.org/10.2166/wcc.2021.244.
18. Elahi E, Li G, Han X, Zhu W, Liu Y, Cheng A, et al. Decoupling livestock and poultry pollution emissions from industrial development: a step towards reducing environmental emissions. *J Environ Manag* 2024;**350**:119654. https://doi.org/10.1016/j.jenvman.2023.119654.
19. IPCC,. Climate change 2022: impacts, adaptation and vulnerability. In *Contribution of working group II to the sixth assessment report of the intergovernmental panel on climate change;* Pörtner H.-O, Roberts DC, Tignor M, Poloczanska ES, Mintenbeck K, Alegría A, Eds.; Cambridge University Press: Cambridge, UK. and New York, NY, USA, 2022, pp. 3056. https://doi.org/10.1017/9781009325844. 3068 pp.
20. Chanapathi T, Thatikonda S. Investigating the impact of climate and land-use land cover changes on hydrological predictions over the Krishna river basin under present and future scenarios. *Sci Total Environ* 2020;**721**:137736. https://doi.org/10.1016/j.scitotenv.2020.137736.
21. Srivastava A, Chinnasamy P. Understanding Declining Storage Capacity of Tank Cascade System of Madurai: Potential for Better Water Management for Rural, Peri-Urban, and Urban Catchments. In *Rural Technology Development and Delivery. Design Science and Innovation;* Chigullapalli S, Susha Lekshmi SU, Deshpande AP, Eds.; Springer: Singapore, 2023, https://doi.org/10.1007/978-981-19-2312-8_14
22. Srivastava A, Maity R, Desai VR. Investigating Spatio-Temporal Trends and Anomalies in Long-Term Meteorological Variables to Determine If Maharashtra is an Emerging Warming State in India. In *Civil Engineering for Multi-Hazard Risk Reduction. IACESD 2023;* Sreekeshava KS, Kolathayar S, Vinod Chandra Menon N, Eds.; Springer: Singapore, 2024, https://doi.org/10.1007/978-981-99-9610-0_25
23. Gosling SN, Arnell NW. A global assessment of the impact of climate change on water scarcity. *Climatic Change* 2016;**134**:371–385. https://doi.org/10.1007/s10584-013-0853-x.
24. Kendall MG. *Rank correlation methods;* Charles Griffin: London, UK, 1975, .
25. Mann HB. Nonparametric tests against trend. *Econometrica* 1945;**13**:245–259.
26. Surendran U, Anagha B, Gopinath G, Joseph EJ. Long-term rainfall analysis towards detection of meteorological drought over Kozhikode District of Kerala. *J Clim Change* 2019;**5**(2):29–34. https://doi.org/10.3233/JCC190010.
27. Hare W. Assessment of knowledge on impacts of climate change - contribution to the specification of article 2 of the UNFCCC: impacts on ecosystems, food production, water and socio-economic systems, Berlin, text; 2003. https://www.wbgu.de/wbgu_sn2003_ex01.pdf.
28. Srivastava A, Maity R, Desai VR. Synthesising causal loop between environmental and compound droughts: A systems-driven framework for climate-resilient governance in monsoon-sensitive agroecosystems. *Next Sustainability* 2025;**6**:100159. https://doi.org/10.1016/j.nxsust.2025.100159.
29. Maity R, Srivastava A, Sarkar S, Khan MI. Revolutionizing the future of hydrological science: Impact of machine learning and deep learning amidst emerging explainable AI and transfer learning. *Applied Computing and Geosciences* 2024;**24**:100206. https://doi.org/10.1016/j.acags.2024.100206.

CHAPTER THIRTEEN

Determination of reservoir water quality parameters in a sugar cane plantation using remote sensing algorithms

Godknows I.J. Mugabe[a] **and Never Mujere**[b,*]
[a]Zimbawe National Geospatial Space Agency, University of Zimbabwe, Harare, Zimbabwe
[b]Department of Geography, Geospatial Sciences and Earth Observation, University of Zimbabwe, Harare, Zimbabwe
*Corresponding author. e-mail address: nemuj@yahoo.co.uk

Contents

Abstract

Climate change is a cross-cutting issue affecting water quality and quantity. Together with human activities, climate variability significantly drives the spatial and temporal dynamics of reservoir water quality in croplands. Nevertheless, little research has been conducted to determine the seasonal distribution of water pollution in sugar cane farmlands. This study assesses the spatial and seasonal variations of chlorophyll-a, turbidity, nitrates, and phosphates in Mteri Reservoir in Zimbabwe. The reservoir lies in a sugar cane *(Saccharum officinarum L.)* estate in the southeastern part of the country. Remotely sensed water quality data were collected from Landsat 8 Operational Land

Advances in Chemical Pollution, Environmental Management and Protection, Volume 15
ISSN 2468-9289, https://doi.org/10.1016/bs.apmp.2026.01.001

Imager imagery. In situ wet season and dry season data were collected during fieldwork. A statistical analysis was conducted to determine the relationship between in situ and remotely sensed water quality data. Results show significant ($p < 0.05$) and strong positive correlations between remote sensing and in situ measurements during the dry season (turbidity- $r = 0.880$), (nitrates- $r = 0.946$), (phosphates- $r_s = 0.744$). Wet season correlations are weak and negative (turbidity- $r = -0.150$), (nitrates- $r = -0.143$), (phosphates- $r = -0.329$). Levels of water pollution were higher in the dry seasons as compared to the wet seasons. In the dry season, there is weak dilution of pollutants due to reduced water levels. Overall, this research demonstrates that remote sensing is a quick and effective approach for estimating water quality in water bodies in intensively cropped fields, especially during dry seasons. Findings help stakeholders to employ appropriate targeted water management intervention strategies, such as precision farming and bioremediation, to mitigate the adverse effects of pollutants, especially during dry seasons.

Keywords: In situ, Remote sensing, Spatial variations, Water quality, Mteri Reservoir, Zimbabwe.

1. Introduction

The dynamics of water quality in sugar cane cropland reservoirs emanating from artificial inputs such as fertilizers, pesticides, and exhaust fuels cause a myriad of socio-economic and biophysical adverse effects. Thus, there is a need to monitor water pollutants in order to employ appropriate and targeted management strategies in farmlands. Nevertheless, there are a few studies conducted to assess the spatial and temporal variation of water quality parameters in irrigated areas. Point or diffuse sources of pollution include industrial wastewater treatment facilities. Non-point sources of pollutants include urban stormwater flows and agricultural[1] the aim of this study is to assess the spatial and temporal variation of a reservoir in a low-lying and highly irrigated area using remote sensing. Few studies have attempted to assess agricultural pollution in Zimbabwe, but little work was conducted in sugar cane fields.[2] Information on the spatial variation of water quality parameters is highly relevant for regional water resource management. Understanding the variations of irrigation water is important for policy and decision making in the quest for efficient and sustainable agriculture.

The irrigation sector produces about 40 % of the global food production. It draws almost 70 % of the global freshwater resources to irrigate 324 million hectares, which contributes 40 % of the total cultivated land. Due to persistent water shortages, Sub-Saharan Africa has the lowest portion of irrigated land of about 3 % compared to other regions. The region has high prevalence rates of malnutrition. About one-fifth of the land with irrigation potential in Sub-Saharan Africa has been equipped. In Zimbabwe, the irrigation potential is approximately 365,624 ha, considering the available renewable water resources. The development and expansion of irrigation in intensive agricultural areas has a significant impact on the water quality of aquatic ecosystems.[3]

Sugar cane (*Saccharum officinarum L.*) production is a major user of global freshwater resources because water use is fundamental to both the growing and processing stages. The main irrigation techniques used in sugar cane farming are sprinkler and flood systems. Flood irrigation is a highly common technique used at large-scale commercial farms of Zimbabwe because of its cost-effectiveness.[4] These systems leach nutrients from the soil into water bodies.

2. Approaches to estimating water pollution

Water quality is traditionally assessed using grab sampling techniques. They involve the onsite collection of water samples and analyzing them in situ or in laboratories within 24 h.[5] These point-based methods are costly, labor-intensive, and time-consuming. They are inadequate to offer information on the spatial coverage of sources of pollution.[6] Traditional methods of water quality assessment often under-sample the spatial and temporal variability needed for the accurate monitoring and management of reservoir resources. Given these critical shortcomings of traditional water quality measurements, remote sensing (RS) offers complementary techniques.

Satellite remote sensing (SRS) is useful in monitoring inland water quality based on correlations between the ratio of reflectance in different parts of the electromagnetic spectrum and the water quality parameters.[7–11] The active optical constituents of water that interact with light and alter the electromagnetic spectrum reflect solar radiation from waterbodies can be measured using remote sensing.

Satellite remote sensing (SRS) offers synoptic, repetitive, consistent, less time-consuming, cost-effective, and comprehensive spatio-temporal views. It offers data acquisition at broad time scales and for a broad spectral range. Remote-sensing-based approaches are considered superior to traditional methods in terms of data acquisition. Free access multispectral satellite datasets provide useful information for decision support and improved environmental monitoring systems (Magome et al., 2002, 2003). Application of geotechnologies, i.e., remote sensing, GIS, and GPS, to monitor or estimate reservoir water quality produces maps to visualise, understand, and interpret. However, geospatial analysis requires experienced personnel to work with spatial data. Some GIS software is expensive to acquire and requires heavy processing computers.[12]

This study quantifies the spatial and temporal variations of water chlorophyll (Chl-a), phosphates, nitrates, and turbidity in Mteri Reservoir in Zimbabwe using satellite remote sensing during the dry and wet seasons. The reservoir lies in a sugar cane plantation. This study area was selected because it covers a wide range of typical Zimbabwean land uses, which are constantly changing. Due to the high volumes of return flow, nutrient concentration from sugar cane fields was assumed to be high and shows seasonal variations. The drivers of water quality dynamics are the type of land use, with expansion of agriculture being the major cause of stream water pollution, and also urban growth.[13] Land use and land cover dynamics have a significant effect on the spatio-temporal variation of diffuse pollutants in water bodies. Different types of crop nutrients from the soil have a significant effect on the variation of water quality parameters. The assessment of the spatial and temporal variation in water quality in the South Eastern Lowveld will contribute to the establishment of key pollutant parameters associated with intensive agriculture in the region.

3. Materials and methods

3.1 Location and description of study area

Mteri Reservoir (Latitude 21° 10′ 0″ S to 21 ° 10′ 60″ S, Longitude 31 ° 33′ 0″ to 31 ° 60′ 0″ E) lies in Hippo Valley Sugar Estate in Chiredzi district in Zimbabwe (Fig. 1). It has a full storage capacity of 75,000 megalitres and occupies an area of approximately 975 ha. Sugar milling companies in the Lowveld and institutions such as EMA rely on point-based techniques in water pollution monitoring[14] that do not have a spatial and temporal coverage. Parameters such as Chlorophyll-a are not measured at these institutions, which is a vital component in water quality studies because it determines the biophysical status and eutrophication level of a water body.

Mteri Reservoir is at an altitude of 326 m in a hot and semi-arid region, which receives a mean annual rainfall of 450 mm. Almost 90 % of the precipitation is received from mid-November to early April. The temperatures

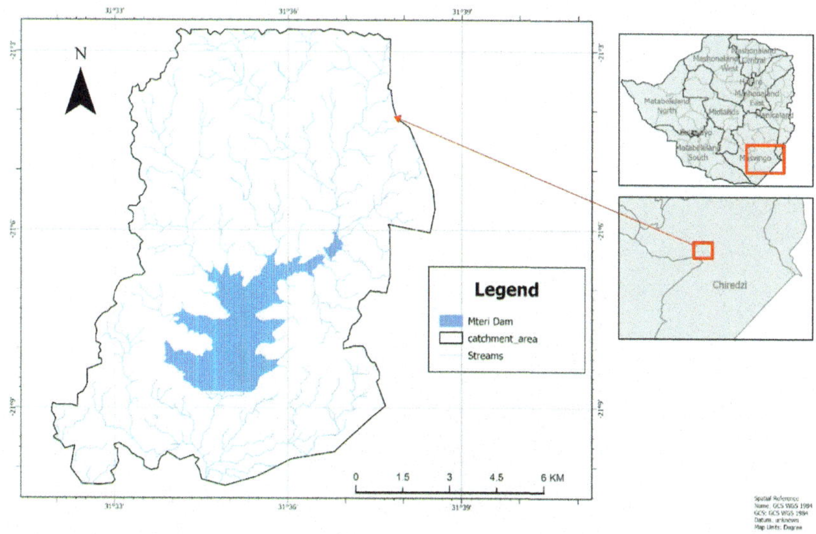

Fig. 1 Location of the Mteri reservoir in Zimbabwe and the water sampling sites. Map lines delineate study areas and do not necessarily depict accepted national boundaries.

range from 8.1 °C in July to 50 °C in January, with a mean of 24 °C to 36 °C. Daily maximum temperature can be as high as 50 °C during hot spells, making the area water-stressed and unfavorable for rainfed cropping activities. Droughts are frequent, and the annual average evaporative demand is ~1500 mm. The concurrent effect of low rainfall and high evaporation resulted in the development of irrigation for crop production.

Mteri Reservoir is fed by canals and streams that flow from the sugar cane plantations. Land use and land cover within the vicinity of the reservoir comprise shrub lands, open grasslands, rain-fed and irrigated croplands (mainly sugar cane), settlements, and effluent treatment plants (Fig. 2).

Forest, shrubland, and grassland cover approximately 29% of the reservoir catchment. Agriculture takes approximately 42% whilst the remaining 28% is for built-up area (Table 1). The settlements consist of flats and gable-type rectangular-shaped houses with asbestos roofs and are linear along major dust roads used in industrial transportation of haulage. There is extensive farming, and other socio-economic activities include recreation and tourism. It is therefore apparent that rainfall and runoff have an impact on the socio-economy of the people living in the catchment. Fig. 3 shows the proportions of land use and land cover in the Mteri reservoir catchment.

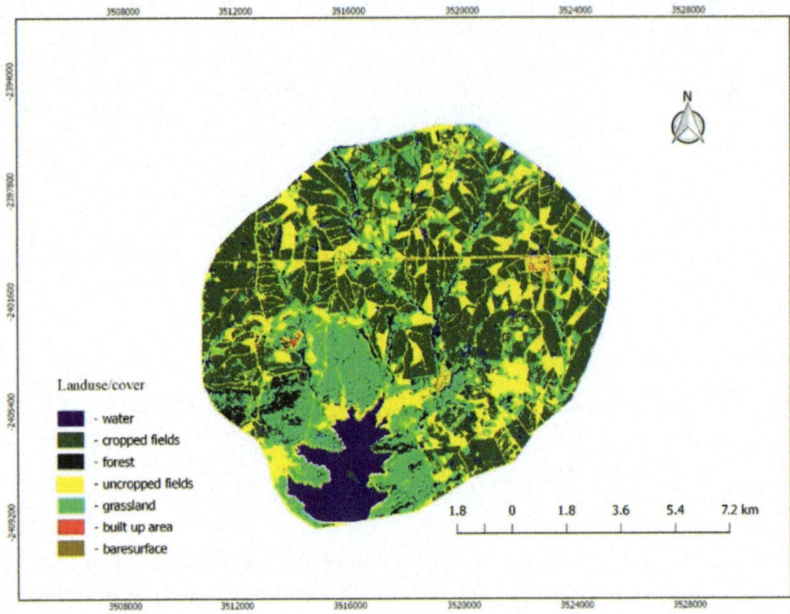

Fig. 2 Land use and land cover in the Mteri Reservoir catchment.

Table 1 Sampling locations, their coordinates, and altitude.

Class	Area (ha)	Percentage (%)
Water	735.48	6.02
Cropped fields	446.21	35.28
Forest	599.76	4.7
Uncropped fields	938.34	7.4
Grassland	2042.1	16.1
Built up area	3597.21	28.4
Bare surface	272.25	2.1
Total	8631.35	100

Mteri Reservoir is surrounded by granite and gneiss rocks and rocks of Karoo sediments. The reservoir catchment lies on paragenic soils called the triangle e-series, developed from the Jurassic basalt, and are moderately shallow to shallow and dark reddish-grained. They are light-textured soils

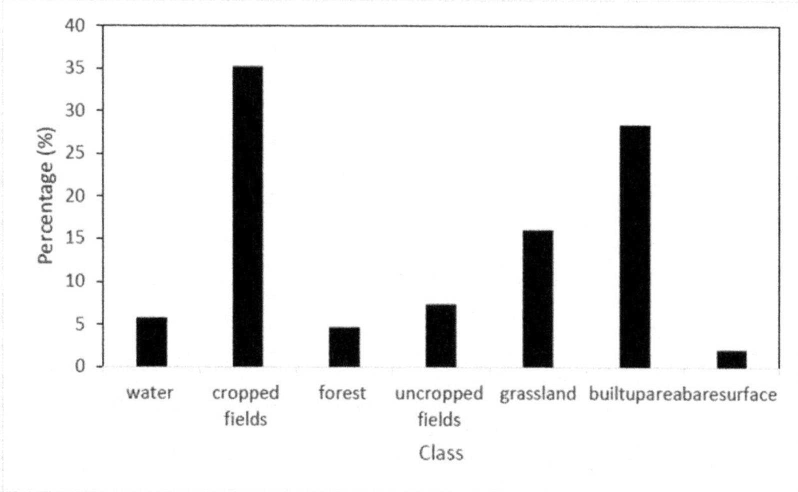

Fig. 3 Land use and land cover proportions in the Mteri reservoir catchment.

ranging from 0.02 to 0.2 mm and fairly flat areas, basalt weathers to produce deep to shallow self-churning black clay soils that are lightly calcareous with soil color varying with the nature of the parent sediments.

3.2 In situ water sample collection

Water quality data were from obtained from the Environmental Management Agency (EMA). The organisation routinely monitors water quality at selected sites in the country. The study area is one of EMA's water quality monitoring points (Table 1).

3.3 Remote sensing of water quality

Cloud-free (less than 0.1 % cloud cover, path/row: 169/75) open-source Landsat 8 OLI images with a 30 m spatial resolution spanning from 2013 to 2019 were downloaded from the USGS website: http://glovis.usgs.gov. The study site was covered by one image scene with less than 0.1 % cloud cover. The satellite images were also split at yearly intervals to gain better temporal variation in the water quality parameters projected into the Universal Transverse Mercator (UTM) coordinate system. Table 2 shows the Landsat image path, raw type, date, cloud cover, and the band used in the estimation of water quality parameters. The metadata file in the image was used to extract sensor calibration information, such as solar zenith angle and Earth–Sun distance. Fig. 4 shows an overview of the pre-processing of the Landsat image.

Table 2 Landsat satellite imagery date used in the study.

Image dates for the dry season	**Image dates for the wet season**	**Spectral bands used**
30 May 2013	6^{t} November 2013	1,2,3,4,5,7
18^{t} June 2014	9 November 2014	1,2,3,4,5,7
20 May 2015	12 November 2015	1,2,3,4,5,7
6 May 2016	30 November 2016	1,2,3,4,5,7
25 May 2017	16 October 2017	1,2,3,4,5,7
29 June 2018	20 November 2018	1,2,3,4,5,7
	24 February 2019	1,2,3,4,5,7

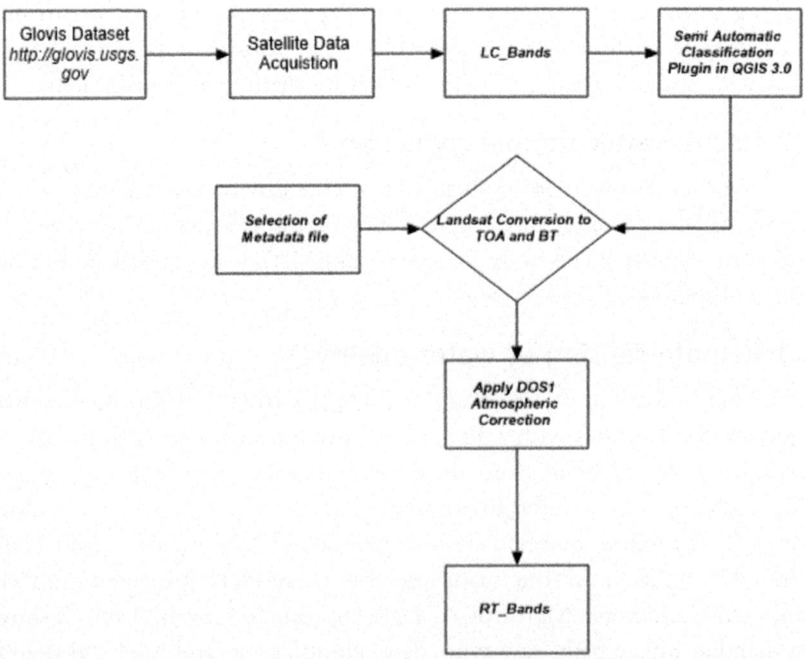

Fig. 4 Pre-processing of landsat 8 OLI images [Note that: DOS1 refers to dark object subtraction, LC_Bands – Digital Number Images, RT_Bands – Reflectance Images].

The conversion of Landsat digital numbers to top of the atmosphere reflectance and brightness was done using the dimensionless equation:[10]

$$\rho TOA = \pi L_\lambda d^2 / ESUN\lambda Cos\Theta s \quad (1)$$

Where: L_λ is the spectral radiance at the sensor, d is the Earth–Sun distance in astronomical units, ESUNλ is the mean solar exoatmospheric irradiance for each band, and Cos Θ_s is the solar zenith angle in degrees.

Estimation of the water quality parameters was done for the dry and wet seasons to capture the variability of the pollutants. Table 3 shows the algorithms used for estimating water quality parameters from Landsat images.

3.4 Test for relationships

Water quality data were tested for normality using the Kolmogorov-Smirnov (KS) test. All data followed a normal distribution ($p > 0.05$).

Table 3 Algorithms for the retrieval of water quality parameters from Landsat images.

Sampling point	X	Y	Altitude (m)
SS1	352349	7663199	362
SS2	353079	7663304	370
SS3	352751	7663126	390
SS4	352858	7663334	371
SS5	352956	7663615	409

NB. *B1 and B7 refer to Bands 1–7 of Landsat 8 OLI.*

Table 4 Algorithms used to extract water quality paramters.

Parameter	Algorithm	Units
Chlorophyll-a	Chl-a = (486*NDVI+44)*−0.001 Where: NDVI = (B5-B4)/(B5+B4)	ug/l
Turbidity	Tb = 5.0819*(B1/B3)^−1.0125	NTU
Phosphates	Ln (TP) = 4.334-4.594(B1/B2)+1.103(B4/B7)	mg/l
Nitrates	TN = 1.047532-54.928(B2)−46.2947(B3) + 120.8943 (B4)−19.223(B5)	mg/l

Thus nitrates: $p = 0.290$, $df = 7$, $n = 7$, phosphates: $p = 0.296$, $df = 7$, $n = 7$ and turbidity: p value $= 0.305$, $df = 6$, $n = 7$. Pearson's correlation analysis was performed to establish the strength and direction of the relationship between remotely sensed water quality variables, as dependent variables, and field-based water quality variables, as independent variables Table 4.

4. Results and discussion

4.1 Variation of Chlorophyll-a

Chl-a varied from −0.261 μg/l to 0.254 μg/l. The reservoir's edges show relatively lower concentration of Chl-a, ranging between −0.433 μg/l and −0.261 μg/l in all of the months throughout the years. The Chl-a content in the reservoir is more uniform in cool dry periods, whilst in the hot period it shows a more heterogeneous pattern. The concentration was high from 2017 to 2018 in the dry season, ranging between 0.082 and 0.254 μg/l, compared to the wet season of the same period, when the concentration ranged between −0.090 μg/l and 0.082 μg/l. Fig. 5 shows the monthly distribution of chlorophyll-a from 2013 to 2018.

4.2 Turbidity

The results show that Mteri Reservoir is more turbid in the cool season, with NTU ranging above 5.5. In the hot season, turbidity is more varied, with the years 2015 and 2016 reaching maximum NTU (Fig. 6). The years 2017 and 2018 of the hot season show that Mteri Reservoir was less turbid below EMA standards (5 NTU), ranging between 2.5 and 3.5 NTU. The edges of the reservoir are highly turbid across the period, and the dispersal of higher concentration occurs starting from the centre of the water body towards the edges for the years 2013–2014 and 2017–2018. However, in 2015 and 2016, higher concentration of turbidity was noted above 5.5 NTU.

Fig. 6 shows the mean concentration of turbidity in the Mteri Reservoir. The highest mean concentration of turbidity is observed between January and April, ranging between 10 NTU and 25 NTU. A sharp decrease between late-April and mid-May occurred to 1.15 NTU. Between late-May and December, the mean concentration shows a steady increase, ranging from 2 NTU to 5.62 NTU.

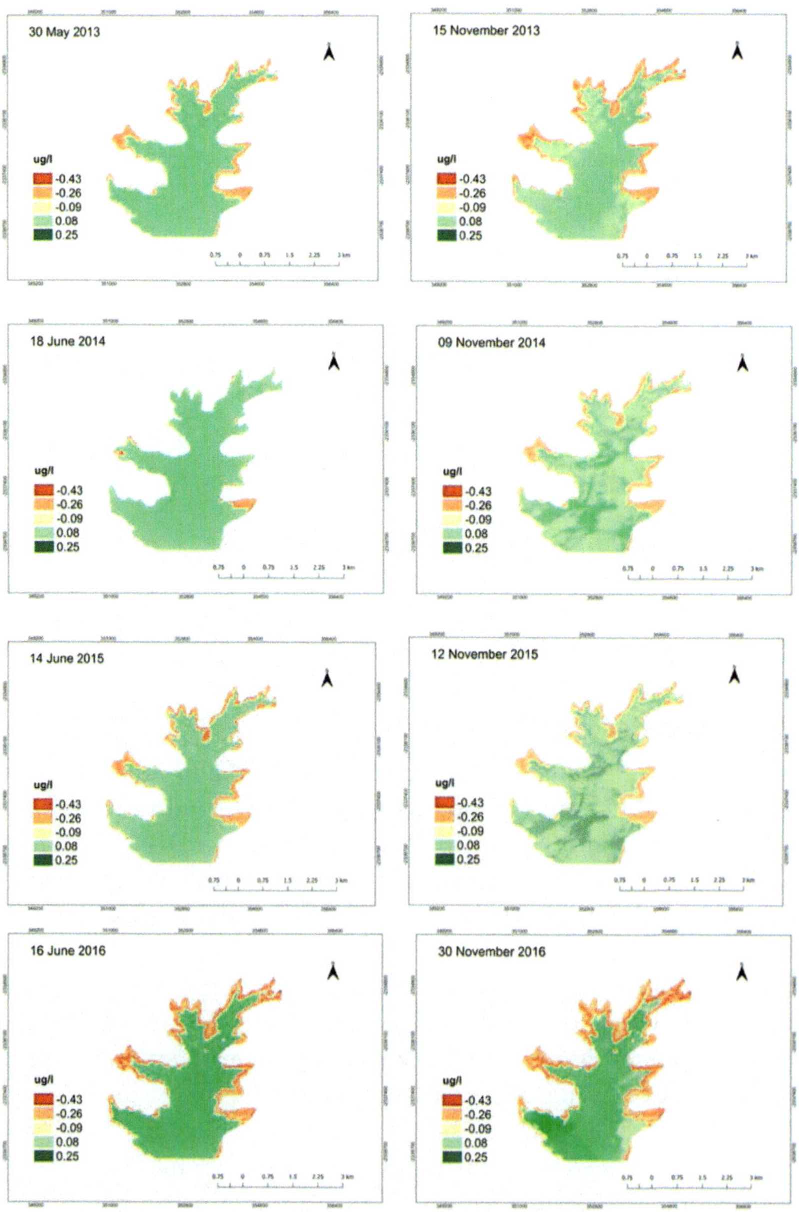

Fig. 5 Monthly distribution of chlorophyll-a from 2013 to 2018.

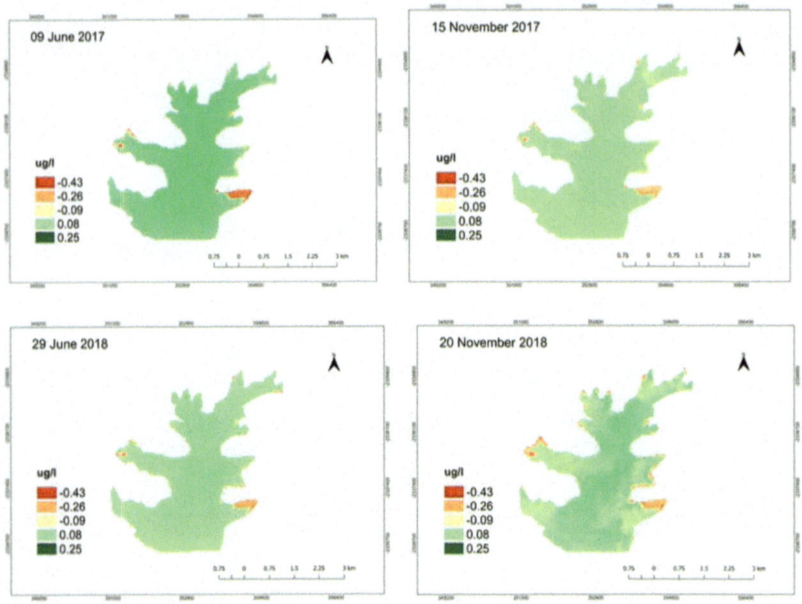

Fig. 5 *(continued)*

4.3 Phosphates

The results show that phosphates were increasing up to June 2016 in the cool season, with more variation between 2.5 mg/l and surpassing maximum levels at 5.5 mg/l according to the EMA standards (Fig. 7). The hot season shows a non-uniform variation in concentration ranging between 0.50 mg/l and 5.5 mg/l. There was a notable decrease in pollution across the reservoir surface from 2013 to 2018, from about 2.75 mg/l to 3.88 mg/l over time. The annual cycle of the diatoms also showed a decrease between 2017 and 2018 in the November months, ranging between 2.5 mg/l and 4.5 mg/l, then a sharp increase in 2018, with the reservoir surface having a greater concentration up to 5.5 mg/l.

Fig. 7 also shows that the highest mean concentration was observed between January and May, ranging from 0.03 to 0.10 mg/l, and a lower concentration in May and October. The concentration then spikes upward between November and December.

4.4 Nitrates

In both the wet and dry seasons, nitrate levels ranged between 3.28 mg/l and 5.21 mg/l (Fig. 8). However, in the hot season, concentration

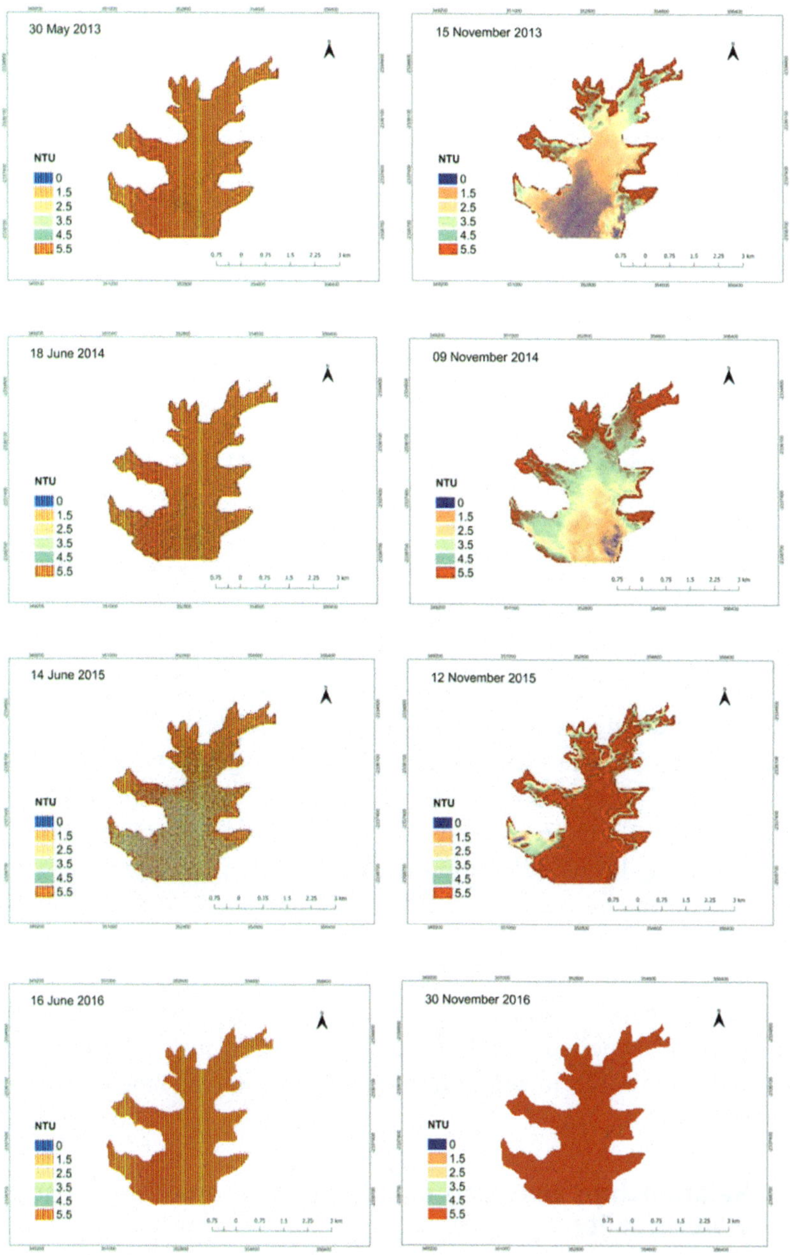

Fig. 6 Variations of turbidity from 2013 to 2018.

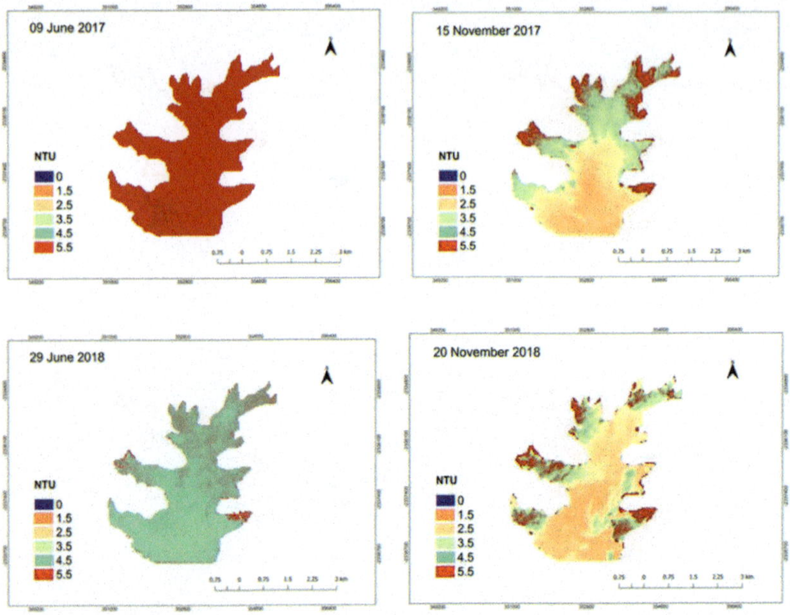

Fig. 6 *(continued)*

increased, reaching up to 5.21 mg/l from 2015 to 2018, below the EMA standards of <10. The concentration generally ranged between 3.38 mg/l and 7.05 mg/l in all seasons from 2013 to 2018. However, concentration increased in 2019 in the reservoir source in the northern parts, reaching up to 10.70 mg/l. The concentration is more varied in the northern parts of the reservoir and becomes lower (3.38 mg/l) moving towards the dam wall on the southern part.

Fig. 8 also shows the monthly mean concentration of nitrates. Higher mean concentration of nitrates is observed in January, April, and December for the 7-year period, reaching as high as 0.50 mg/l. February and December show the lowest concentrations of nitrates, ranging between 0.20 mg/l and 0.27 mg/l.

4.5 Relationships between measured and remotely sensed water quality

Fig. 9 shows that field measurements in the dry season (a), ($r = 0.946$, $p < 0.05$, $n = 7$) have a strong positive correlation with the remotely sensed measurements. A negative relationship was observed in the hot season ($r = -0.143$, $p < 0.05$, $n = 7$).

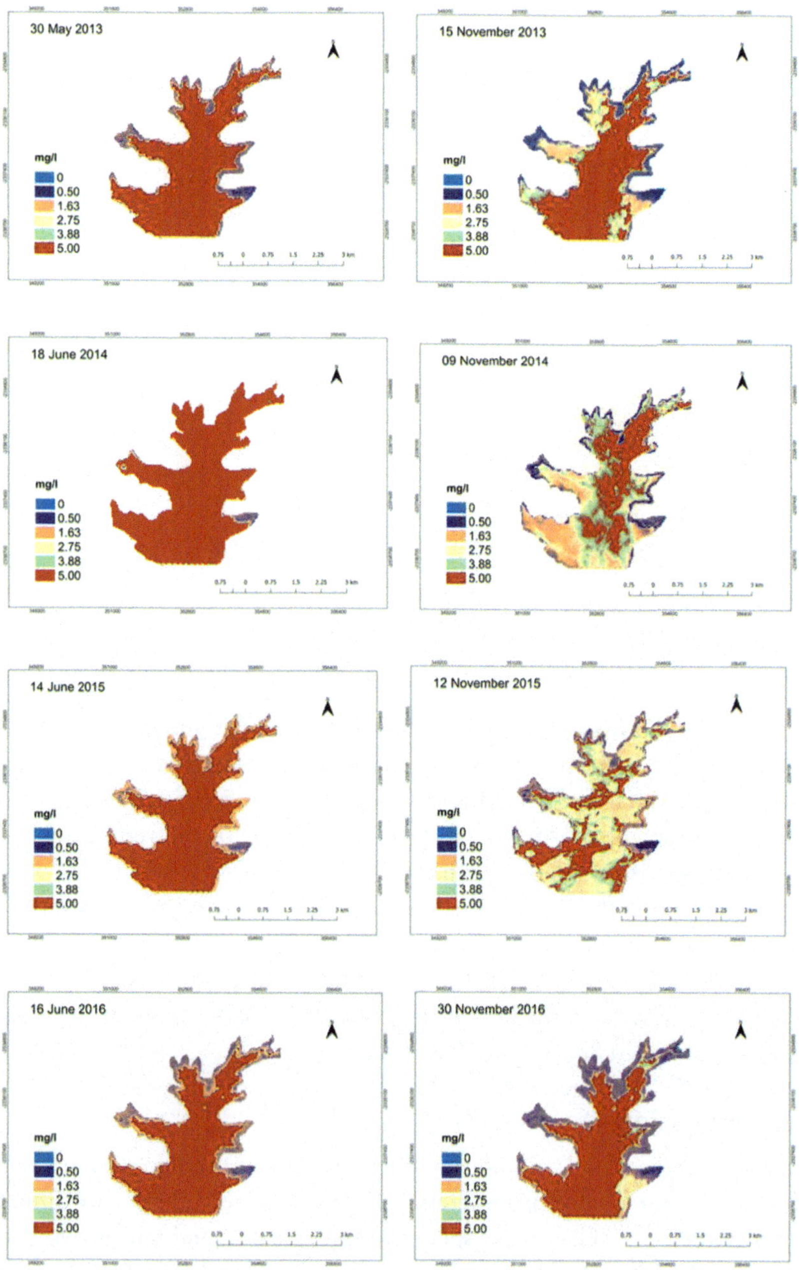

Fig. 7 Monthly distribution maps of phosphates from 2013 to 2018.

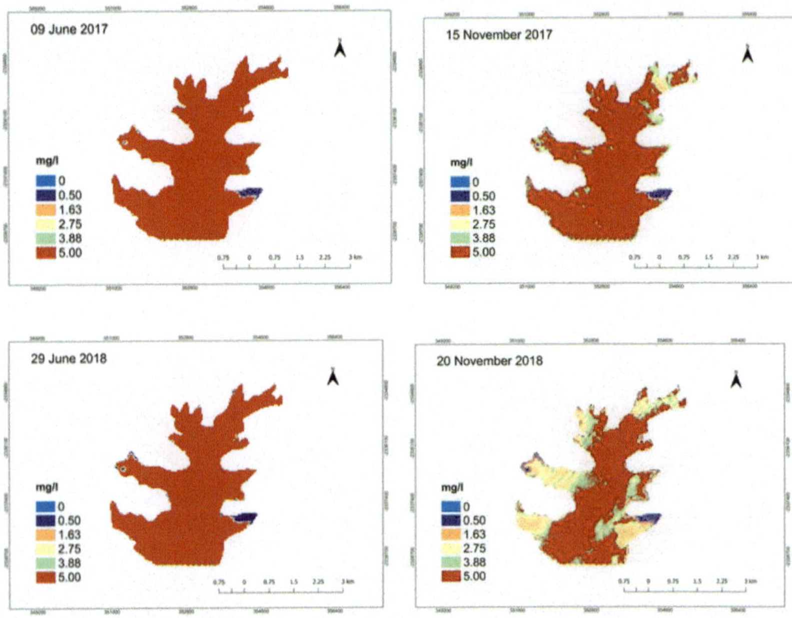

Fig. 7 *(continued)*

Fig. 10 shows that field phosphate measurements ($r = -0.329$, $p < 0.05$, $n = 7$) are negatively correlated with the remotely sensed measurements in the hot season. The results show a strong correlation between field measurements ($r = 0.744$, $p < 0.05$, $n = 7$) with the remotely sensed measurements in the cool season.

Fig. 11 shows that field-measured turbidity is negatively correlated with the remotely sensed measurements ($r = -0.150$, $p < 0.05$, $n = 7$), in the hot season. Results show that field-measured turbidity has a very strong positive correlation with the remotely sensed measurements in the winter (dry) season ($r = 0.88$, $p < 0.05$, $n = 7$).

4.6 Discussion

Previous studies show significant correlations between in situ measurements of water quality parameters and remotely sensed data in different water bodies.[15–19] The visible spectrum (blue, green, and red) provides the best correlations and also the near infrared spectral ranges. Therefore, Landsat conversion to TOA reflectance and brightness temperature was performed to capture the accuracy and precision of the quantitative data obtained using remote sensing.

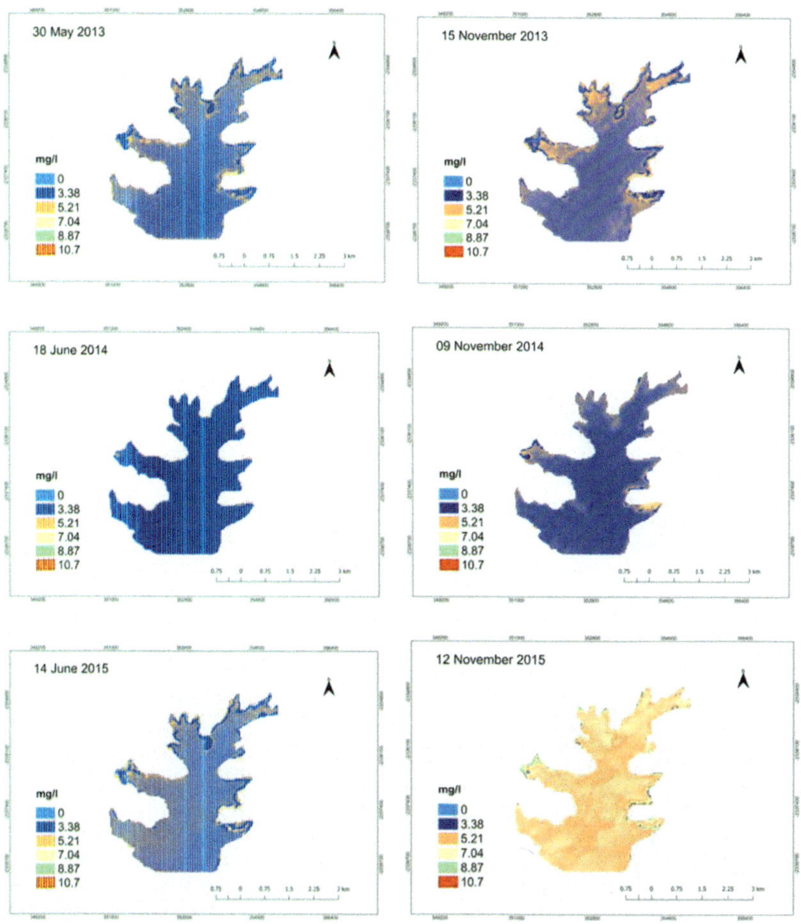

Fig. 8 Monthly distribution of nitrates from 2013 to 2019.

Studies by Lim et al. and Masocha et al.[10] showed that NIR-Red band ratios, which are Bands 4 and 5, showed greater correlation with the measured in situ chlorophyll ($R = 0.85$, $p < 0.05$). Therefore, NIR-Red bands 4 and 5 were used to estimate chlorophyll using NDVI as a proxy by normalising with coefficients. Remote sensing of turbidity has been found to be correlated to the red spectrum of the satellite bands for low to moderate values.[20]

The spatio-temporal variation in chlorophyll-a concentration within the reservoir is more likely to be explained by the variation in the land use activities within the agricultural region and less by the season. The variation

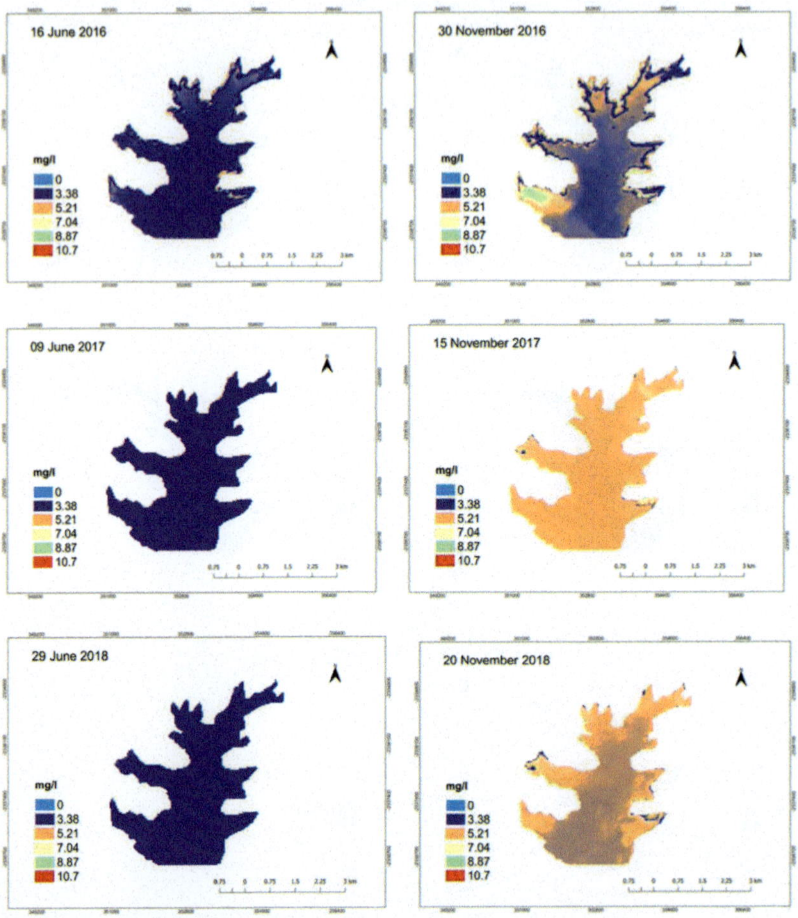

Fig. 8 (*continued*)

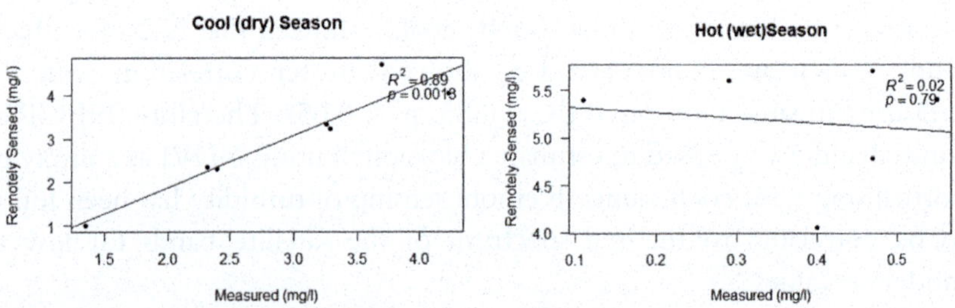

Fig. 9 Relationships between field-measured and remotely sensed nitrates.

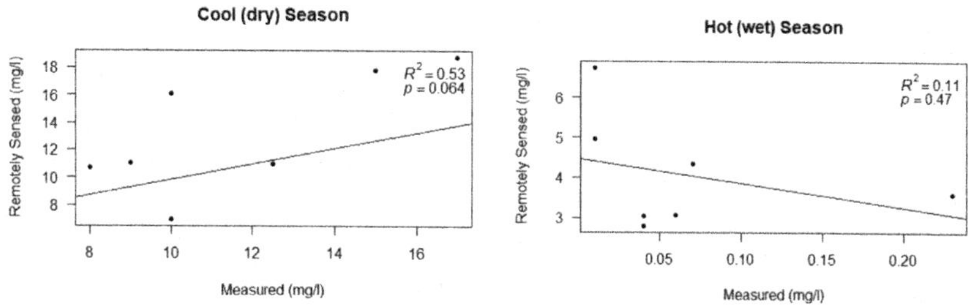

Fig. 10 Relationships between field-measured and remotely sensed phosphates.

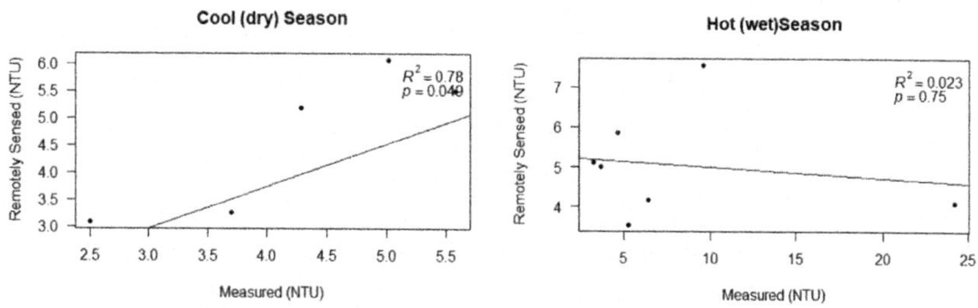

Fig. 11 Relationships between measured and estimated turbidity.

of Chl-a concentration is most likely ascribed to the land conversion to new plantation fields, which increases inorganic fertilizer application. The variation in the land use for sugar cane plantation affects the chlorophyll content in the lake. With more fields opened up for sugar cane growing, more surface water is lost through irrigation, flowing its way to the lake. Therefore, this causes greater net primary productivity in the reservoir, promoting chlorophyll to grow through the inflow of leached dissolved fertilizers into the reservoir, reducing the effect of seasonality.

In land lakes, there is a high temporal water level fluctuation; therefore, there is constant turnover, allowing oxygen to recharge below the photic zone, increasing surface nutrients.[21,22] This explains the high fluctuation in the concentration of Chlorophyll-a in both seasons. The continuous supply of agricultural runoff throughout the year allows nutrient enrichment in the reservoir waters, further explaining the similar pattern of change in both seasons. The differences in the chlorophyll-a concentration at the edges of the water body show the variation in the water levels, explained by the shallow water and even sand surface, resulting in negative values of chlorophyll.

There was a considerable difference between the summer (wet) and winter (dry) seasons in turbidity units and total phosphate concentration in mg/l within the reservoir. The reservoir is more turbid in the winter (dry) season compared to the summer (wet) season, where oscillating patterns are observed. This could be explained by the reverse effect of the rain season, where more water is available to dissolve solids in the reservoir. Whereas in the winter season, less water is available to dissolve the materials from the agricultural region, with 90 % of the surface water recharge being agricultural runoff from sugar cane plantations.

Results show that phosphates are also more varied in summer because discharge variations are dependent on the weather and seasonal variations. Turinayo revealed that in River Musamya in Uganda, higher concentrations of total phosphates are due to the nitrogenous (nitrogen, asparagine, glutamine, aspartic) and phosphate (phosphorus acid) matter found in sugar cane juice. These compounds come along with raw sugar cane from the estate fields and are processed into molasses to be used as wastewater during the post-sugar process. Van Der Laan et al. found that in the northern sugarcane-producing regions of South Africa, the flush phenomenon is a major cause for higher concentrations of nutrients where loadings occur during the first few storm events of a season. Inorganic fertilizer Diammonium phosphate (DAP) and Urea, Monoammonium phosphate (MAP), and Muriate of Potash (MOP) application occurs at the onset of the dry season between June and September under intense irrigation during replanting.

Nitrate variation had similar variations in both the dry and wet seasons. This can be explained by lower application rates of nitrate fertilizers and more of phosphate-rich fertilizers, which promote the growth of crops. The variation is more homogenous in both seasons compared to the other months, with the months of 2015 and 2017 of the hot seasons and 2018 of the cool seasons being excluded from the trend due to the high values.

Based on the measured and remotely sensed water quality parameters, the results show significant positive relationships in the cool season (Turbidity, $r = 0.880$), (Nitrates, $r = 0.946$), and Phosphates, $r = 0.744$. Whereas in the hot season, negative relationships were observed between measured and remotely sensed water quality parameters (Turbidity, $r = -0.150$), (Nitrates, $r = -0.143$), and Phosphates, $r = -0.329$. The seasonal difference between the measured and the remotely sensed data can be explained by the effect of land use and land cover dynamics in the reservoir catchment, where discharge and recharge rates are mainly determined by the availability of water, and also the irrigation return flows from the agricultural fields.

Previous studies show that developing a climate-resilient watershed is an important adaptation strategy to cope with water stress under changing climatic scenarios.[23] In the 21st century, climate change has emerged as a significant challenge to agriculture, freshwater resources, and the food security of billions of people in the world. Water-smart agricultural technologies integrate traditional and innovative practices, technologies, and services that are relevant for a particular location to adapt to climate change and variability. The study findings highlight key interactions between land use, seasonal variations, and water quality in sugarcane-dominated regions. These insights can significantly contribute to climate-smart water management strategies by improving monitoring and guiding targeted interventions. The strong correlation between in situ measurements and remotely sensed data (especially in the cool season) suggests that remote sensing can serve as a cost-effective and continuous monitoring tool. Using satellite-derived NDVI-based chlorophyll estimates, managers can track nutrient loading trends and assess pollution hotspots in near real-time. The integration of Landsat TOA reflectance with water quality parameters enables predictive modeling, improving early warning systems for nutrient enrichment events. The study confirms that phosphates and nitrates are significantly influenced by agricultural runoff, particularly from sugarcane estates. The negative correlation between remotely sensed and measured water quality parameters during the dry season suggests that the concentration of pollutants increases due to low water availability and reduced dilution. The first storm events flush accumulated nutrients into the reservoir, leading to episodic spikes in nutrient levels (flush phenomenon). Given that land use changes (e.g., expansion of sugarcane plantations) contribute to nutrient loading, policies should encourage Land use planning regulations to prevent excessive expansion into ecologically sensitive areas. The findings provide strong evidence for integrating remote sensing into water quality management in agricultural reservoirs. By combining precision agriculture, sustainable irrigation, and seasonal nutrient control, sugarcane-dominated regions can minimize pollution risks, ensuring long-term water security in the face of climate change.

Climate-smart water management in agricultural reservoirs is essential for enhancing resilience in Chiredzi District, a semi-arid region heavily reliant on irrigated agriculture supported by reservoirs such as Manjirenji and Lake Mutirikwi. Climate change has led to unpredictable rainfall patterns and shorter rainy seasons, resulting in erratic reservoir inflows and increased pressure on irrigation systems, particularly for sugarcane estates and outgrower schemes. This situation calls for dynamic, forecast-based reservoir operations using seasonal and real-time climate data. As irrigation demand grows due to unreliable rainfall, there is a pressing need to adopt water-efficient technologies like drip irrigation and to reduce conveyance losses through canal lining or piping. Additionally, sedimentation caused by catchment degradation from deforestation and overgrazing is reducing reservoir capacity, highlighting the need for integrated watershed management practices such as reforestation and gully reclamation. With water scarcity comes the risk of conflict over access, especially between commercial estates, smallholders, and conservation areas like Gonarezhou. Strengthening governance structures through inclusive water user associations can help ensure equitable distribution and conflict resolution. A significant challenge remains the lack of hydrological data and monitoring systems for many small dams. Ultimately, effective climate-smart water management in Chiredzi will depend on coordinated policy efforts, stakeholder collaboration, and sustainable use of both reservoir infrastructure and surrounding ecosystems to support agricultural productivity and community livelihoods in a changing climate.

The findings have important implications for improving water quality monitoring and targeted management practices, especially in sugarcane-dominated regions like Chiredzi, where pollution risks increase during the dry season. As water availability diminishes in the dry months, the concentration of pollutants such as agrochemicals, fertilizers, and sediment runoff rises significantly due to reduced dilution capacity in reservoirs and canals. These findings emphasize the need for a seasonal approach to water quality monitoring, with increased surveillance during the dry season when pollutant loads are likely to peak.

In sugarcane-growing areas, the use of fertilizers and pesticides is intensive, and runoff from fields can introduce nutrients like nitrates and phosphates into water bodies, promoting algal blooms and degrading aquatic ecosystems. Targeted monitoring using remote sensing technologies, combined with in-situ testing, can help detect changes in water color, turbidity, and temperature that signal contamination. This real-time data can inform adaptive reservoir management, allowing water managers to adjust flows, flush systems when needed, or restrict irrigation at critical times.

The findings also support the implementation of buffer zones, such as vegetated strips or constructed wetlands, around irrigation canals and reservoirs to filter runoff before it reaches open water. Moreover, water quality data can guide farmer education programs to promote the responsible use of agrochemicals and soil conservation techniques. Establishing community-led water monitoring committees can enhance local stewardship and create feedback loops between farmers, water authorities, and environmental regulators.

In summary, the findings advocate for a proactive, data-driven approach to managing water quality in sugarcane-dominated landscapes, where pollution pressures are highest during the dry season. Integrating environmental monitoring with farm-level management and policy interventions can help safeguard water resources while sustaining agricultural productivity.

5. Conclusions

This study utilized mathematical algorithms for water quality monitoring in an inland reservoir using Landsat 8 OLI satellite imagery from 2013 to 2018. Intensive sugar cane agriculture was the main driver of water quality dynamics for Mteri Reservoir. All the water quality parameters varied spatially and temporally in the reservoir. Based on the research findings, remote sensing algorithms can be effective in determining Chl-a, phosphates, nitrates, and turbidity in reservoirs. The use of remote sensing aids in timely targeted interventions in problems affecting water resources. The findings underscore the urgent need for climate-smart, data-driven water management in sugarcane-dominated regions like Chiredzi, especially during the dry season when pollution risks intensify. Reduced water availability increases the concentration of agrochemical pollutants, making seasonal water quality monitoring critical. Integrating remote sensing with in-situ measurements enables real-time tracking of water quality, guiding adaptive management actions such as controlled flushing and irrigation scheduling. Implementing buffer zones, promoting responsible agrochemical use, and engaging local communities through water monitoring committees can enhance pollution control. Ultimately, a coordinated approach combining technology, policy, and stakeholder participation is essential for protecting water resources and sustaining agricultural productivity under changing climatic conditions.

Declaration of generative AI and AI-assisted technologies in the writing process

Authors declare non-use of generative AI in scientific writing upon submission of this chapter.

References

1. Tafangenyasha C. *The influence of point source and non-point source pollution in the Runde river, Zimbabwe;* Department of National Parks and Wildlife Management: Harare, Zimbabwe, 2017.
2. Tafangenyasha C, Dzinomwa T. Land-use impacts on river water quality in Lowveld sand river systems in south-east Zimbabwe; 2005.
3. Liu W, Chan W. Assessing the influence of nonpoint source and discharge changes on water quality in a tidal river estuary using a three-dimensional model; 2014. https://doi.org/10.3390/environments1020157.
4. Nhiwatiwa T, et al. Impact of irrigation based sugarcane cultivation on the Chiredzi and Runde Rivers quality, Zimbabwe. *Sci Total Environ* 2017. https://doi.org/10.1016/j.scitotenv.2017.02.155.
5. EMA: Environmental Management Agency – Water testing services, available at: http://www.ema.co.zw/index.php/2014–06-1203–49-33/2014–06-12–03-50–38/ema-services/50-water-testing-services.html (last access: 22 April 2015), 2012.
6. Gumindoga W, Rientjes Htm, Shekede Md, Rwasoka Dt, Nhapi I, Haile At. Hydrological impacts of urbanisation of two catchments in Harare, Zimbabwe. *Rem. Sens* 2014;**6**(12):12544–12574. https://doi.org/10.3390/rs61212544.
7. Chawira M, Dube T, Gumindoga W. Remote sensing based water quality monitoring in Chivero and Manyame lakes of Zimbabwe. *Phys. Chem. Earth* 2013. https://doi.org/10.1016/j.pce.2013.09.003.
8. Kibena J, Nhapi I, Gumindoga W. Assessing the relationship between water quality parameters and changes in landuse patterns in the Upper Manyame River, Zimbabwe. *Phys. Chem. Earth Parts A/B/C* 2014;**67–69**(0):153–163.
9. Dlamini S, et al. Assessing the feasibility of integrating remote sensing and in-situ measurements in monitoring water quality status of Reservoir Chivero, Zimbabwe. *Physics and Chemistry of the Earth. Elsevier Ltd.* 2016. https://doi.org/10.1016/j.pce.2016.04.004.
10. Masocha M, Murwira A, Magadza CHD. Remote sensing of surface water quality in relation to catchment condition catchment condition in Zimbabwe; 2017. https://doi.org/10.1016/j.pce.2017.02.013.
11. Gumindoga W, Dube T, Chawira M. An integration of GIS and remote sensing with field measurements in monitoring water quality in ReservoirMutirikwi. *Masvingo Province PROPOSAL Submitted to Masvingo City Council* 2018.
12. Li YL, Liu K, Li L, Xu ZX. Relationship of land use/cover on water quality in the Liao River basin, China. *Procedia Environ. Sci.* 2012;**13**(0):1484–1493.
13. Hossain S. Impact of land use change on stream water quality: a review of modelling approaches. India. *J Res Eng Appl Sci* 2017.
14. EMA: Ambient water quality monitoring; 2014, available at: https://www.herald.co.zw/ambient-water-quality-monitoring/ (last access: 20 March 2015), 2014.
15. Dlamini S. *Evaluation of the water quality status in reservoir Chivero, Zimbabwe;* University of Zimbabwe, 2012.
16. Kapalanga TS. Faculty of engineering assessment and development of remote sensing based algorithms for water quality monitoring in olushandja reservoir, north-central namibia algorithms for water quality monitoring in; 2015.
17. Shafique NA, Fulk F, Autrey BC, Flotemersch J. *Hyperspectral remote sensing of water quality parameters for large rivers in the ohio river Basin;* Environmental Protection Agency: USA, 2002.
18. Abdelmalik KW. Role of statistical remote sensing for Inland water quality parameters prediction. *Department of Geology, Faculty of Science, Ain Shams University – Cairo, Egypt* 2018.

19. Muchini R, Gumindoga W, Togarepi S, Masarira T. Near real time water quality monitoring of Chivero and Manyame lakes of Zimbabwe. Wallingford, UK. *Int Assoc Hydrol Sci* 2018.
20. Dogliotti KG, Ruddickb B, Nechadb D, Doxaranc E, Knaeps,. *A single algorithm to retrieve turbidity from remotely-sensed data in all coastal and estuarine watersI. Instituto de Astronomía y Física del Espacio (IAFE);* CONICET/UBA: Argentina, 2014.
21. Dalu T, Moyo S, Clegg BW, Nhiwatiwa T. Stratification, diel and seasonal energy transfers of Malilangwe Reservoir in the south-eastern Lowveld of Zimbabwe. *ReservoirReserv ResManage* 2013;**18**:285–296.
22. Dalu T, Thackeray Z, Leuci R, Clegg B, Chari LD, Nhiwatiwa T. First results on bathymetry, stratification and physicochemical limnology of a small tropical African reservoir (Malilangwe, Zimbabwe). *Water SA* 2013;**39**:119–130.
23. Bhattacharyya P, Pathak H, Pal S. *Water management for climate-smart agriculture. Climate smart agriculture. Green energy and technology;* Springer: Singapore, 2020. https://doi.org/10.1007/978-981-15-9132-7_5

CHAPTER FOURTEEN

Advancements in LiDAR technology for precision water resource management

Abdulhalim Musa Abubakar[a,*,1], Siraj Hussain[b], Saroj Raj Kafle[c], Abubakar Yusuf Waziri[a], Ahmad Abubakar Mustapha[d], and Eze Cordelia Nneka[e]

[a]Department of Chemical Engineering, Faculty of Engineering, Modibbo Adama University, Yola, Adamawa State, Nigeria
[b]Department of Geological Engineering, Balochistan University of Technology, Engineering and Management Sciences (BUITEMS), Quetta, Pakistan
[c]Department of Chemical Engineering, Chungbuk National University, Cheongju, South Korea
[d]School of Computer Science and Engineering, Vellore Institute of Technology, Andhra Pradesh (VIT-AP) University, Andhra Pradesh, India
[e]Department of Chemical Engineering, Faculty of Engineering, Chukwuemeka Odumegwu Ojukwu University (COOU), Uli, Anambra State, Nigeria
*Corresponding author. e-mail address: abdulhalim@mau.edu.ng

Contents

Abstract

Progressions in Light Detection and Ranging (LiDAR) technology have ushered in a new era of precision water resource management, offering an unprecedented level of detail and accuracy. This review explores the transformative role of LiDAR in hydrological studies, highlighting its capacity to generate high-resolution topographical maps and three-dimensional (3D) models of water bodies and surrounding

[1] Senior Contributor

Advances in Chemical Pollution, Environmental Management and Protection, Volume 15
ISSN 2468-9289, https://doi.org/10.1016/bs.apmp.2025.11.013

landscapes. Through a series of case studies and practical applications, the narrative demonstrates how LiDAR data facilitates the meticulous monitoring of watershed dynamics, floodplain mapping, and the assessment of aquatic habitats. Besides, the technical nuances of LiDAR, from data acquisition through airborne and terrestrial platforms to sophisticated data processing techniques, was delved into. It emphasizes the significance of LiDAR-derived digital elevation models (DEMs) in simulating hydrological processes and enhancing the predictive accuracy of flood models. Over and above that, the integration of LiDAR with other remote sensing technologies and Geographic Information Systems (GIS) is examined, illustrating a holistic approach to water resource management. Challenges associated with LiDAR deployment, such as cost, data volume, and the necessity for specialized skills, are candidly discussed. Innovations aimed at overcoming these obstacles are also presented, showcasing the continuous evolution of LiDAR technology in the face of climate change.

Keywords: LiDAR, Water resource management, Floodplain mapping, Hydrological modeling, Geographic information systems.

1. Introduction

Light Detection and Ranging (LiDAR) systems are mainly manufactured in technologically advanced regions such as the United States, Europe (notably Germany and France), China, and Japan by specialized geospatial and sensor companies. These regions dominate due to strong electronics industries, advanced optics manufacturing, and sustained investment in aerospace, autonomous systems, and remote sensing technologies. Improvements in remote sensing technologies have greatly remodeled environmental monitoring and management practices.[1,2,3] LiDAR technology, in particular, has surfaced as a decisive tool in precision water resource management,[4] as it offers high-resolution spatial data that enhances the understanding of water systems.[5] Previous research has highlighted the effectiveness of LiDAR in mapping floodplains,[6] river morphology assessment, and the monitoring of wetland ecosystems.[7] Studies have demonstrated its capability to provide detailed elevation models, which are essential for flood risk assessments and habitat analysis. Recent developments later pushed the boundaries of LiDAR applications, by integrating it with other remote sensing technologies and advanced analytical tools. To give an instance, integration with satellite imagery and Geographic Information Systems (GIS) now allowed for more comprehensive water resource mapping and management.[8,9] Although, existing research has established the foundational applications of LiDAR in environmental monitoring, there remains a need to explore its full potential in assessing water quality, climate change impacts adaptation, and the optimization of water resource management strategies.[10] It is already being used in South Africa for flood and erosion assessment using drones and airborne systems, and in aerial survey projects like terrain mapping along the Niger River in West Africa, which supports catchment and river studies. Emerging applications also include integration with digital twins (DT) for river basins such as the Limpopo, where high-resolution LiDAR and multispectral data help model surface features for planning and decision-making. Other tools include satellite optical remote sensing (e.g., Landsat, Sentinel), radar systems e.g., Synthetic Aperture Radar (SAR), and ground-based Global Positioning System (GPS)/total station surveys, which are useful for broad monitoring but often lack fine elevation detail or are affected by cloud cover. LiDAR is superior because it provides very high-resolution, accurate 3D elevation data, can penetrate vegetation canopies, and enables precise floodplain, drainage, and water-infrastructure analysis. Hence, the objectives of this study include exploring advancements in LiDAR technology, evaluating its impact on water quality assessments, and assessing its role in climate change adaptation for water resources. Moreover, integrating LiDAR with emerging technologies and addressing challenges associated with its application will be examined. The current examination aims at providing a comprehensive understanding of how LiDAR can enhance precision in water resource management and guide future research and practical applications in this field.

2. Technical fundamentals of LiDAR

LiDAR is a remote sensing method that utilizes laser light to measure variable distances to the earth. It operates by emitting laser pulses from a sensor towards the ground and records the time it takes for the light to bounce back after hitting an object or surface. Fundamental components of a LiDAR system include (Fig. 1): a laser,[11] a scanner,[10] and a specialized GPS receiver.[12] When laser pulses are emitted, the scanner directs the laser across the target area, and the GPS receiver accurately records the position and movement of the LiDAR sensor. Operation of LiDAR relies on precise timing and positioning. Each laser pulse travels at the speed of light, and the time delay between emission and return is measured to calculate the distance traveled. It is known as time-of-flight measurement, because the process enables the creation of detailed 3D models of the scanned

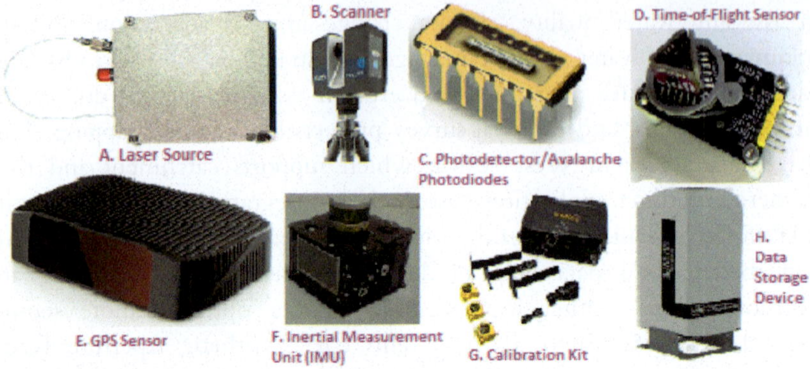

Fig. 1 Components of LiDAR systems.

environment.[13,14] Modern LiDAR systems usually emit hundreds of thousands of pulses per second, so as to boost data density and resolution.

A critical aspect of LiDAR technology is its ability to penetrate vegetation and provide high-resolution topographic data even in densely forested areas.[15–17] Achieving this capability involves multiple returns, where the sensor records the reflected signals from different surfaces, such as the top of the canopy[18,19] and the ground. It is particularly useful in water resource management, as understanding the terrain beneath vegetation is crucial for accurate modeling and analysis.[20] Data collected by LiDAR sensors are processed to generate various types of output, including DEMs,[17,21] digital surface models (DSMs),[22] and point clouds.[23] DEMs represent the bare ground surface without any objects,[24,25] whereas DSMs include all the objects on the surface, such as buildings and vegetation.[26] Point clouds are collections of points that represent the 3D positions of objects in the scanned area, which can be used for further analysis and visualization.[14,15] LiDAR systems can be deployed from aircraft, drones,[27] and ground-based vehicular platforms,[15,25] each offering different advantages depending on the application.

3. Applications in water resource management

LiDAR technology is an influential tool in water resource management,[22]due to its ability to offer high-resolution, accurate topographic and bathymetric data,[28,29] Floodplain mapping and flood risk assessment are primary applications. Creating detailed DEMs with LiDAR will help identify flood-prone areas according to Kilpatrick,[24] further enhancing planning and mitigation strategies. High-resolution data also supports the development of hydraulic models that predict flood behavior under various scenarios. Watershed monitoring and management benefit immensely from LiDAR's precision.[18] Accurate mapping of watershed boundaries, drainage patterns, and stream networks is essential for effective water resource planning.[30] LiDAR data identify potential erosion zones and sediment transport paths, which are critical for maintaining water quality and ecosystem health. Over and above that, LiDAR-derived models assist in designing and implementing best management practices to reduce erosion and sedimentation.

River and stream management applications leverage LiDAR for detailed information on channel morphology and riparian vegetation.[31] It is mentioned in Buytaert et al.[32] that understanding channel dynamics and vegetation patterns is momentous for maintaining healthy aquatic habitats and ensuring sustainable water resources. LiDAR data facilitate the assessment of streambank stability and the identification of areas requiring restoration or protection, in order to support efforts made atenhancing habitat quality, improving water quality, and maintaining biodiversity. Reservoir and dam management also reap significant benefits from LiDAR technology.[33] Accurate topographic data are essential during the design and maintenance of these structures. LiDAR surveys provide detailed information on reservoir capacity, sedimentation rates, and potential hazards, aiding in optimizing reservoir operations, ensuring structural safety, and extending the lifespan of critical infrastructure.[34]

Coastal[35] and estuarine management applications use LiDAR to monitor highly dynamic areas susceptible to erosion, sea-level rise, and storm surges.[36] High-resolution coastal maps generated from LiDAR data monitors shoreline changes, assess coastal erosion, and plan for climate change impacts. In order to develop coastal protection measures and manage coastal ecosystems effectively, those maps are essential.[37] Irrigation management optimization is another application of LiDAR-derived data.[38] Accurate elevation models assist in designing efficient irrigation systems by identifying optimal water distribution paths and minimizing water loss.[39] Precision agriculture,[9,40,41] benefits from LiDAR by ensuring that water resources are used efficiently, which is vital in arid and semi-arid regions where water scarcity poses real concern, as it also affects crops.[42,43] Groundwater management also sees benefits from LiDAR.[44] Simply because, it provide exhaustive surface elevation data helps in digesting the relationship between surface water[45,46] and groundwater. Though, it is essential for recharge studies, aquifer mapping, and sustainable groundwater extraction practices, accurate topographic data will assist during the identification of potential recharge areas and the effective management of groundwater resources.

3.1 Impact of LiDAR on water quality assessment

LiDAR technology chiefly augment water quality assessment (WQA) by providing detailed and accurate data that traditional methods often miss. One of its key contributions is in mapping and monitoring watershed areas.[33] Accurate topographic data from LiDAR seldom identify runoff from agricultural lands,[47] industrial sites, urban areas, as well as other sources of pollution.[48] Knowing precisely where pollutants originate gives room for targeted interventions to prevent contaminants from reaching water bodies. Another critical area where LiDAR impacts water quality is in the assessment of erosion and sedimentation patterns. Sediment is a major pollutant in many water bodies, as it affects aquatic habitats and water quality. LiDAR provides high-resolution elevation data[49] that helps identify erosion hotspots and sediment transport pathways. This information is vital for the development of effective soil conservation practices and sediment management strategies,[50] and the reduction in the amount of sediment that enters rivers, lakes, and reservoirs.[51] Riparian zone management benefits greatly from LiDAR technology. Riparian zones, the interface between land and a river or stream, play an observable role in maintaining water quality. They act as buffers that filter out pollutants before they reach the water. LiDAR data helps map these zones with high precision, so that better management practices to preserve or restore these natural filters are enabled.[52] As it is, properly managed riparian zones reduce nutrient loads and improve overall water quality.

LiDAR also aids in the detection and monitoring of algal blooms, which are prime concerns for water quality. Excess nutrients, particularly nitrogen and phosphorus, often lead to harmful algal blooms that deplete oxygen in water and produce toxins. High-resolution LiDAR data then identify areas contributing to nutrient loading, such as agricultural fields with heavy fertilizer use[53] or wastewater discharge sites. With this info, measures can be implemented to control nutrient inputs, thereby mitigating algal blooms. Wetland mapping and monitoring is another area were LiDAR impacts WQA. Wetlands are natural water purifiers capable of trapping sediments and absorbing pollutants. Accurate elevation data from

LiDAR most times, enables the precise delineation of wetlands, monitors changes over time, and assess the health of these crucial ecosystems.[19] Protecting and restoring wetlands based on LiDAR data contributes to improved water quality, due to the fact that it enhances their natural filtration functions.

Urban water quality management also benefits from LiDAR technology.[34] Urban areas contribute voluminous amounts of pollutants to water bodies through stormwater runoff. LiDAR data then supports the design and implementation of green infrastructure, such as bioswales and rain gardens, which mitigate the impact of stormwater runoff. Detailed topographic information is another way to ensure whether these systems are optimally placed and effective in filtering pollutants before they reach water bodies.[28] In agricultural settings, LiDAR's role is in precision farming practices that benefit water quality.[38,40] The provision of detailed terrain models enables LiDAR to aid in the design of efficient irrigation systems[42] that minimize water and fertilizer use. To improve water quality, reducing excess irrigation and fertilizer runoff directly decreases the amount of pollutants entering water bodies. LiDAR's role in floodplain mapping also indirectly benefits water quality. Accurate floodplain maps help manage land use in flood-prone areas, and prevent the development that could increase pollution risks during flooding events. Managing floodplains effectively is said to reduce the likelihood of pollutants being washed into water bodies during floods.

3.2 Role of LiDAR in climate change adaptation for water resources

Adapting water resources to climate change requires innovative technologies, and LiDAR is heavily involve in this effort. High-resolution topographic data from LiDAR systems may identify vulnerable areas prone to flooding, for better planning and mitigation. Accurate elevation models still aid in predicting flood zones,[54,55] in order to facilitate the ability to protect communities and infrastructure. Increased precipitation and extreme weather events, which are hallmarks of climate change, necessitate robust watershed management. LiDAR provides essential data for mapping watersheds, analyzing drainage patterns, and modeling water flow.[24,54] Such information is vital when designing interventions to manage runoff, reduce erosion, and prevent sedimentation in water bodies. Coastal areas facing rising sea levels and increased storm intensity, benefit sizeably from LiDAR. Coastal erosion, storm surge, and sea-level rise can be monitored and modeled accurately, informing strategies for shoreline protection and habitat restoration. Coastal managers may decide to use this data to develop effective responses to mitigate the impacts of climate change on coastal water resources. Agricultural practices must also adapt to changing climate conditions, and precision farming supported by LiDAR data is key.[56] Terrain models guide efficient irrigation system design and optimizes water use and waste reduction. In areas facing increased drought risk,[57,58] it helps conserve water resources and ensures sustainable agricultural practices.[39,47]

Groundwater resources, often impacted by climate variability, can be better managed using LiDAR. Surface elevation data aids in identifying recharge areas[52] and understanding the interaction between surface water and groundwater. According to Ibrahim et al.,[59] this is central for sustainable groundwater management, so that aquifers are replenished and not over-extracted. Urban areas, with their unique challenges, also see significant benefits. Stormwater management systems, essential for coping with increased rainfall and runoff emphasized in Doost et al.,[57] are sometimes optimized using precise topographic data. Green infrastructure, such as permeable pavements and green roofs, can be strategically placed to enhance stormwater absorption and reduce urban flooding. Wetlands, acting as natural buffers, are critical in climate adaptation. Accurate mapping and monitoring of wetlands ensure their preservation and restoration.[6] Healthy wetlands absorb floodwaters, filter pollutants, and provide habitat for wildlife, all crucial functions in a changing climate.

4. Integration with other technologies

Coalescing LiDAR with other advanced technologies naturally foster its effectiveness in precision water resource management.[60] Again, combining LiDAR data with GIS trigger some sophisticated spatial analyses and visualizations. GIS platforms can overlay LiDAR-derived elevation models with other spatial data, such as land use and vegetation cover, to provide a comprehensive understanding of water resources.[45,56,61] Remote sensing technologies complement LiDAR by offering broader coverage and different data types.[1] Satellites[62,63] and aerial[64] imagery provide large-scale views of water bodies, vegetation health, and land surface changes. When paired with LiDAR's high-resolution data, the aforementioned technologies create a robust framework for monitoring and managing water resources effectively. Unmanned Aerial Vehicles (UAVs) highlighted by

Ballabh et al.[40] and Mishra & Rai,[64] are commonly referred to as drones, and are increasingly being used to deploy LiDAR systems in hard-to-reach areas. Drones offer flexibility and cost-effectiveness, as well as frequent and targeted data collection.[65,66] The combination gives real-time monitoring and rapid response to changing environmental conditions. Sharma et al.[34] and Wada et al.[67] found that the integration with hydrological models undeniably supports water resource management. LiDAR data feeds into various models to simulate water flow, predict flooding events, and manage water distribution. In that case, accurate topographic information ensures that models are precise, for decision-making and planning.

Cutting-edge data processing and machine learning (ML) algorithms[68] further expand the capabilities of LiDAR. ML techniques analyze vast amounts of LiDAR data to identify patterns and predict trends.[69] This enables more proactive and efficient management of water resources, to address issues before they escalate. Incorporating Internet of Things (IoT) devices creates a network of sensors that continuously collect environmental data.[70] IoT sensors monitor water quality, temperature, and flow rates parameters.[71] When integrated with LiDAR data, this information provides a real-time, dynamic picture of water resource conditions, further simplifying the ability to manage them adaptively. Collaboration with cloud computing platforms[72] facilitates the storage, processing, and sharing of large LiDAR datasets. Cloud-based systems enable seamless access to data for stakeholders,[16] to promote collaborative water resource management. Real-time data processing in the cloud also ensures timely analysis and dissemination of information. Solar-powered and autonomous systems extend the reach of LiDAR in remote and off-grid locations. Several or similar systems can operate independently, collecting data over long periods without human intervention. It ensures continuous monitoring of water resources, even in areas that are difficult to access,

5. Challenges and solutions

Several setbacks may be encountered in this respect, as shown in Table 1. However, a well-maintained equipment operates more reliably to minimize downtime and associated costs.[80] Combining LiDAR with other remote sensing methods, such as aerial imagery,[43] further improve accuracy in challenging environments.

Table 1 Some identified demerits of LiDAR technologies and their solutions.

No.	challenges	Solutions	Literature source
1.	High Initial Costs	Utilizing government grants, partnerships, and phased implementation to spread out expenses.	LIDAR[73]
2.	Data Processing Complexity	Employing advanced software, machine learning, and cloud computing for efficient data handling.	Kanga[16]
3.	Limited Expertise	Providing training programs, workshops, and collaborating with academic institutions to build expertise.	Hopkinson et al.,[74,75]
4.	Weather Dependency	Scheduling surveys during optimal weather conditions and using drones for flexibility.	Debell et al.,[21,75]
5.	Data Integration	Developing standardized protocols and using interoperable systems to ensure seamless data integration.	Dowman[76]
6.	Equipment Maintenance	Implementing regular maintenance schedules and investing in durable, high-quality equipment.	SICK[77]
7.	Accuracy in Dense Vegetation	Using multi-return LiDAR technology and combining with other remote sensing methods for improved accuracy.	Sharma et al.,[35]
8.	Legal and Regulatory Issues	Ensuring compliance with local regulations through thorough research and legal consultation.	Crane et al.,[78,79]

Consistent data formats and standards facilitate the combination of different types of data, which enhances the overall analysis and decision-making processes. Furthermore, according to Maurya et al.,[81] drones provide additional flexibility for data collection during short windows of favorable weather, thus reducing delays and improving data reliability. It is foretold that creating a well-trained workforce ensures that LiDAR technology is used effectively and to its full potential in managing water resources. As mentioned in Table 1, spreading out costs over time and leveraging external funding sources make the technology more accessible to various water management organizations.

6. Next-generation developments and technological revolutions

LiDAR know-how has seriously transformed remote sensing by delivering high-resolution,[49] precise spatial data crucial for water resource management. Future advancements in LiDAR technology promise to further multiply its utility in this field. One key area of innovation is the improvement of sensor capabilities, with developments in laser and detector technologies expected to provide even finer spatial resolution[82] and greater accuracy. This progress will enhance the precision of mapping river networks, wetlands, and floodplains, as mentioned by Kumar et al.,[58] and later leads to more accurate modeling and management of water resources.[59] On top of that, integrating LiDAR with multi-spectral and hyperspectral imaging will offer deeper insights into vegetation health and water quality,[4,83] also enabling more comprehensive assessments of aquatic ecosystems.[84]

The future of LiDAR also lies in its integration with other technologies. Fusing LiDAR data with satellite imagery and aerial photography will create more accurate land and water resource maps, while advanced GIS will facilitate the creation of detailed 3D models of watersheds[61] and floodplains. The two models will support better decision-making in water management, including flood risk assessment and infrastructure planning. Moreover, innovations in real-time data acquisition and processing will allow for continuous monitoring of water bodies, and provide up-to-date information on water levels,[62] flow rates, and landscape changes. Cloud computing platforms will further enhance this capability by enabling efficient processing and accessibility of large volumes of LiDAR data, where timely and informed decision-making is facilitated.

Developments in UAV[31,46] and mobile LiDAR systems will expand data collection possibilities in diverse and challenging environments. UAVs equipped with LiDAR sensors will be able to access remote or hazardous areas (e.g., landslide location) that traditional methods might miss,[13,20,85,] while mobile LiDAR systems mounted on vehicles or boats will capture detailed data along river corridors and other linear features. Monitoring extensive regions and evaluating the effects of infrastructure on water resources are key advantages of these systems.[34,86] At length, enhanced data analytics and modeling techniques will be pivotal in extracting meaningful insights from LiDAR data. ML and artificial intelligence (AI) will automate feature identification, land cover classification, and trend prediction, for improved flood risk assessments and water quality forecasts.[65,69,87] Advanced modeling tools will also enable the simulation of various climate change impacts and land use changes scenarios, in order to support proactive management strategies and better preparedness for future challenges. Sit et al.[88] and Dixon & Uddameri[8] critically reviewed the role of artificial neural network (ANN) modelling and fuzzy set theory, respectively, in water resources and hydrology.

6.1 Potential applications of LiDAR in climate-smart water management

LiDAR technology has great potential in climate-smart water management.[89] First, it provides high-resolution topographic data that help in mapping watersheds, monitoring water bodies, and assessing flood risks. Using LiDAR in view of researchers, is an opportunity to be able to analyze river flow patterns and identify areas prone to erosion for better planning for flood control and water conservation. Secondly, LiDAR supports groundwater management.[90] It helps detect changes in land elevation, which can indicate groundwater depletion. Identification of suitable locations for water storage and recharge projects is assisted by this technology. In agriculture, LiDAR improves irrigation planning by mapping soil moisture levels and terrain slopes.[91,92] As a result, it allows farmers to use water efficiently, for the purpose of reducing waste and enhancing crop yield. LiDAR is valuable in monitoring the effects of climate change on water resources. It helps track changes in glacier volume, river paths, and coastal erosion. Related data obtained supports climate adaptation strategies and provide accurate predictions of water availability.

7. Case studies

LiDAR technology has been instrumental in enhancing water resource management through detailed spatial data and improved modeling techniques. Table 2 display some of the illustrative case studies that demonstrate the application and benefits of advanced LiDAR systems in this field.

Table 2 Case studies on advanced LiDAR technology for precision water resource management.

Elements	Case study				
	A	B	C	D	E
Title	Flood Risk Assessment in the Mississippi River Basin	Monitoring Wetland Changes in the Everglades National Park	Riverbank Erosion Control in the Rhine River	Urban Water Management in Singapore	Watershed Management in the Amazon Basin
Objective	Assess flood risk and create accurate floodplain maps.	Track changes in wetland areas and vegetation over time.	Evaluate and control erosion along riverbanks to protect infrastructure.	Optimize water management and stormwater runoff in a rapidly urbanizing city.	Manage and conserve watershed areas to prevent deforestation and maintain biodiversity.
Technology Used	High-resolution airborne LiDAR and satellite imagery integration.	UAV-based LiDAR and multi-spectral imaging.	Mobile LiDAR systems mounted on boats and aerial platforms.	Ground-based and UAV-based LiDAR combined with real-time monitoring systems.	Airborne LiDAR and satellite remote sensing combined with GIS.
Data Collected	Elevation data, floodplain delineation, vegetation mapping, and hydrological modeling.	Detailed elevation changes, vegetation health, water levels, and land cover classification.	Riverbank topography, erosion rates, sediment transport, and vegetation mapping.	Urban drainage patterns, water quality, flood risk areas, and infrastructure impacts.	Watershed boundaries, forest cover changes, soil erosion rates, and river flow patterns.

(continued)

Table 2 Case studies on advanced LiDAR technology for precision water resource management. (*cont'd*)

Elements	Case study				
	A	**B**	**C**	**D**	**E**
Key Findings	Identified high-risk flood zones and improved floodplain management strategies.	Detected significant changes in wetland extent and vegetation health, informing conservation efforts.	Successfully mitigated erosion and protected critical infrastructure through targeted interventions.	Enhanced stormwater management practices and reduced urban flooding incidents.	Provided insights for better watershed conservation practices and reduced deforestation.
Impact	Improved flood preparedness and emergency response, reducing flood damage.	Enabled more effective wetland preservation and management strategies.	Strengthened riverbank stability and safeguarded valuable infrastructure.	Improved urban water management and resilience to extreme weather events.	Advanced conservation measures and better management of natural resources.
Future Directions	Expand to include more dynamic hydrological models and integrate with real-time data sources.	Utilize higher resolution LiDAR and further integrate with habitat modeling tools.	Implement additional erosion control measures based on detailed LiDAR data.	Develop more sophisticated models incorporating climate change projections and urban growth.	Increase collaboration with local communities and integrate indigenous knowledge in management strategies.
Source	Rango, Anderson[93]	Hong, Wdowinski[94,95]	Krapesch et al.,[96]	Banerjee et al.,[97,98]	Braga et al.,[99,100]

Evidently, the case studies in Table 2 illustrate the diverse applications of advanced LiDAR technology in water resource management. It also highlight its role in improving flood risk assessments, monitoring environmental changes, and enhancing urban[4] and watershed management.[5] Continuous evolution of LiDAR technology and its integration with other tools promise further advancements in these areas.

8. Conclusion

In summary, advancements in LiDAR technology have profoundly influenced the field of water resource management, offering unprecedented levels of detail and accuracy in spatial data collection. Through the exploration of its technical fundamentals, diverse applications, and integration with other technologies, it becomes clear that LiDAR is pivotal in enhancing water quality assessments and adapting to climate change impacts. The ability to generate high-resolution elevation models and monitor dynamic changes in water systems has improved flood risk management, informed conservation efforts, and optimized urban water management practices. While challenges such as data processing and integration remain, ongoing innovations promise to address these issues and expand LiDAR's capabilities. Future research should continue to focus on refining LiDAR technologies and exploring new applications to address emerging water management challenges. As the field progresses, the integration of LiDAR with other advanced tools and methodologies will undoubtedly lead to more effective and sustainable management of our precious water resources.

Declaration of generative AI and AI-assisted technologies in the writing process

This work has been prepared with the assistance of generative AI in the writing process, specifically for drafting and language refinement. However, the final content has been extensively reviewed, edited, and verified by the authors to ensure accuracy, originality, and compliance with academic standards.

References

1. Leckie DG. Advances in remote sensing technologies for forest surveys and management. *Can J For Res* 1990;**20**:464–483. https://cdnsciencepub.com.
2. Pandey PC. In *Advances in remote sensing for natural resource;* Pandey PC, Sharma LK, Eds.; John Wiley & Sons Ltd, 2021, https://lccn.loc.gov/2020026498.

3. Srivastava A, Jain S. Remote Sensing for Environment Assessment: Multispectral, Hyperspectral, and Thermal Imaging Applications. In *Remote Sensing for Environmental Monitoring;* Saritha V, Pande CB, Singh R, Shahid M, Eds.; Springer: Singapore, 2025, https://doi.org/10.1007/978-981-96-5546-5_1
4. Lakra D, Singh SK, Gupta SK, Kanga S. Enhancing integrated resource management through remote sensing and GIS. *J Geogr Cartography* 2024;**7**(1):1–13. https://doi.org/10.24294/jgc.v7i1.426.
5. Ogren BM. Precision conservation in the Zumbro River watershed using LiDAR and Digital Terrain analysis to identify critical areas associated with water resource impairment in agricultural landscapes. *Pap Resour Anal* 2012;**14**:1–15. http://www.gis.smumn.edu.
6. Duan W, Maskey S, Chaffe PLB, Luo P, He B, Wu Y, et al. Recent advancement in remote sensing technology for hydrology analysis and water resources management. *Remote Sens* 2021;**13**(1097):1–4. https://doi.org/10.3390/rs13061097.
7. Jeziorska J. UAS for wetland mapping and hydrological modeling. *Remote Sens* 2019;**11**(1997):1–39. https://doi.org/10.3390/rs11171997.
8. Dixon B, Uddameri V. *GIS and geocomputation for water resource science and engineering;* John Wiley & Sons, Ltd, 2016, https://doi.org/10.1002/9781118826171
9. Katkani D, Babbar A, Mishra VK, Trivedi A, Tiwari S, Kumawat RK. A review on applications and utility of remote sensing and Geographic Information Systems in agriculture and natural resource management. *Int J Environ Clim Change* 2022;**12**(4):1–18. https://doi.org/10.9734/IJECC/2022/v12i430651.
10. Javaid A, Mahmood N, Mehmood MQ. Review: optimizing LiDAR technology for enhanced 3D remote sensing. Proceedings Volume 13024, Optical instrument science, technology and applications III; SPIE optical systems design [13024OP 2024] Strasbourg, France; 2024, p. 13024. https://doi.org/10.1/1/17/12.3022156.
11. Large ARG. In *Laser scanning for the environmental sciences;* Heritage GL, Large ARG, Eds.; Blackwell Publishing Ltd, 2009, https://doi.org/10.1002/9781444311952.fmatter
12. White AM, Gardner WP, Borsa AA, Argus DF, Martens HR. A review of GNSS/GPS in hydrogeodesy: hydrologic loading applications and their implications for water resource research. *Water Resour Res* 2022;**58**(7):1–8. https://doi.org/10.1029/2022WR032078.
13. Jaboyedoff M, Oppikofer T, Abella A, Derron M-H, Loye A, Metzger R, et al. Use of LIDAR in landslide investigations: a review. *Nat Hazards* 2012;**61**:5–28. https://doi.org/10.1007/s11069-010-9634-2.
14. Sampson CC, Fewtrell TJ, Duncan A, Shaad K, Horritt MS, Bates PD. Use of terrestrial laser scanning data to drive decimetric resolution urban inundation models. *Adv Water Resour* 2012;**41**:1–17. https://doi.org/10.1016/j.advwatres.2012.02.010.
15. Dassot M, Constant T, Fournier M. The use of terrestrial LiDAR technology in forest science: application fields, benefits and challenges. *Ann For Sci* 2011;**68**:959–974. https://doi.org/10.1007/s13595-011-0102-2.
16. Kanga S. Advancements in remote sensing tools for forestry analysis. *Sustain Forestry* 2023;**6**(1):1–24. https://doi.org/10.24294/sf.v6i1.2269.
17. Akay AE, Oğuz H, Karas IR, Aruga K. Using LiDAR technology in forestry activities. *Environ Monit Assess Assess* 2009;**151**:117–125. https://doi.org/10.1007/s10661-008-0254-1.
18. Hudak AT, Evans JS, Smith AMS. LiDAR utility for natural resource managers. *Remote Sens* 2009;**1**:934–951. https://doi.org/10.3390/rs1040934.
19. Lefsky MA, Cohen WB, Parker GG, Harding DJ. Lidar remote sensing for ecosystem studies. *BioScience* 2002;**52**(1):19–30. https://doi.org/10.1641/0006-3568(2002)052[0019:LRSFES]2.0.CO;2.

20. Jiao Z. The application of remote sensing techniques in ecological environment monitoring. *Highlights Science, Eng Technol* 2024;**81**:449–455. https://doi.org/10.54097/7dqegz64.
21. Liu X. Airborne LiDAR for DEM generation: some critical issues. *Prog Phys Geogr* 2008;**32**(1):31–49. https://doi.org/10.1177/0309133308089496.
22. Debell L, Anderson K, Brazier RE, King N, Jones L. Water resource management at catchment scales using lightweight UAVs: current capabilities and future perspectives. *J Unmanned Veh Syst* 2016;**30**:7–30. http://www.nrcresearchpress.com/juvs.
23. Melchior P, Ledig C. Advanced algorithms for real-time detection of floating objects in water bodies using AI and neural networks. *J Environ Sci Technol (JEST)* 2024;**3**(1): 913–927. https://jest.com.pk/index.php/jest/article/view/161.
24. Kilpatrick GR. *DEM generation and hydrologic modelling using LiDAR data* [Degree of Bachelor of Spatial Science (Surveying) Dissetation, Faculty of Health, Engineering and Sciences, University of Southern Queensland]; 2015. https://sear.unisq.edu.au/29263.
25. Bellian JA, Kerans C, Jennette DC. Digital outcrop models: applications of terrestrial scanning LIDAR technology in statigraphic modeling. *J Sediment Res* 2005;**75**(2): 166–176. https://doi.org/10.2110/jsr.2005.013.
26. Acharya BS, Bhandari M, Bandini F, Pizarro A, Perks M, Joshi DR, et al. Unmanned aerial vehicles in hydrology and water management: applications, challenges, and perspectives. *Water Resour Res* 2021;**57**:1–33. https://doi.org/10.1029/2021WR029925.
27. Glennie CL, Carter WE, Shrestha RL, Dietrich WE. Geodetic imaging with airborne LiDAR: the Earth's surface revealed. *Rep Prog Phys* 2013;**76**(086801):1–086824. https://doi.org/10.1088/0034-4885/76/8/086801.
28. Höfle B, Rutzinger M. Topographic airborne LiDAR in geomorphology: a technological perspective. *Z Für Geomorphologie* 2015;**55**(2):1–29. https://doi.org/10.1127/0372-8854/2011/0055S2-0043.
29. Pricope NG, Bashit S. Emerging trends in topobathymetric LiDAR technology and mapping. *Int J Remote Sens* 2023;**44**(24):7706–7731. https://doi.org/10.1080/01431161.2023.2287564.
30. Jain SK, Singh VP. Water resources systems planning and management. In *Developments in water science;* Jain SK, Singh VP, Eds.; 2nd ed. Elsevier, 2003, pp. 503–858. https://www.sciencedirect.com/bookseries/developments-in-water-science/vol/51/suppl/C.
31. Peko M, Milic DC, Vidakovic I. Unmanned aerial vehicle mapping of river flow for water resources management. *32nd International Conference on Organization and Technology of Maintenance [OTO 2023]*; 2024, p. 154–163. https://doi.org/10.1007/978-3-031-51494-4_14.
32. Buytaert W, Zulkafl Z, Grainger S, Acosta L, Alemie TC, Bastiaensen J, et al. Citizen science in hydrology and water resources: opportunities for knowledge generation, ecosystem service management, and sustainable development. *Front Earth Sci* 2014;**2**:1–21. https://doi.org/10.3389/feart.2014.00026.
33. Park D, You H. A digital twin dam and watershed management platform. *Water* 2023;**15**(2106):1–19. https://doi.org/10.3390/w15112106.
34. Molua CO, Ataman JO. Applications of Lidar technology in urban geophysical surveys. *Int J Inf Technol Computer Eng (IJITC)* 2024;**4**(4):46–58. https://doi.org/10.55529/ijitc.44.47.58.
35. Sharma A, Rout C, Garg V. The emerging role of LIDAR remote sensing for hydrological application: opportunities and constraints. *National Conference on Hydraulics, Water Resources, Coastal and Environmental Engineering HYDRO 2010 at MMEC, Mullana*; 2010 December 16–18, 2010, p. 461–467. https://www.researchgate.net/publication/284032528.
36. Chust G, Galparsoro I, Borja Á., Franco J, Uriarte A. Coastal and estuarine habitat mapping, using LIDAR height and intensity and multi-spectral imagery. *Estuarine, Coast Shelf Sci* 2008;**78**(4):633–643. https://doi.org/10.1016/j.ecss.2008.02.003.

37. Masria A. Bridging coastal challenges: the role of remote sensing and future research. *Regional Stud Mar Sci* 2018;**73**:103502. https://doi.org/10.1016/j.rsma.2024.103502.
38. Adinarayana S, Raju MG, Srirangam DP, Prasad DS, Kumar MR, Veesan SB. Enhancing resource management in precision farming through AI-based irrigation optimization. In *How mach learn is innovating today's world: a concise technical guide;* Dey A, Nayak S, Kumar R, Mohant SN, Eds.; Wiley, 2024, https://doi.org/10.1002/9781394214167.ch1
39. Wanniarachchi S, Sarukkalige R. A review on evapotranspiration estimation in agricultural water management: past, present, and future. *Hydrology* 2022;**9**(123):1–12. https://doi.org/10.3390/hydrology9070123.
40. Ballabh J, Bhatt A, Joshi R, Prakash S, Prasad R. Advancements in precision agriculture technologies for enhancing crop yields. *J Surv Fish Sci* 2022;**8**(3):403–408. https://doi.org/10.53555/sfs,v8i3.2399.
41. Singh V. Advances in precision agriculture technologies for sustainable crop production. *J Sci Res Rep (JSRR)* 2024;**30**(2):61–71. https://doi.org/10.9734/JSRR/2024/v30i21844.
42. Debangshi U, Sadhukhan A. Crop water stress monitoring through precision technologies: a review. *Pharma Innov J* 2023;**12**(11):803–808. https://doi.org/10.22271/tpi.2023.v12.i11Sk.24052.
43. Ahmad U, Alvino A, Marino S. A review of crop water stress assessment using remote sensing. *Remote Sens* 2021;**13**(4155):1–26. https://doi.org/10.3390/rs13204155.
44. Shaikh M, Birajdar F. Advancements in remote sensing and GIS for sustainable groundwater monitoring: applications, challenges, and future directions. *Int J Res Engineering, Sci Manag (IJRESM* 2024;**7**(3):16–24. https://www.ijresm.com.
45. Mashala MJ, Dube T, Mudereri BT, Ayisi KK, Ramudzuli MR. A systematic review on advancements in remote sensing for assessing and monitoring land use and land cover changes impacts on surface water resources in semi-arid tropical environments. *Remote Sens* 2023;**15**(3926):1–29. https://doi.org/10.3390/rs15163926.
46. Sigopi M, Shoko C, Dube T. Advancements in remote sensing technologies for accurate monitoring and management of surface water resources in Africa: an overview, limitations, and future directions. *Geocarto Int* 2024;**39**(1):1–33. https://doi.org/10.1080/10106049.2024.2347935.
47. Kumar S, Meena RS, Sheoran S, Jangir CK, Jhariya MK, Banerjee A, et al. Remote sensing for agriculture and resource management. *Nat Resour Conserv Adv Sustain* 2022:91–135. https://doi.org/10.1016/B978-0-12-822976-7.00012-0.
48. Selvarajan S, Mohamed A, White T, Boodram N. Assessment of geospatial technologies for natural resource management in Florida. *Geospatial Technol* 2009:242–249. https://academic.oup.com/jof/article/107/5/242/4598899.
49. Li Y, Dang B, Zhang Y, Du Z. Water body classification from high-resolution optical remote sensing imagery: achievements and perspectives. *ISPRS J Photogrammetry Remote Sens* 2018;**187**:306–327. https://doi.org/10.1016/j.isprsjprs.2022.03.013.
50. Abdulraheem MI, Zhang W, Li S, Moshayedi AJ, Farooque AA, Hu J. Advancement of remote sensing for soil measurements and applications: a comprehensive review. *Sustainability* 2023;**15**(15444):1–32. https://doi.org/10.3390/su152115444.
51. Marcé R, George G, Buscarinu P, Deidda M, Dunalska J, De Eyto EDE, et al. Automatic high frequency monitoring for improved lake and reservoir management. *Environ Sci & Technol* 2016;**50**:10780–10794. https://doi.org/10.1021/acs.est.6b01604.
52. McKean J, Nagel D, Tonina D, Bailey P, Wright CW, Bohn C, et al. Remote sensing of channels and Riparian Zones with a narrow-beam aquatic-terrestrial LIDAR. *Remote Sens* 2009;**1**:1065–1096. https://doi.org/10.3390/rs1041065.

53. Galzki JC, Birr AS, Mulla DJ. Identifying critical agricultural areas with three-meter LiDAR elevation data for precision conservation. *J Soil Water Conserv* 2011;**66**(6):423–430. https://doi.org/10.2489/jswc.66.6.423.
54. Garbrecht BJ, Martz LW. Digital elevation model issues water Resour modeling; 1992. p. 1 14. https://proceedings.esri.com/library/userconf/proc99.
55. Heidemann HK. *Lidar base specification. Collection and delineation of spatial data;* USGS Publications, 2018, pp. 114. https://doi.org/10.3133/TM11B4
56. Marques P, Padua L, Sousa JJ, Fernandes-Silva A. Advancements in remote sensing imagery applications for precision management in olive growing: a systematic review. *Remote Sens* 2024;**16**(1324):1–41. https://doi.org/10.3390/rs16081324.
57. Doost ZH, Alsuwaiyan M, Yaseen ZM. Runoff management based water harvesting for better water resources sustainability: a comprehensive review. *Knowl Eng Sci* 2024;**5**(1):1–45. https://doi.org/10.51526/kbes.2024.5.1.1-45.
58. Kumar V, Sharma KV, Pham QB, Yadav SM. Advancements in drought using remote sensing: assessing progress, overcoming challenges, and exploring future opportunities. *Theor Appl Climatol* 2024;**155**:4251–4288. https://doi.org/10.1007/s00704-024-04914-w.
59. Ibrahim A, Wayayok A, Shafri HZM, Toridi NM. Remote sensing technologies for unlocking new groundwater insights: a comprehensive review. *J Hydrol X* 2024;**23**(100175):1–22. https://doi.org/10.1016/j.hydroa.2024.100175.
60. Pandey A, Singh G. Overview of geospatial technologies for land and water resources management. *Geospatial Technol LWater Resour Manag* 2021:1–16. https://doi.org/10.1007/978-3-030-90479-1_1.
61. Martin PH, Leboeuf EJ, Dobbins JP, Daniel EB, Abkowitz MD. Interfacing GIS with water resource models: a state-of-the-art review. *J Am Water Resour Assoc (JAWRA)* 2006;**41**(6):1471–1487. https://doi.org/10.1111/j.1752-1688.2005.tb03813.x.
62. Kossieris S, Tsiakos V, Tsimiklis G, Amditis A. Inland water level monitoring from satellite observations: a scoping review of current advances and future opportunities. *Remote Sens* 2024;**16**(1181):1–37. https://doi.org/10.3390/rs16071181.
63. Kumar S, Imen S, Sridharan VK, Gupta A, Mcdonald W, Ramirez-Avila JJ, et al. Perceived barriers and advances in integrating earth observations with water resources modeling. *Remote Sens Applications: Soc Environ* 2024;**33**(101119):1–19. https://doi.org/10.1016/j.rsase.2023.101119.
64. Mishra PK, Rai A. Role of unmanned aerial systems for natural resource management. *J Indian Soc Remote Sens* 2020:1–9. https://doi.org/10.1007/s12524-020-01230-4.
65. Quamar M, Al-Ramadan B, Khan K, Shafiullah, El Ferik S. Advancements and applications of drone-integrated geographic information system technology—A review. *Remote Sens* 2023;**15**(5039):1–35. https://doi.org/10.3390/rs15205039.
66. Wada Y, Bierkens MFP, De Roo A, Dirmeyer PA, Famiglietti JS, Hanasaki N, et al. Human–water interface in hydrological modelling: current status and future directions. *Hydrol Earth Syst Syst* 2017;**21**:4169–4193. https://doi.org/10.5194/hess-21-4169-2017.
67. Srivastava A, Jain S, Maity R, Desai VR. In *Current Directions in Water Scarcity Research;* Zakwan M, Wahid A, Niazkar M, Chatterjee U, Eds.; Elsevier, 2022, pp. 17–35. https://doi.org/10.1016/B978-0-323-91910-4.00002-9
68. Shen, C. A trans-disciplinary review of deep learning research and its relevance for water resources scientists; 2018. p. 1–81. https://doi.org/10.1029/2018WR022643.
69. Erfani SMH. Developing a vision-based framework for measuring and monitoring water resource systems using computer vision and deep learning techniques. In *Degree of doctor of philosophy in civil and environmental engineering college of engineering and computing;* Goharian E, Imran J, Qian Y, Tong Y, Vail A, Eds.; University of South Carolina, 2023, https://scholarcommons.sc.edu/etd/7444.

70. Iqbal U, Bin Riaz MZ, Barthelemy J, Perez P, Idrees MB. The last two decades of computer vision technologies in water resource management: a bibliometric analysis. *Water Environ J* 2023:1–17. https://doi.org/10.1111/wej.12845.
71. Patel A, Kushwaha NL, Ajaykumar K, Naorem A. Review of artificial intelligence and internet of things technologies in land and water management research during 1991-2021: a bibliometric analysis. *Eng Appl Artif Intell* 2023;**123**. https://doi.org/10.1016/j.engappai.2023.106335.
72. Chen J, Chen S, Fu R, Li D, Jiang H, Wang C, et al. Remote sensing big data for water environment monitoring: current status, challenges, and future prospects. *Earth's Future* 2022;**10**:1–33. https://doi.org/10.1029/2021EF002289.
73. LIDAR. *Lidar vs Radar: differences, advantages and disadvantages;* Global GPS Systems, 2024, https://globalgpssystems.com/lidar/lidar-vs-radar-differences-advantages-and-disadvantages/.
74. Hopkinson C, Popescu S, Flood M, Maher R. A survey on the need for airborne Lidar training. *Photogrammetric Eng & Remote Sens* 2007;**73**(5):537–546. https://doi.org/10.14358/PERS.73.5.537.
75. Zhou J. A review of LiDAR sensor technologies for perception in automated driving. *Academic J Sci Technol* 2022;**3**(3):255–261. https://doi.org/10.54097/ajst.v3i3.2993.
76. Dowman IJ. Integration of LiDAR and IFSAR for mapping. *Invited Paper Commission II, WGII/2*; 2004. p. 1–12. https://www.researchgate.net/publication/32894099.
77. SICK. *Maintenance of LiDAR sensors.* Verification and optimization: maintenance; 2024. https://www.sick.com/ch/en/catalog/service/verification.
78. Crane DA, Logue KD, Pilz BC. A survey of legal issues arising from the deployment of autonomous and connected vehicles. *Mich Telecommun Technol Law Rev* 2017;**23**(2): 193–224 https://repository.law.umich.edu.mttlr/vol23/iss2/1.
79. Guarnera C. The ethics of drone LiDAR: a guide to responsible data collection. *Blue Falcon Aer* 2023. https://www.bluefalconaerial.com/the-ethics-of-drone-lidar-a-guide-to.
80. Mao F, Khamis K, Clark J, Krause S, Buytaert W, Ochoa-Tocachi BF, et al. Moving beyond the technology: a socio-technical roadmap for low-cost water sensor network applications. *Environ Sci & Technol* 2020;**54**:9145–9158. https://doi.org/10.1021/acs.est.9b07125.
81. Maurya NK, Tripathi AK, Chauhan A, Pandey PC, Lamin S. Recent advancement and role of drones in forest monitoring. *Adv Remote Sens Monit* 2022:2024. https://doi.org/10.1002/9781119788157.ch11.
82. Andresen CG, Schultz-Fellenz ES. Change detection applications in the earth sciences using UAS-based sensing: a review and future opportunities. *Drones* 2023;**7**(258):1–28. https://doi.org/10.3390/drones7040258.
83. Eitel JUH, Höfle B, Vierling LA, Abellán A, Asner GP, Deems JS, et al. Beyond 3-D: the new spectrum of lidar applications for earth and ecological sciences. *Remote Sens Environ* 2016;**186**:372–392. https://doi.org/10.1016/j.rse.2016.08.018.
84. Bhadra AB, Rayhan F, Rime MHK, Jahan S, Bhuiyan EA, Urmi SD. Paradigm-shifting the subaquatic frontier: advancements and prospects of sensor-based underwater positioning systems. *SSRN* 2023:1–22. https://doi.org/10.2139/ssrn.453187.
85. Singh, K. Current trends in River Bathymetry using UAV-borne technology to inform E-flow assessments in Southern Africa; 2023. https://on.cgiar.org/digital.
86. Boretti A. A perspective on single-photon LiDAR systems. *Microw Optical Technol Lett* 2024;**66**(1). https://doi.org/10.1002/mop.33918.
87. Yan WY, Shaker A, El-Ashmawy N. Urban land cover classification using airborne LiDAR data: a review. *Remote Sens Environ* 2015;**158**:295–310. https://doi.org/10.1016/j.rse.2014.11.001.
88. Sit M, Demiray BZ, Xiang Z, Ewing GJ, Sermet Y, Demir I. A comprehensive review of deep learning applications in hydrology and water resources. *Water Sci & Technol* 2020;**82**(12):2635–2670. https://doi.org/10.2166/wst.2020.369.

89. Jha G, Debangshi U, Palla S, Nazrul F, Dey S, Dutta W, et al. Precision agriculture technologies for climate-resiliency and water resource management. In *Navigating the nexus. Water science and technology library;* Nanda A, Gupta PK, Gupta V, Jha PK, Dubey SK, Eds.; Springer: Cham, 2025, pp. 351–381. https://doi.org/10.1007/978-3-031-76532-2_15
90. García-López S, Vélez-Nicolás M, Zarandona-Palacio P, Curcio AC, Ruiz-Ortiz V, Barbero L. UAV-borne LiDAR revolutionizing groundwater level mapping. *Sci Total Environ* 2023;**859**(Part 1). https://doi.org/10.1016/j.scitotenv.2022.160272.
91. McCoy MD, Asner GP, Graves MW. Airborne lidar survey of irrigated agricultural landscapes: an application of the slope contrast method. *J Archaeological Sci* 2011;**38**(9):2141–2154. https://doi.org/10.1016/j.jas.2011.02.033.
92. Pak J, Son H, Il. 3D LiDAR-based semantic SLAM for intelligent irrigation using UAV. *IEEE J Sel Top Appl Earth Observations Remote Sens* 2025:1–15. https://doi.org/10.1109/JSTARS.2025.3547717.
93. Rango A, Anderson AT. Flood hazard studies in the Mississippi River basin using remote sensing. *J Am Water Resour Assoc (JAWRA)* 2007;**10**(5):1060–1081. https://doi.org/10.1111/j.1752-1688.1974.tb00625.x.
94. Hong S-H, Wdowinski S. A review on monitoring the Everglades Wetlands in the Southern Florida using space-based synthetic aperture radar (SAR) observations. *Korean J Remote Sens* 2017;**33**(4):377–390. https://doi.org/10.7780/kjrs.2017.33.4.4.
95. Liao H, Wdowinski S, Li S. Regional-scale hydrological monitoring of wetlands with Sentinel-1 InSAR observations: case study of the South Florida Everglades. *Remote Sens Environ* 2020;**251**(9). https://doi.org/10.1016/j.rse.2020.112051.
96. Krapesch M, Klösch M, Brinke WT, Habersack H. The Rhine catchment: a review of sediment-related knowledge, monitoring, and a future research perspective. *Water* 2024;**16**(8) https://doi.org/10.3390/w16081121.
97. Ding N, Hamel P, Cherqui F, Bertrand-Krajewski J-L, Zhu Q. Water sensitive urban design performance in Singapore: monitoring with lowcost sensors. *Singap Int Water Week – SIWW, PUB* 2023:1–4. https://hal.science/hal-04133144.
98. Banerjee C, Bhaduri A, Saraswat C. Digitalization in urban water governance: case study of Bengaluru and Singapore. *Front Environ Sci* 2022:1–12. https://doi.org/10.3389/fenvs.2022.816824.
99. Braga D, Aragão LEOC, Anderson LO, Dutra, DJ, Cabral BF, Dalagnol R. Quantifying selective logging intensity through airborne LiDAR data in an Amazon rainforest: study case at Jamari National Forest. In: Vinhas L, da F, Feitosa F, Monteiro AMV (Eds.), Conference Proceedings XXIV [GEOINFO 2023] at São José dos Campos, SP - Brazil, vol. 24; 2023, p. 256–267. https://www.researchgate.net/publication/378796771.
100. Tejada G, Görgens EB, Espírito-Santo FDB, Cantinho RZ, Ometto JP. Evaluating spatial coverage of data on the aboveground biomass in undisturbed forests in the Brazilian Amazon. *Carbon Balance Manag* 2019;**14**(11):1–18. https://doi.org/10.1186/s13021-019-0126-8.

CHAPTER FIFTEEN

Integrating remote sensing and machine learning in Indian aquaculture water management: Lessons from two decades

G. Sri Bala[a,b] and T. Vamsi Nagaraju[a,b,*]
[a]Department of Civil Engineering, SRKR Engineering College, Bhimavaram, India
[b]Centre for Sustainable Inland Aquaculture, SRKR Engineering College, Bhimavaram, India
*Corresponding author. e-mail address: varshith.varma@gmail.com

Contents

Abstract

Indian inland aquaculture is the leading contributor to the sector in South Asia. Water is a critical factor, and its availability depends on the balance between demand and supply at any given time and location. The ongoing expansion of inland aquaculture in South India's coastal regions, farming practices, and climate change have disrupted the balance between water quality and demand, leading to increased water insecurity in nearby rural communities. Over the past two decades, the use of remote sensing and artificial intelligence (AI) tools for assessing water quality and balance has grown significantly. This chapter reviews the scientific and technological advancements in remote sensing and machine learning for inland aquaculture water management. It explores how these advanced integrated tools meet current and future research and innovation needs in the inland aquaculture zones and estuaries. The review highlights the diverse sources of chlorophyll formation, salinity intrusion, water balance, and

Advances in Chemical Pollution, Environmental Management and Protection, Volume 15
ISSN 2468-9289, https://doi.org/10.1016/bs.apmp.2025.11.011

land dynamics and how indicators have utilized remote sensing and machine learning to identify temporal and geospatial attributes, scales, and metrics. Metrics such as land cover, vegetation indices, and water indices have been widely used in various case studies, playing a crucial role in these assessments. Promoting knowledge graphs could help integrate various parameters and provide a comprehensive understanding of water scarcity. Enhancing inland aquaculture water management can facilitate the reuse of pond water across multiple crop cycles, reduce eutrophication in adjacent water bodies and estuaries, and further promotes aquaculture sustainability.

Keywords: Inland aquaculture, Circular economy, Remote sensing, Machine learning, Sustainable aquaculture.

1. Introduction

Global aquaculture production 2022 reached approximately 130.9 million tonnes, an estimated USD 312.8 billion.[1] This output represented about 59 % of the combined global fisheries and aquaculture production (see Fig. 1).[2] Inland aquaculture was the primary contributor, accounting for 62.6 % of farmed aquatic animal yields, while marine and coastal aquaculture contributed 37.4 %.[2] For the first time, aquaculture's production of aquatic animals surpassed capture fisheries, reaching 94.4 million tonnes, 51 % of the global total and a significant 57 % of the aquatic production intended for human consumption.[3] This growth reflects aquaculture's critical role in meeting global seafood demands, with

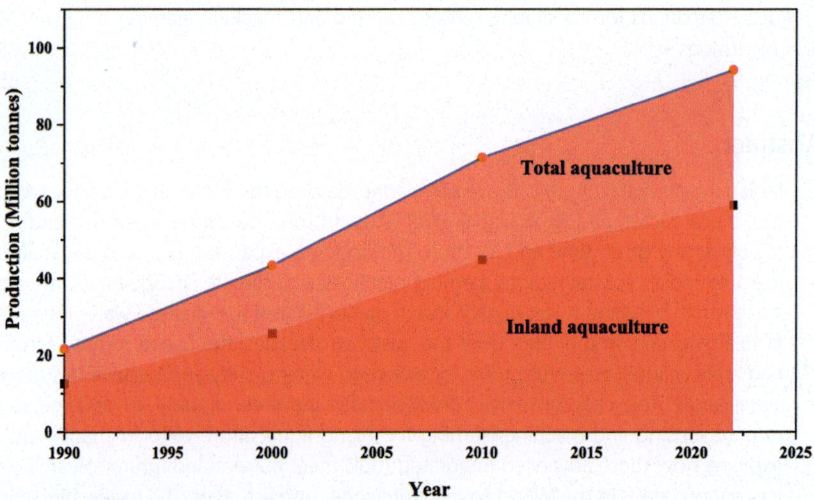

Fig. 1 Global inland aquaculture production.

production trends indicating an increasing reliance on sustainable farming practices to support food security. Inland aquaculture plays a pivotal role in meeting the growing global demand for aquatic food, offering a sustainable source of protein and other nutrients.[4,5] However, this industry faces significant challenges, particularly in the realms of water and health management. Effective feed and energy management are critical to overcome these challenges, and mismanagement in these areas can have severe consequences.[6] For maximum growth, fish and other animals require a well-balanced diet that provides all the necessary critical components. However, these nutrient requirements of a species are influenced by numerous factors such as species, age, sex, environment, various stages of larvae development, and varying health issues.[7] All varieties of fish species at least require a minimum of 35–45 nutrients for their healthy growth.[8] To achieve the most efficient growth and minimize environmental damage, it is essential to have a thorough comprehension of the feeding habits and nutrient requirements of the species.[9] The environmental impact of shore-based and open-water farming systems is significantly influenced by the low quality of feed and ineffective feeding strategies. Any surplus nutrients that are not used by the fish or shrimp are discharged into the environment and must either be absorbed or accumulated in soil which is a serious consequence affecting ecosystem.[10,11] As an outcome for this bio-accumulation in intensive aquaculture, the excessive feed results in the generation of significant quantity of organic waste discharges.[12] Accordingly, this waste exists in two forms: particulate matter, primarily consisting of uneaten food; and soluble compounds, such as excreta. These waste products further contribute to an increase in concentration of biochemical oxygen demand (BOD), nitrates, and phosphates in soil.[13] Still, over usage of fertilizers or manure in ponds also lead to a rise in the amount of nutrients, which in turn stimulates the growth of phytoplankton.[14]

Approximately half of the overall nitrogen and carbon is expelled through the gills and dissolves into the water, whereas more than half of the total phosphorus is discharged as solid particles and settles on the bottom of ponds' bed.[15] However, utilizing untreated organic waste and raw livestock manure in fish ponds heightens the likelihood of disease transmission to the fish. Few more studies are conducted on elevated levels of faecal matter in waste discharges.[16,17] Subsequently, decreased faecal integrity is also seen after replacing fish meal with plant protein sources.[18] Therefore, the inconsistent texture of the faeces and the small particles present in it pose a challenge in eliminating the faecal matter from the water, leading to elevated levels of total suspended solids in both the system and the discharged water.[19] Further, the inclusion of plant ingredients in the diet negatively affects nutrient digestibility and increases the formation of faecal waste in rainbow trout when they are fed to apparent satiation.[20] Besides, it was noted that fat digestion was specifically impaired when the animals were fed to the point of satiation. This demonstrates the significance of the feeding level in relation to the production and types of faecal waste.[19]

An illness in fish and shrimp is a condition in which the regular physiological functioning of different organs is changed, resulting in long-lasting symptoms.[21] The increase in fish infections is directly associated with the existence of different pathogenic bacteria in water and the absence of necessary nutritional components in prepared fish feed.[22] Typical signs of nutritional illnesses include reduced fertility, slower rate of growth, loss of appetite, greater vulnerability to diseases, poor movement, and abnormalities in tissue structure.[23,24] Finally, death can potentially occur under severe circumstances. Likewise, in the context of fish farming, inadequate nutrition in fish might be initially identified by decreased growth and spawning. Persistent starvation can elevate the death rate in fish populations.[25] Fingerlings, fry, and brood stock are especially susceptible to infections due to insufficient qualitative feed.[26] Above and beyond, the implementation of appropriate pond management practices, such as daily feed input, proper stocking density, and fertilization, exerts a substantial influence on the fish, and shrimp growth respectively.[27] On the contrary, the inadequate technical expertise, pumping expenses, and conflicting water demands typically result in poor water quality in ponds used for small-scale aquaculture. Overall, sustainable practices guarantee the maintenance of ecological equilibrium and the long-term ability to survive.[28]

Recently, AI technologies, such as machine learning and deep learning,[29,30] have demonstrated significant potential in the analysis of extensive datasets gathered from fish farms.[31] Through the utilization of AI techniques, fish farmers can acquire vital knowledge on fish growth patterns, feeding habits, and the impact of environmental factors on fish well-being.[31] These techniques have the capability to identify and anticipate abnormalities, illnesses, and symptoms of stress, allowing for preventative measures to address and minimize negative outcomes.[31,32] However, it is crucial to engage in ongoing research, promote data sharing, and foster collaboration among academics, industry stakeholders, and policymakers to fully utilise the capabilities of AI and achieve a sustainable aquaculture market. On the other hand, remote sensing has become essential for monitoring water bodies, assessing flood impacts and water quality, and evaluating water scarcity.[33–36] The Sentinel-2 satellite, with its moderate spatial resolutions, has proven particularly effective for regional water body mapping.[37] Its freely available multispectral imagery and frequent revisit period have made Sentinel-2 instrumental in tracking and analysing water resources.[37] Compared to the Landsat series, Sentinel-2 offers advantages with higher spatial resolution, additional spectral bands, shorter revisit intervals, and a wider swath, all enhancing its utility in water resource studies.[38] Lower-resolution sensors, like the Moderate Resolution Imaging Spectroradiometer (MODIS), are also accessible for water studies.[38] However, they are best suited for monitoring large bodies of water, such as lakes and reservoirs, rather than smaller or more fragmented regions. The water index method remains one of the most effective approaches for water body extraction, excelling in differentiating pure and mixed water pixels over conventional pixel-based and object-based classifications, especially in complex environments.[39–42]

This chapter will explore how these remote sensing technologies and artificial intelligence can advance aquaculture water management, optimizing sustainability practices. A discussion on AI integration, case studies, and emerging trends aims to shed light on future research pathways that enhance sustainable aquaculture and resource management.

2. Indian inland aquaculture

Shrimp farming is prevalent in coastal lagoons, deltas, estuaries, and transition zones next to mangrove forests in India, the world's second-largest aquaculture producer.[43] The mainland of India has a 5422 km coastline, and small-scale farmers mostly grow shrimp in 102 districts throughout nine coastal states (Fig. 2). Regarding aquaculture area and production, the various Indian states differ significantly from one another in terms of current data and in terms of development and heritage. India has had 1.4 million hectares of brackish water area accessible for coastal aquaculture since 2000.[44] It was anticipated that the eastern provinces, including West Bengal, Andhra Pradesh, and Tamil Nadu, would expand brackish water aquaculture on a bigger scale.[45] The physical circumstances may also be a contributing factor. With several lagoons and sizable rivers pouring into the sea in broad deltas, the eastern coastal plain is far more expansive than the narrow western plain. The Tapi, Godavari, Cauvery,

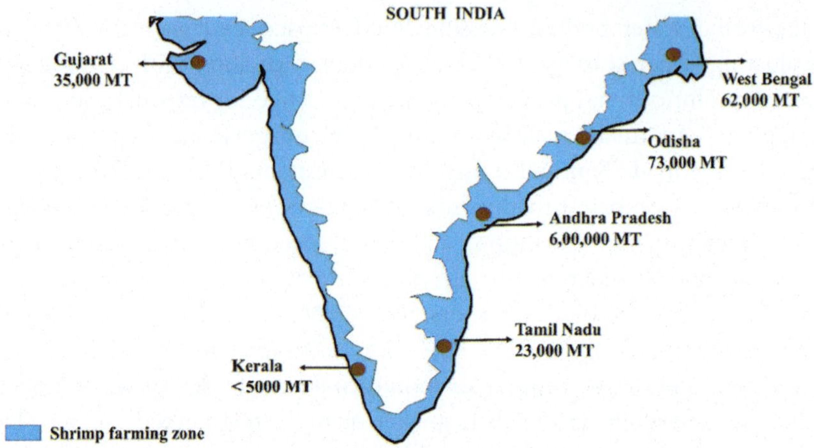

Fig. 2 South India inland aquaculture shrimp farming. Map lines delineate study areas and do not necessarily depict accepted national boundaries.

and Brahmani deltas are a few more deltaic areas that exhibit significant aquaculture concentrations.[46–48] On the other hand, Gujarat's Gulf of Kutch is the only significant coastal aquaculture area on the western coast, and estuaries and mangrove forests primarily surround it. Shrimp ponds are most widely distributed among metropolitan areas in the Godavari districts of Andhra Pradesh.[49] Because of economic reforms, Andhra Pradesh, for instance, became the primary producer of brackish water shrimps and now accounts for almost 1108 km^2 of India's total aquaculture area, which is by far the most significant percentage of ponds in the nation.[49]

Traditionally, tiger shrimp farming in coastal India utilized minimal feed, drawing primarily from the water body's natural productivity, particularly in the early stages of aquaculture growth.[50] In these early systems, feed often included pellet-based supplements, sometimes cultivated on fishpond dykes, alongside organic fertilization to stimulate primary production in the ponds.[50] This approach gradually transformed as government policies began to support aquaculture expansion. The Indian government relaxed regulations on land use for aquaculture, leading to a significant increase in land allocated for shrimp and fish farming. This policy shift contributed to a notable expansion in aquaculture practices, particularly in regions like the Upputeru catchment area of Andhra Pradesh, where intensive inland aquaculture is now a prominent feature, as illustrated in Fig. 3.

The development of inland aquaculture in India from 1980 to 2020 reflects a significant shift in techniques and technology, aligned with

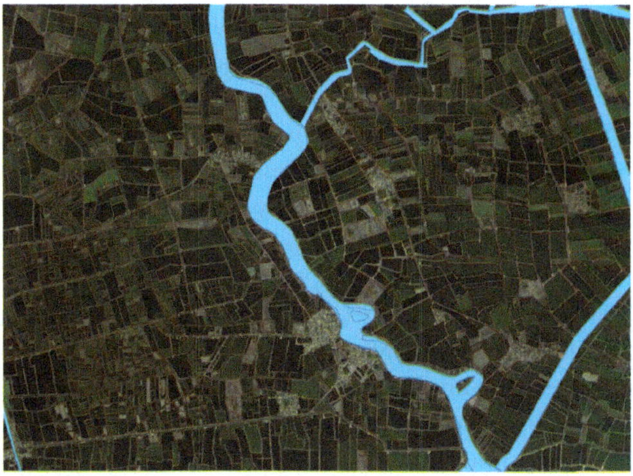

Fig. 3 Satellite image of inland aquaculture in Andhra Pradesh.

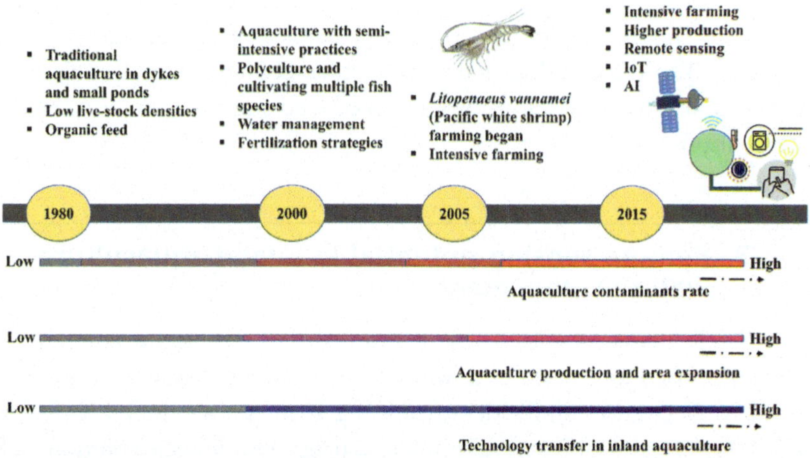

Fig. 4 Transformation of Indian inland aquaculture.

increasing demands and advancements (Fig. 4).[50] During the 1980s, aquaculture in India was based on traditional, extensive farming methods, primarily cultivating carp in ponds using natural food sources and basic, manual management practices. Traditional practices limited productivity with minimal intervention for water quality. By the early 2000s, aquaculture began shifting towards semi-intensive practices, emphasizing polyculture and cultivating multiple fish species, with added water management and fertilization strategies to boost yield.[50] This era also introduced fundamental monitoring techniques, although comprehensive tools still needed to be widely applied. A central turning point arrived in 2005 with introducing *Litopenaeus vannamei* (Pacific white shrimp) farming in India, marking a move toward intensive aquaculture practices.[50] This approach required stringent water quality, biosecurity, and feed management measures to accommodate higher-density stocking, especially in areas near coastal and inland water bodies.

From 2015 onward, internet of things (IoT) based systems were increasingly adopted, providing real-time monitoring of critical parameters like pH, dissolved oxygen, and salinity. These systems helped reduce labor demands and improved disease management by allowing immediate responses to environmental changes.[50] Towards the end of the decade, inland aquaculture in India began leveraging machine learning and remote sensing for further advancements. Machine learning applications enabled predictive analytics, optimizing harvest periods and growth tracking, while remote sensing tools such as Sentinel-2 and Landsat satellites facilitated large-scale monitoring of water quality, habitat mapping, and pollutant sources.[50] These technologies not only enhanced efficiency but also supported sustainable expansion through improved site selection and targeted interventions.

3. Remote sensing potential in inland aquaculture water management

Satellite images with 10 m or more spatial resolutions is frequently used for aquaculture pond detection.[51] For example, fish ponds in coastal areas have been successfully identified using 10-meter Sentinel-1 SAR data. However, decades-old satellite photos, usually with lesser resolutions, like those from the Landsat series, are necessary for long-term monitoring of inland aquaculture ponds.[51] Although methods for identifying coastal aquaculture ponds using Landsat data are available, they often perform less well for inland aquaculture ponds because of many important issues. First, inland lakes and ponds are often smaller than their coastal counterparts. Second, the visual scene is complicated by the fact that different types of land cover often encircle inland water bodies. Aquaculture ponds and lake surfaces can seem comparable in size since many inland aquaculture ponds close to lakes were created from previous coastal areas. It can be challenging to differentiate aquaculture ponds from natural lakes using multispectral satellite imagery due to the diverse ecology of inland lakes and ponds, particularly when examining long-term changes.

3.1 Remote sensing in inland aquaculture resources management

Combining remote sensing with machine learning is a practical approach to enhance water quality prediction in aquaculture.[52] Remote sensing enables widespread monitoring of key water quality indicators, which is particularly valuable when in-person monitoring is constrained by budget or logistical issues. Machine learning models can process and analyse the complex data obtained through remote sensing to produce accurate water quality forecasts.[53] Though primarily experimental in aquaculture settings, remote sensing paired with machine learning has shown potential for water quality prediction in larger water bodies, such as lakes. Cross-mission data integration, supported by machine learning algorithms, exemplifies the power of multi-sensor satellite data for consistent, daily water quality monitoring.[54] Techniques such as back-propagation neural networks allow the quantification of various surface water quality indicators from imagery, such as Landsat-8 data. The amalgamation of remote sensing and machine learning not only extends the scope of water quality monitoring but also significantly improves prediction accuracy. For instance, the development of multiscale modelling systems and neural networks has enabled the effective analysis of satellite data, leading to comprehensive environmental assessments and proactive aquaculture management.[55]

Google Earth has enabled access to high-resolution satellite images from as early as 2006, allowing for detailed land-use and land-cover (LULC) types analysis.[56] These historical images serve as valuable references in the LULC classification process, making it possible to detect and analyse the relationships between specific characteristics of satellite imagery and various LULC categories. In coastal regions of India, where agricultural practices and aquaculture are significant, changes in LULC have been mapped to guide sustainable and optimized farming practices. This analysis has been instrumental in identifying transformations in land use over time, supporting better decision-making for coastal agriculture and aquaculture development. The insights gained from LULC changes aid in adapting best practices to enhance productivity while preserving local ecosystems. Fig. 5 shows the change detection of LULC classification of the delta of river Godavari in 2024. It is evident that, compared with previous studies in this region, there is a drastic increase in inland aquaculture of 826.5 km^2.[57]

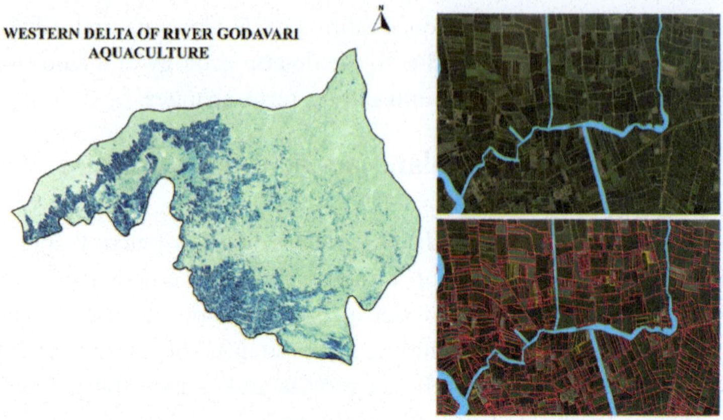

Fig. 5 Land use and land cover of western delta of river Godavari in 2024.

Table 1 Application of remote sensing in Indian inland water quality monitoring.

Location	**Data utilized**	**Parameters analysed**	**References**
South India	Sentinel-1 synthetic-aperture radar (SAR) data at 10-m spatial resolution.	Aquaculture ponds and their density.	Prasad et al.[59]
Gujarat	MODIS-Aqua images.	Turbidity and chlorophyll concentration.	Kulshreshtha and Shanumugam[60]
Orissa	Sentinel-3 satellite data.	Chlorophyll-a concentration.	Kolluru et al.[61]
Andhra Pradesh	Sentinel-2 satellite data and MODIS.	Turbidity and water quality index	Nagaraju et al.[62]
Kerala	Linear Imagine Self Scanning System (LISS III FCC Print)	Site selection for brackish water ponds	Gupta et al.[63]
Maharashtra	Multi-temporal satellite imagery	Water spread area	Anand et al.[64]

3.2 Remote sensing in inland water quality

Real-time monitoring systems integrating remote sensing and machine learning technologies represent substantial advancements in aquaculture management. These systems leverage satellite imagery and unmanned aerial vehicle (UAV) data to collect accurate information on critical environmental factors such as water quality, temperature, and chlorophyll distribution at aquaculture regions.[58] These systems can analyse data streams, predict outcomes, and optimize aquaculture operations by applying machine learning techniques like neural networks and decision trees. Predictive models based on fish behaviour and growth patterns support adaptive feed distribution, while the comprehensive analysis of historical data plays a crucial role in detecting early signs of diseases or environmental stressors, enabling rapid response to mitigate potential impacts. This approach not only improves operational efficiency but also fosters sustainability by reducing resource usage and minimizing environmental impacts. Table 1 shows the application of remote sensing in the inland aquaculture water quality in the Indian conditions.

4. AI in inland aquaculture management

4.1 AI in water quality management

Due to rapid exponential increase seen in freshwater aquacultures, demanding prerequisite is essential to prioritize the conservation and management of freshwater resources. To prevent hindering growth in miscellaneous species, it is imperative to ensure that the future demands of the water industry do not compromise the security of freshwater supplies for inhabitants. Further, the effluent generated from water usage in aquaculture is also a significant issue binding with the yield that impacts ecosystem badly if not handled properly. Artificial ponds with brackish and fresh water inland cultures are mainly accountable for massive waste water effluents.[65] This water is considered not benign to the environment as it contains high amounts of organic pollutants, nutrients such as proteins, hormones and biomasses.[65] Ensuring the success of aquaculture operations influenced by various complex elements relies heavily on the implementation of efficient water quality monitoring. Monitoring water quality is an essential concern, especially in the fish farming business and AI is increasingly employed to assist in this domain. AI has the capacity to identify trends and irregularities that may suggest issues with the water quality. This can assist farmers in implementing remedial measures prior to any detrimental impact on the fish and enables farmers to promptly address any fluctuations in water quality, hence minimizing the likelihood of fish mortality and other adverse consequences. Fig. 6 illustrates the AI process for aquaculture water quality monitoring, beginning with data collection, then data pre-processing and partitioning, and concluding with applying the refined data to aquaculture operations. This workflow underscores the systematic approach necessary to transform raw data into actionable insights that support sustainable aquaculture management.

4.2 AI in nutrient management

An optimal diet, rich in nutrients, that is easily consumed and digested, is crucial for promoting the growth and development of aquaculture species.[66] Most of the recently hatched fish and shrimp rely on live food as a critical primary diet. Copepods, artemia and rotifers are widely regarded as the standard diet in terms of nutrition for a variety of fish especially in marine aquaculture. Also, fish demands vital amino acids found in proteins to facilitate development, tissue regeneration, overall well-being, and reproduction. Besides, the performance of fish is influenced by the quality of protein. Fish meal, although of superior quality, is nevertheless quite costly. On the other hand, more affordable protein sources can be utilized as long as they fulfil the necessary amino acid needs of the fish. Phospholipids, such as soybean lecithin, is also necessary for crustaceans,

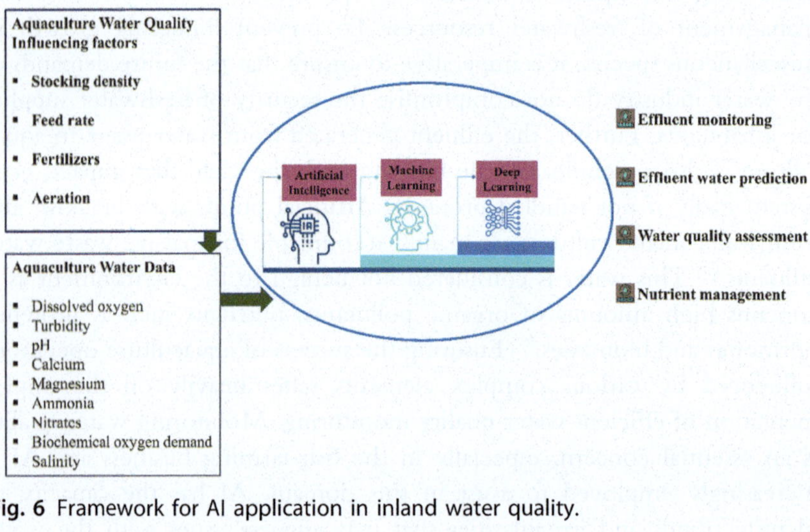

Fig. 6 Framework for AI application in inland water quality.

mollusks, and perhaps young fish. The requirements of fatty acid varies among species. Herbivorous and omnivorous fish often flourish on plant oils rich in shorter-chain fatty acids, but carnivorous species normally necessitate longer-chain fatty acids, exemplified by those found in marine fish oil.

In the interim, the escalation of multi-omics has given rise to the concept of 'precision nutrition' in the field of medicine and clinical research. Meanwhile, precise nutrition in aquaculture has been gaining significant interest. Practical implementation of precise nutritional regulation aims to exactly deliver all necessary nutrients to the fish, considering the specific species, evelopmental stage, and breeding environment.[67] This approach helps to minimize environmental impact and enhance financial gains. The assessment of precision nutritional intervention must encompass the various features such as precision feeding, precision metabolism, precision output, and precision support simultaneously.[68]

5. Economic and environmental benefits of AI and remote sensing in aquaculture

Aquaculture has significant economic advantages and hence it stimulates employment opportunities and ensures a dependable food source besides contributing to the conservation of animals and the maintenance of functional waterfront areas. Fish is a nutritious and environmental-friendly protein source that can support global efforts to maintain a balanced diet. Despite the fact, the aquaculture production is prominent for global food security and achieving sustainable development goals (SDGs), further it has significant resource consumption and adverse environmental impacts too. These include excessive usage of water and energy, emission of greenhouse gases (GHGs), eutrophication, and degradation of aquatic, benthic, and coastal habitats and ecosystems. Still, problems such as species invasion, changes in biodiversity, loss of mangrove habitats, and the excessive usage of resources derived from fish are also quite common.[69]

Developing and sustaining the aquaculture business is a complex endeavour that requires meticulous adjustments to get optimal results. The intelligent aquaculture industry relies on advanced technology and efficient algorithms. In this scenario, the utilization of sensing and machine learning algorithms is crucial as they facilitate prompt decision-making to mitigate hazards and enable autonomous real-time monitoring of the environment.

6. Challenges and opportunities in climate-smart aquaculture water management

A key aspect of future-proof aquaculture systems against escalating climate stressors is climate-smart aquaculture (see Fig. 7). According to recent studies, shrimp growth rates can be slowed by 15–20 % and their susceptibility to disease can increase tenfold with each 1 °C increase in water temperature over ideal values.[70] Alarming diurnal pH swings and dissolved oxygen crashes throughout the summer months were found by water unbalance profiling across South Indian brackish water systems. These conditions are directly related to the 30–40 % crop losses that farmers claim. The climate-smart aquaculture framework can meet these biophysical thresholds by using precision water management and Internet of Things (IoT) sensors to forecast hypoxic episodes five hours in advance with 98 % accuracy. The technological conundrum in the application of climate-smart aquaculture offers both tremendous potential and daunting obstacles. Through machine learning-driven oxygenation schedules, smart aerator pilot deployments frequently lowered energy utilization by 40–50 %; yet, the digital divide is still quite evident.[71] Sensor alerts were unable to be sent during crucial monsoon periods when cellular networks

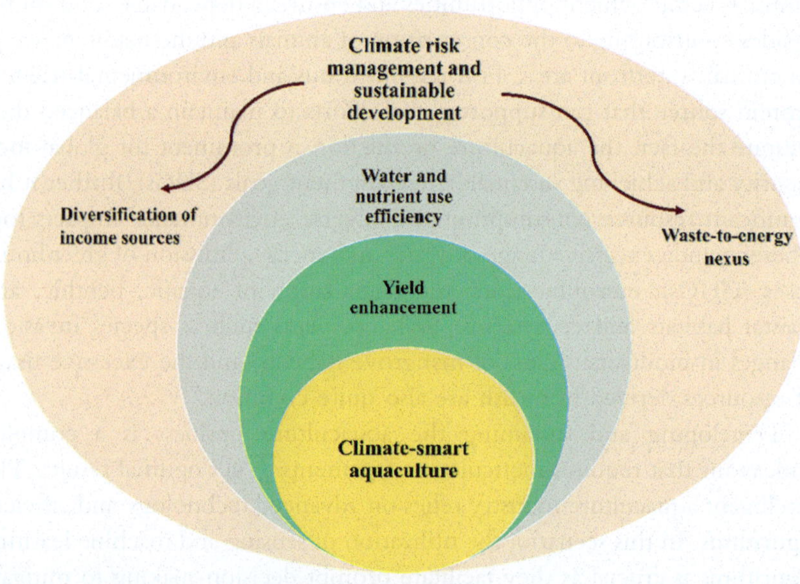

Fig. 7 Climate-smart aquaculture opportunities.

were degraded, according to field tests conducted in South Asian nations.[71] When water quality issues occur during these periods of significant precipitation, this connectivity gap is very worrisome. Aquaculture governance needs to undergo a socio-technical revolution in order to scale climate-smart aquaculture. This multifaceted strategy provides a roadmap for climate adaptation that is both realistically achievable and supported by science.

7. Future directions for integrating remote sensing and machine learning

The application of computer vision and AI to aquaculture offers a revolutionary technique to improve fish output, a vital source of protein for the world's diet. Maintaining ideal water quality and early illness identification is crucial to averting large mortality events as semi-intensive and intensive aquaculture becomes more common. With over 90% accuracy, recent developments in real-time monitoring systems driven by deep learning algorithms have shown remarkable efficacy in forecasting disease outbreaks (such as Vibrio infections in shrimp).[72] Losses can now be significantly decreased by using AI-driven picture analysis of fish activity and gill discoloration to identify stress responses 48 h before clinical signs manifest; furthermore, by identifying chlorophyll-a, turbidity, and thermal anomalies, essential markers of aquaculture health, multispectral and hyperspectral remote sensing from next-generation satellites permits extensive water quality studies.[73] Nevertheless, there are also issues with applying these models for small-scale inland ponds, where limits in spatial resolution and cloud cover make it challenging to conduct continuous monitoring.

High-resolution aquaculture surveillance is being revolutionized by emerging autonomous sensing platforms, such as remotely operated vehicles (ROV), autonomous underwater vehicles (AUV), and small unoccupied aircraft systems (SUAS). In order to maximize feed efficiency in offshore aquaculture, sUAS-mounted hyperspectral sensors (400–1000 nm) have effectively evaluated tissue nitrogen concentration and kelp canopy biomass.[74] Meanwhile, side-scan sonar and deep learning-classified underwater imaging can map submerged aquaculture equipment (such as cages and longlines) and identify biofouling with error rates.[75] Despite these developments, there are still important gaps: (1) due to cloud obstruction, satellites are less effective for inland ponds, requiring gap-filling with radar-based (Sentinel-1) or UAV-assisted methods; (2) water column light attenuation makes it difficult for commercial SUAS sensors to quantify subsurface kelp; and (3) the cost of AUV deployment is still too high for small farms, underscoring the need for inexpensive ROV with edge AI processing.

Future studies must focus on multi-sensor data fusion in order to close these gaps. They should use Google Earth Engine's cloud computing to harmonize data from satellites, drones, and in-situ sensors. By combining satellite data with radiative transfer models, physics-informed neural networks could enhance forecasts of kelp biomass. Swarm robotics, which uses fleets of inexpensive AUV, may also make it possible to monitor offshore farms continuously while lowering operating expenses. Real-time decision-making without reliance on cloud connectivity is possible with edge AI deployment on SUAS and ROVs, which is essential for remote aquaculture locations. Next-generation remote sensing and machine learning systems can provide a scalable, precision aquaculture framework by tackling these technological and analytical issues, guaranteeing sustainable fish production in the face of climate change and rising global demand.

8. Conclusions

This chapter explores the essential role that remote sensing and machine learning play in advancing sustainable aquaculture. Given the rising global population, leveraging technology has become vital to inland aquaculture to safeguard ecosystems and boost production. Both remote sensing and AI, independently and combined, serve as valuable tools for monitoring, managing, and optimizing aquaculture practices. They address significant challenges in resource management, environmental stewardship, and productivity enhancement. Remote sensing provides extensive and timely insights into aquatic ecosystems, even in remote or challenging areas, while AI offers robust data analysis and predictive modelling capabilities that enhance aquaculture management. These advancements enable precise decision-making on feeding schedules, disease prevention, and market projections. Integrating remote sensing with AI applications creates a robust synergy that fosters efficient and sustainable aquaculture, contributing to better resource allocation, minimized ecological impacts, and improved aquatic ecosystem health.

As the demand for seafood continues to rise, investing in further research and development to refine and expand these technologies across diverse aquaculture systems is critical. The aquaculture industry can create a more sustainable and resilient future by embracing these advancements. In summary, the progression of inland aquaculture largely hinges on adopting innovative techniques, which should be prioritized to ensure a balanced approach to industry growth and environmental preservation.

References

1. Slobodskova VV, Dovzhenko NV, Kukla SP, Chelomin VP, Mazur AA. Express diagnosis and prediction of remote mass mortality of scallop mizuhopecten yessoensis in mariculture farms using biomarkers. *J Mar Sci Eng* 2024;**12**(7):1151.
2. Verdegem M, Buschmann AH, Latt UW, Dalsgaard AJ, Lovatelli A. The contribution of aquaculture systems to global aquaculture production. *J World Aquaculture Soc* 2023;**54**(2):206–250.
3. Ahmad AL, Chin JY, Harun MH, Low SC. Environmental impacts and imperative technologies towards sustainable treatment of aquaculture wastewater: A review. *J Water Process Eng* 2022;**46**:102553.
4. Fiorella KJ, Okronipa H, Baker K, Heilpern S. Contemporary aquaculture: implications for human nutrition. *Curr OpBiotechnol* 2021;**70**:83–90.
5. Boyd CE, D'Abramo LR, Glencross BD, Huyben DC, Juarez LM, Lockwood GS, et al. Achieving sustainable aquaculture: historical and current perspectives and future needs and challenges. *J World Aquac Soc* 2020;**51**(3):578–633.
6. Majeed Y, Khan MU, Waseem M, Zahid U, Mahmood F, Majeed F, et al. Renewable energy as an alternative source for energy management in agriculture. *Energy Rep* 2023;**10**:344–359.
7. Finke MD, Oonincx D. *Insects as food for insectivores. Mass production of beneficial organisms;* Academic Press, 2023; pp. 511–540.
8. Colombo SM, Roy K, Mraz J, Wan AH, Davies SJ, Tibbetts SM, et al. Towards achieving circularity and sustainability in feeds for farmed blue foods. *Rev Aquaculture* 2023;**15**(3):1115–1141.
9. Li H, Cui Z, Cui H, Bai Y, Yin Z, Qu K. A review of influencing factors on a recirculating aquaculture system: environmental conditions, feeding strategies, and disinfection methods. *J World Aquaculture Soc* 2023;**54**(3):566–602.
10. Mullungal MN, Peediyakkathodi S, Bibi S, Ratheesh Kumar CS, Abu-Dieyeh M. *Nutrient contamination in marine environment. Contaminated land and water: remediation and management;* Springer International Publishing: Cham, 2024; pp. 15–33.
11. Tabrett S, Ramsay I, Paterson B, Burford MA. A review of the benefits and limitations of waste nutrient treatment in aquaculture pond facilities. *Rev Aquaculture* 2024.
12. Naz S, Arshad M, Majeed S, Maqaddas S, Habib SS, Kesbiç OS, et al. Assessing heavy metal contamination in commonly used fertilizers for polyculture fish ponds and its implications for human health: a comprehensive investigation. *Biol Trace Elem Res* 2024:1–7.
13. Vivaldi GA, Salcedo FP, Tallou A. *Application biol Effl inmicroirrigation crop prod.* Elsevier Science, 2024; pp. 393–410.
14. Vasseghian Y, Nadagouda MM, Aminabhavi TM. Biochar-enhanced bioremediation of eutrophic waters impacted by algal blooms. *J Environ Manag* 2024;**367**:122044.

15. Kumari S, Dwivedi S, Khan ME, Nayanam S, Dhasmana A, Malik S. *The challenges of wastewater and wastewater management. Advanced and innovative approaches of environmental biotechnology in industrial wastewater treatment;* Springer Nature Singapore: Singapore, 2023; pp. 99–121.
16. Shanks OC, Diedrich A, Sivaganesan M, Willis JR, Sharifi A. Quantitative fecal source characterization of urban municipal storm sewer system outfall 'Wet'and 'Dry'weather discharges. *Water Res* 2024:121857.
17. Minier P, Esculier F, Tassin B, Chatzis K. Can sewerage be considered safe management of human feces?*City Environ Interact* 2023;**19**:100107.
18. Horstmann P, Maas RM, de Boer XV, de Jong TM, Staessen TW, Kokou F, et al. Effect of dietary protein source and ingredient grinding size on fish performance, faecal waste production and characteristics of yellowtail kingfish (Seriola lalandi) fed restrictively and to apparent satiation. *Aquaculture* 2023;**562**:738875.
19. Singh PK, Kumar U, Kumar I, Dwivedi A, Singh P, Mishra S, et al. Critical review on toxic contaminants in surface water ecosystem: sources, monitoring, and its impact on human health. *Environ Sci Pollut Res* 2024:1–35.
20. Zhang Y, Maas RM, Horstmann P, Prakash S, Staessen TW, Kals J, et al. Effect of carbohydrate source, feeding level (restricted vs. satiation) and their combination on nutrient digestibility, bile acid balance, faecal waste production and characteristics of yellowtail kingfish (Seriola lalandi). *Aquaculture Rep* 2024;**36**:102179.
21. Haridevamuthu B, Raj D, Arshad A, Arockiaraj J. Comprehensive review of Argulus infestations in aquaculture: biological impacts and advanced management strategies. *Fish & Shellfish Immunol.* 2024:109851.
22. Senthamarai MD, Rajan MR, Bharathi PV. Current risks of microbial infections in fish and their prevention methods: a review. *Microb pathogenesis* 2023;**185**:106400.
23. Styne DM. *Disorders of Growth. Pediatric endocrinology: a clinical handbook;* Springer International Publishing: Cham, 2023; pp. 55–119.
24. Singh J, Kottarath SK, Bano F, Farhan M, Tripathi AK, Srivastava Y. *Hormonal manifestations of endocrine diseases. Clinical applications of biomolecules in disease diagnosis: a comprehensive guide to biochemistry and metabolism;* Springer Nature Singapore: Singapore, 2024; pp. 257–287.
25. Araujo GS, Silva JW, Cotas J, Pereira L. Fish farming techniques: current situation and trends. *J Mar Sci Eng* 2022;**10**(11):1598.
26. Yue GH, Wang L, Sun F, Yang ZT, Wong J, Wen YF, et al. Improving growth, omega-3 contents, and disease resistance of Asian seabass: status of a 20-year family-based breeding program. *Rev Fish Biol Fish* 2024;**34**(1):91–110.
27. Bauer RT. *Fisheries and aquaculture. Shrimps: their diversity, intriguing adaptations and varied lifestyles;* Springer International Publishing: Cham, 2023; pp. 583–655.
28. Hariram NP, Mekha KB, Suganthan V, Sudhakar K. Sustainalism: an integrated socio-economic-environmental model to address sustainable development and sustainability. *Sustainability* 2023;**15**(13):10682.
29. Srivastava A, Maity R. Assessing the potential of AI–ML in urban climate change adaptation and sustainable development. *Sustainability* 2023;**15**(23):16461. https://doi.org/10.3390/su152316461.
30. Maity R, Srivastava A, Sarkar S, Khan MI. Revolutionizing the future of hydrological science: Impact of machine learning and deep learning amidst emerging explainable AI and transfer learning. *Applied Computing and Geosciences* 2024;**24**:100206. https://doi.org/10.1016/j.acags.2024.100206.
31. Ubina NA, Lan HY, Cheng SC, Chang CC, Lin SS, Zhang KX, et al. Digital twin-based intelligent fish farming with artificial intelligence Internet of Things (AIoT). *Smart Agric Technol* 2023;**5**:100285.

32. Mandal A, Ghosh AR. Role of artificial intelligence (AI) in fish growth and health status monitoring: a review on sustainable aquaculture. *Aquaculture Int* 2024;**32**(3): 2791–2820.
33. Lausch A, Bannehr L, Berger SA, Borg E, Bumberger J, Hacker JM, et al. Monitoring water diversity and water quality with remote sensing and traits. *Remote Sens* 2024;**16**(13):2425.
34. Priyanka N, Sheeja JP, Bhaskaran G. Remote sensing in water resource monitoring a comprehensive review on monitoring the quality and quantity of the water resources using remote sensing applications. *Adv Water Resour Plan Sustainability* 2023:1–22.
35. Gani MA, Sajib AM, Siddik MA, Moniruzzaman M. Assessing the impact of land use and land cover on river water quality using water quality index and remote sensing techniques. *Environ Monit Assess* 2023;**195**(4):449.
36. Srivastava A, Jain S. Remote Sensing for Environment Assessment: Multispectral, Hyperspectral, and Thermal Imaging Applications. In *Remote Sensing for Environmental Monitoring;* Saritha, V., Pande, C. B., Singh, R., Shahid, M., Eds.; Springer: Singapore, 2025https://doi.org/10.1007/978-981-96-5546-5_1
37. Latwal A, Rehana S, Rajan KS. Detection and mapping of water and chlorophyll-a spread using Sentinel-2 satellite imagery for water quality assessment of inland water bodies. *Environ Monit Assess* 2023;**195**(11):1304.
38. Adjovu GE, Stephen H, James D, Ahmad S. Overview of the application of remote sensing in effective monitoring of water quality parameters. *Remote Sens* 2023;**15**(7): 1938.
39. Huang W, Yu A, Xu Q, Sun Q, Guo W, Ji S, et al. Sea ice extraction via remote sensing imagery: algorithms, datasets, applications and challenges. *Remote Sens* 2024;**16**(5):842.
40. Cheng G, Huang Y, Li X, Lyu S, Xu Z, Zhao H, et al. Change detection methods for remote sensing in the last decade: a comprehensive review. *Remote Sens* 2024;**16**(13):2355.
41. Srivastava A, Chinnasamy P. Investigating impact of land-use and land cover changes on hydro-ecological balance using GIS: insights from IIT Bombay, India. *SN Appl. Sci.* 2021;**3**:343. https://doi.org/10.1007/s42452-021-04328-7.
42. Srivastava A, Chinnasamy P. Watershed development interventions for rural water safety, security, and sustainability in semi-arid region of Western-India. *Environ Dev Sustain* 2024;**26**:18231–18265. https://doi.org/10.1007/s10668-023-03387-7.
43. Basu A, Bhattacharjee A. *A tale of crab collectors and fatteners: negotiating climate change in indian sundarbans. Climate crisis: adaptive approaches and sustainability;* Springer Nature Switzerland: Cham, 2024; pp. 511–527.
44. Sidiq MJ, Ganie PA, Sarma D. *Freshwater aquaculture assuring the food and livelihood security in india: a decadal perspective (2011–2022). Aquaculture and conservation of inland coldwater fishes;* Springer Nature Singapore: Singapore, 2024; pp. 101–125.
45. Mondal M, Guchhait R, Mustak S. *Analysis of ecological vulnerability behind the land conversion from agriculture to aquaculture of purba Medinipur District in West Bengal, India. Advanced remote sensing for urban and landscape ecology;* Springer Nature Singapore: Singapore, 2023; pp. 251–281.
46. Rahman S, Rahman S, Ahsan A, Eslamian S. *Climate change impacts on water resources and sediment yield in South Asia. Handbook of climate change impacts on river basin management;* CRC Press, 2024; pp. 193–228.
47. Bulian F, Alessandri L, Attema P, Sevink J. Bronze age to Roman period salt production in the coastal areas of peninsular Italy: palaeoenvironments, production methods and archaeological evidence. *Quaternary Sci Rev* 2024;**344**:108930.
48. Ogunleye RO, Oluwakuse MM, Ajidahun J, Ale TO, Oluwakuse A. Sedimentologic characteristics and correlations of coastal plain sand and alluvium deposits within Lagos environs southwestern Nigeria. *Arab J Geosci* 2024;**17**(6):1–7.

49. Nagaraju TV, Sunil BM, Chaudhary B, Prasad CD, Gobinath R. Prediction of ammonia contaminants in the aquaculture ponds using soft computing coupled with wavelet analysis. *Environ Pollut* 2023;**331**:121924.
50. Gottumukkala SB, Thotakura VN, Gvr SR, Chinta DP, Park R. Balancing aquaculture and estuarine ecosystems: machine learning–based water quality indices for effective management. *Environ Sci Pollut Res* 2024:1–7.
51. Liu C, Xu J, Xu K, Yu J. Mapping large-scale aquaculture ponds in Jiangsu Province, China: an automatic extraction framework based on Sentinel-1 time-series imagery. *Aquaculture* 2024;**581**:740441.
52. Tian S, Guo H, Xu W, Zhu X, Wang B, Zeng Q, et al. Remote sensing retrieval of inland water quality parameters using Sentinel-2 and multiple machine learning algorithms. *Environ Sci Pollut Res* 2023;**30**(7):18617–18630.
53. Chen P, Wang B, Wu Y, Wang Q, Huang Z, Wang C. Urban river water quality monitoring based on self-optimizing machine learning method using multi-source remote sensing data. *Ecol Indic* 2023;**146**:109750.
54. Drogkoula M, Kokkinos K, Samaras N. A comprehensive survey of machine learning methodologies with emphasis in water resources management. *Appl Sci* 2023;**13**(22):12147.
55. Wu W, Wan L. *Coastal ecological and environmental management under multiple anthropogenic pressures: a review of theory and evaluation methods. Current trends in estuarine and coastal dynamics;* Elsevier, 2024; pp. 385–415.
56. Ganjirad M, Bagheri H. Google earth engine-based mapping of land use and land cover for weather forecast models using Landsat 8 imagery. *Ecol Inform* 2024;**80**:102498.
57. Ahmad H, Abdallah M, Jose F, Elzain HE, Bhuyan MS, Shoemaker DJ, et al. Evaluation and mapping of predicted future land use changes using hybrid models in a coastal area. *Ecol Inform* 2023;**78**:102324.
58. Chen G, Wang Y, Gu X, Chen T, Liu X, Lv W, et al. Estimating water quality parameters of freshwater aquaculture ponds using UAV-based multispectral images. *Agric Water Manag* 2024;**304**:109088.
59. Prasad KA, Ottinger M, Wei C, Leinenkugel P. Assessment of coastal aquaculture for India from Sentinel-1 SAR time series. *Remote Sens* 2019;**11**(3):357.
60. Kulshreshtha A, Shanmugam P. Estimation of underwater visibility in coastal and inland waters using remote sensing data. *Environ Monit Assess* 2017;**189**:1–18.
61. Kolluru S, Gedam SS, Chander S, Sahay A. Development of chlorophyll-a concentration estimation algorithm for turbid productive inland waters in India. *Geocarto Int* 2023:2171143.
62. Nagaraju TV, Malegole SB, Chaudhary B, Ravindran G, Chitturi P, Chinta DP. Novel assessment tools for inland aquaculture in the western Godavari delta region of Andhra Pradesh. *Environ Sci Pollut Res* 2024;**31**(25):36275–36290.
63. Gupta MC, Krishnarajan VP, Nayak S. Brackish water aquaculture site selection in coastal track of cannanore (Kerala) using remote sensing and GIS techniques. *J Indian Soc Remote Sens* 2001;**29**:79–83.
64. Anand A, Kantharajan G, Krishnan P, Abdul Hakeem K, Sai Santosh K, Srinivasa Rao C, et al. Mapping the potential areas for enclosure fish culture in tropical reservoirs: geo-spatial solutions for sustainable aquaculture expansion. *Spat Inf Res* 2019;**27**:733–747.
65. Meerhoff M, Beklioğlu M. *Shallow lakes and ponds. Wetzel's limnology;* Academic Press, 2024; pp. 859–892.
66. Kalaiselvan P, Ranjan A, Nazir MI, Suresh E. Exploring ontogenic development and larval rearing of striped murrel (Channa striatus). *Aquac. Int* 2024:1–44.
67. Isibor PO. *Nanochitosan-based enhancement of fish breeding programs. Nanochitosan-based enhancement of fisheries and aquaculture: aligning with sustainable development goal 14–life below water;* Springer Nature: Cham, 2024; pp. 239–264 (Switzerland).

68. Antwi J. Precision nutrition to improve risk factors of obesity and type 2 diabetes. *Curr Nutr Rep* 2023;**12**(4):679–694.
69. Aransiola SA, Zobeashia SL, Ikhumetse AA, Musa OI, Abioye OP, Ijah UJ, et al. Niger delta mangrove ecosystem: biodiversity, past and present pollution, threat and mitigation. *Regional Stud Mar Sci* 2024:103568.
70. Mugwanya M, Dawood MA, Kimera F, Sewilam H. Anthropogenic temperature fluctuations and their effect on aquaculture: a comprehensive review. *Aquac Fish* 2022;**7**(3):223–243.
71. Das S. *Challenges in predicting extreme weather events over the South Asian region. Extreme natural events: sustainable solutions for developing countries;* Springer Nature Singapore: Singapore, 2022; pp. 51–106.
72. Mishra A, Kim HS, Kumar R, Srivastava V. Advances in Vibrio-related infection management: an integrated technology approach for aquaculture and human health. *Crit Rev Biotechnol* 2024;**44**(8):1610–1637.
73. Mandal A, Ghosh AR. Role of artificial intelligence (AI) in fish growth and health status monitoring: a review on sustainable aquaculture. *Aquac. Int* 2024;**32**(3): 2791–2820.
74. Bell TW, Nidzieko NJ, Siegel DA, Miller RJ, Cavanaugh KC, Nelson NB, et al. The utility of satellites and autonomous remote sensing platforms for monitoring offshore aquaculture farms: a case study for canopy forming kelps. *Front Mar Sci* 2020;**7**:520223.
75. Vo TTE, Ko H, Huh JH, Kim Y. Overview of smart aquaculture system: focusing on applications of machine learning and computer vision. Electronics. 202. 10(22), 2882.

SECTION D

Synthesis and future directions

CHAPTER SIXTEEN

Synthesizing climate-smart water resource management with recently developed technologies (AI-ML, RS-GIS, and LLMs)

Rajib Maity* and **Aman Srivastava**
Department of Civil Engineering, Indian Institute of Technology Kharagpur, Kharagpur, West Bengal, India
*Corresponding author. e-mail address: rajib@civil.iitkgp.ac.in

Contents

Abstract

This study provides a comprehensive synthesis of cutting-edge technologies and their potential to address the increasingly complex challenges from the book 'Climate Smart Water Resource Management.' The synthesis begins with a recap of key themes explored throughout the book, emphasizing the pivotal roles of Artificial Intelligence and Machine Learning (AI-ML), Remote Sensing and Geographic Information Systems (RS-GIS), and Large Language Models (LLMs) in improving decision-making processes, real-time monitoring, and stakeholder engagement in climate-smart water management. By showcasing the convergence of these technologies, it highlights the need for interdisciplinary collaboration to develop effective and adaptive solutions. Recent advancements in AI-ML, RS-GIS, and LLMs are further explored, and actionable strategies for integrating these tools into policy frameworks and governance structures are documented. Strategic recommendations emphasize policy innovation, bridging the research-to-application gap and empowering communities and decision-makers through technology adoption. By reflecting on the journey from traditional methods to smart, data-driven approaches, this synthesis outlines a future vision for climate-resilient water management that is equitable, inclusive, and sustainable. In closing, this chapter calls for continued exploration and commitment to sustainable water

Advances in Chemical Pollution, Environmental Management and Protection, Volume 15
ISSN 2468-9289, https://doi.org/10.1016/bs.apmp.2025.11.009

management practices that are inclined toward technological advancements. It serves as a call to action for researchers, policymakers, and communities to embrace innovation and collaboration as essential to securing water resources for future generations in an increasingly uncertain climate landscape.

Keywords: Climate change adaptation, Climate resilience engineering, Decentralized water management, Predictive ecosystem modelling, Adaptive governance models.

1. Introduction

The global water crisis is a pressing issue, with over 2 billion people lacking access to safe drinking water.[1] Climate change exacerbates this challenge, leading to more frequent and intense droughts and floods.[2,3] Water scarcity affects approximately 40 % of the global population, which is projected to increase as demand outpaces supply.[4] Traditional water management practices commonly struggle to cope with these challenges, leading to inefficient water use, water pollution, and ecosystem degradation.[5–7] To monitor water resources, optimize water use, and improve water quality, climate-smart water resource management provides a more sustainable approach by integrating technology, such as Artificial Intelligence and Machine Learning (AI-ML),[8–15] Remote Sensing and Geographic Information Systems (RS-GIS),[16–21] and Large Language Models (LLMs).[22,23] For instance, AI-powered predictive modelling can help anticipate water shortages and inform decision-making, while remote sensing can track changes in water bodies and vegetation health. By adopting climate-smart water management practices, countries can reduce water consumption by up to 20 % and increase agricultural productivity by up to 15 %.[24] Additionally, these practices can help mitigate the impacts of climate change and protect vulnerable communities. The book has journeyed through these evolving dimensions, beginning with foundational understandings of water resource challenges, progressing through state-of-the-art methodologies, and examining the role of advanced technologies like AI-ML, RS-GIS, and LLMs in shaping the future of water management.

Each technological advancement discussed in this book brings a unique perspective and set of capabilities to the field of water resource management, creating a multi-faceted, data-driven framework for addressing current and emerging challenges. AI-ML is at the forefront of boosting predictive models, which are crucial for precise resource allocation and early warning systems in flood and drought management. For instance, AI-ML models leveraging Deep Learning (DL) techniques and neural networks have demonstrated upwards of 90 % predictive accuracy in water quality monitoring, helping stakeholders proactively manage contamination and conserve freshwater ecosystems.[25] RS-GIS extends the predictive capabilities of AI-ML by providing geospatial data for mapping water bodies,[26] delineating groundwater potential zones,[27] identifying drought-prone areas,[28] estimating suspended sediment yield,[29] agricultural water management,[30] and providing real-time flood monitoring.[31] Advances in satellite imagery and high-resolution data allow for detailed assessments of watershed health and flow patterns, with tools like MODIS, Landsat, and Sentinel-2 satellites providing near-continuous data critical for adaptive management in dynamic climates.[32] Introducing LLMs is revolutionizing how complex data is communicated and interpreted. With NLP-based advancements, LLMs convert intricate scientific data into accessible interpretations, bridging the gap between data scientists, policymakers, and the public.[33] LLMs (if employed) may rapidly process large datasets from AI-ML and RS-GIS outputs, thereby aiding faster policy adjustments and community outreach. Together, these technologies integrate immediate responsiveness with long-term strategic planning, fundamentally transforming traditional water resource paradigms. This interdisciplinary approach marks a critical shift towards a resilient, climate-smart future.

This chapter thus aims to crystallize the book's core themes into actionable recommendations, providing a roadmap for future advancements. By examining the synergies and practical challenges in combining AI-ML, RS-GIS, and LLMs, this chapter provides strategies for implementing a cohesive climate-smart water management framework. Key areas for future innovation are highlighted, such as creating more inclusive, equitable AI systems and expanding LLM functionalities to support community-specific water policies. The AI-ML and RS-GIS innovations explored provide highly effective tools for predictive analytics and resource mapping; however, implementing these at scale requires a multi-sector approach that includes policy alignment, community engagement, and infrastructure readiness. This final chapter thus serves as both a summary of current understandings and a call for continued research, collaboration, and policy development. Together, the knowledge and strategies discussed herein underscore the importance of harnessing the full potential of emerging technologies for sustainable water resource management in a changing climate.

2. Cross-disciplinary integration for water management

This concluding chapter proposes integrating AI-ML, RS-GIS, and LLMs in water resource management to revolutionize the monitoring, prediction, and response to water-related challenges. As discussed, AI-ML algorithms, with their ability to analyze vast datasets of water flow, consumption patterns, and climate projections, can unlock critical understandings that were previously inaccessible. Combining AI-ML with RS-GIS technologies can transform these understandings into actionable knowledge through precise geospatial mapping and environmental monitoring. RS-GIS technologies can provide real-time data on hydrological changes, including water distribution, seasonal fluctuations, and pollution hotspots, which can inform AI-ML models to generate highly localized and context-aware predictions. This integration is particularly useful for early warning systems in drought-prone or flood-risk areas, where precise, data-driven information can support timely decision-making to mitigate risks to populations and infrastructure. Furthermore, LLMs can add another powerful dimension to this integrated framework by improving communication and decision support among diverse stakeholders. By applying NLP capabilities, LLMs can translate complex data outputs from AI-ML and RS-GIS models into clear and accessible language, empowering policymakers, community leaders, and the public to make informed choices. For example, an LLM could summarize real-time flood prediction data and propose actionable measures for stakeholders, from city planners to local emergency responders. This approach particularly benefits community-led water management, where clear communication and inclusive decision-making are crucial. A truly collaborative approach between these technologies can create adaptable, resilient, and user-friendly water management systems. Proposing pilot programs that utilize AI-ML for forecasting, RS-GIS for real-time monitoring, and LLMs for public engagement could serve as a blueprint for other regions facing water stress. Ultimately, the synergy between these technologies may boost data-driven water management and democratize access to information, empowering communities to participate actively in sustainable resource use and conservation. To this end, Table 1 presents a holistic overview of how AI-ML, RS-GIS, and LLMs can synergistically improve climate-smart water resource management by addressing diverse challenges across data analysis, risk assessment, resource allocation, and community engagement. When combined, these

Table 1 Synergistic contributions of AI-ML, RS-GIS, and LLMs in climate-smart water resource management.

Aspect / dimension	AI-ML	RS-GIS	LLMs	Proposed synergistic outcome
Data Analysis and Forecasting	Utilizes machine learning algorithms for high-accuracy water demand forecasts, precipitation models, and drought or flood probability assessments.	Provides spatial data on water sources, weather patterns, and environmental factors in real-time, supporting precise geographical insights.	Translates complex model outputs into user-friendly language, allowing stakeholders to interpret predictions easily.	An accessible, real-time forecast system enabling local governments and communities to prepare for climate-induced water fluctuations.
Risk Mapping and Early Warning	Employs predictive analytics to identify high-risk zones based on historical data and environmental trends.	Maps vulnerable areas with satellite imagery, integrating data on elevation, vegetation, and urban density.	Communicates alerts and emergency instructions, adapting messages to various audiences for clarity.	A multi-tiered early warning system that integrates prediction, spatial insights, and community guidance for rapid response.
Resource Allocation and Efficiency	Optimizes water distribution through smart algorithms that adapt to consumption patterns and climate data.	Monitors water bodies and resources remotely to assess availability and utilization rates.	Provides automated summaries on water availability and distribution plans, aiding decision-makers and local organizations.	An AI-GIS-driven allocation system that maximizes water use efficiency and minimizes wastage, informed by real-time, location-specific data.
Stakeholder Engagement and Education	Adapts models to community-specific needs, creating local forecasts and scenarios.	Offers visual maps that make climate impacts and resource changes tangible for communities.	Generates accessible reports, FAQs, and educational content on water issues and solutions, fostering informed engagement.	Increased community involvement through accessible data visualizations and user-friendly explanations, enabling localized water management strategies.

(continued)

Table 1 Synergistic contributions of AI-ML, RS-GIS, and LLMs in climate-smart water resource management. (*cont'd*)

Aspect / dimension	AI-ML	RS-GIS	LLMs	Proposed synergistic outcome
Policy and Strategic Planning	Supports evidence-based policy by analyzing patterns in water use, efficiency, and climate impacts.	Supplies data-driven insights into geographic disparities, helping prioritize water conservation in sensitive regions.	Summarizes technical reports into policy briefs, aligning findings with policy objectives and regulatory needs.	A policy support system that provides scientific insights alongside simplified summaries for policymakers, facilitating climate-responsive decisions.
Adaptive Management and Resilience	Continuously refines models as new data is received, enhancing resilience against unforeseen climate events.	Provides ongoing environmental monitoring, capturing shifts in water resources and ecosystem health.	Curates best practices and case studies, assisting water managers with insights into adaptive practices worldwide.	A feedback-loop system allowing water managers to adaptively respond to climate and water resource shifts with up-to-date information.
Innovation in Knowledge Dissemination	Customizes analytics and insights to diverse environmental contexts, paving the way for innovative resource management approaches.	Generates interactive geospatial layers for various stakeholders, from scientists to community groups.	Communicates advances in water management technology, generating engagement across scientific and public platforms.	Enhanced awareness of technological solutions, promoting a culture of collaborative innovation in water management practices.

tools create an integrated system that supports informed decision-making, efficient resource allocation, proactive policy planning, and heightened community engagement, ultimately fostering resilient and sustainable water management practices.

3. Technological advances and future directions

Emerging AI-ML innovations are revolutionizing predictive modelling in water resource management, making it possible to anticipate and adapt to environmental fluctuations with unprecedented accuracy. ML techniques, including DL, ensemble methods, and neural networks, provide nuanced analysis and adaptive forecasting capabilities, which are particularly essential in climate-sensitive regions (see Table 2). For instance, models now integrate historical hydrological data, climate projections, and real-time inputs to predict water demand, quality fluctuations, hydrologic processes prediction, and extreme climatic event occurrences.[34–37] Techniques such as transfer learning, where models trained in one climate region are adapted to another, further improve prediction reliability, especially for areas with limited data.[38] With the growing incorporation of reinforcement learning, models can dynamically adjust resource allocations based on changing conditions, supporting real-time decision-making.[39] AI-driven anomaly detection also plays a critical role in flagging unusual water (or air) quality or supply patterns that signal pollution events or infrastructure leaks.[40,41] These AI-ML advancements, with their capacity to simulate diverse climate scenarios, provide resource managers with tools to mitigate risks, optimize resources, and address climate uncertainties proactively, ultimately strengthening resilience against extreme weather patterns.

RS-GIS technologies have already been recognized as essential tools in the precision management of water resources. These are continuously expanding their scope with evolving satellite capabilities and geospatial analytics. High-resolution satellite data from systems like Sentinel and Landsat allow real-time tracking of water bodies, groundwater reserves, soil moisture, and evapotranspiration rates across vast regions. Advanced applications of RS-GIS, such as Synthetic Aperture Radar (SAR) for all-weather imaging and hyperspectral sensors for detailed water quality assessments, provide continuous, accurate monitoring critical for early warning systems and rapid response.[32,42] By integrating data from these systems with machine learning algorithms, it is now possible to detect

Table 2 Technological advances and future directions in climate-smart water resource management.

Dimension	AI-MI innovations for predictive models	Evolving RS-GIS applications in water monitoring	LLM capabilities in knowledge dissemination and stakeholder engagement
Predictive Accuracy and Adaptability	Enhanced climate forecasting models utilizing ensemble and deep learning for precise predictions and risk mitigation. Adaptive models integrate real-time data to respond to immediate environmental changes.	Satellite-based data coupled with AI for accurate, continuous water cycle tracking and rapid anomaly detection is crucial for local and regional climate adaptation strategies.	Rapid synthesis of real-time reports and scientific findings, making technical knowledge accessible to diverse audiences, thus promoting adaptive learning in stakeholder groups.
Resource Allocation and Optimization	Reinforcement learning for dynamic allocation of water resources, aiding in drought management and efficient agricultural water use.	RS-GIS spatial mapping to identify and prioritize areas for water conservation efforts, supporting precise resource allocation in regions with variable water needs.	LLM-driven policy recommendations are based on historical and predictive insights that guide sustainable resource planning and allocation.
Data Integration and Interoperability	AI algorithms combine weather, hydrological, and socioeconomic data to provide a holistic view of water demands and supply risks.	Integration of RS-GIS with other data streams (e.g., IoT sensors, hydrological models) to monitor environmental factors like soil moisture, groundwater levels, and snowmelt.	Cross-referencing insights from RS-GIS and AI-ML data to produce cohesive reports, ensuring a well-rounded view for policymakers and water managers.
Community and Stakeholder Engagement	Predictive models tailored to community-specific needs, helping local leaders understand water scarcity risks and plan accordingly.	GIS-based community engagement tools like interactive maps display water quality and availability in real-time, fostering transparency and trust.	LLM-powered interfaces for public consultations and QandA systems make it easy for communities to ask questions and receive reliable answers on water resource topics.

Policy Support and Compliance	AI-driven analysis to predict policy impacts on water sustainability, enabling data-informed adjustments to climate action plans.	RS-GIS mapping to monitor compliance with water use and pollution standards, assisting policymakers in enforcing regulations.	Automated summarization of legal and regulatory information, enabling policymakers to stay updated on relevant water laws and global best practices.
Educational Outreach and Awareness	AI-enhanced visualization tools, such as scenario-based simulations, for educating communities on water risks and sustainable practices.	RS-GIS-enabled remote learning modules that show how water bodies respond to climate interventions, fostering environmental stewardship.	LLM facilitated workshops and awareness campaigns to explain complex water issues simply, enhancing public knowledge and proactive involvement.
Scalability and Global Application	Transfer learning to adapt AI models across diverse climates and water scarcity regions, allowing global applicability and resilience planning.	Use of RS-GIS to standardize water resource monitoring systems globally, supporting international cooperation on water conservation.	LLM-supported translation and localization of water management resources, fostering international collaboration and knowledge sharing.
Future Research and Development	AI exploration into hybrid predictive models, combining AI-ML with physical models for enhanced climate impact forecasting.	Advancing RS-GIS through partnerships with emerging satellite programs and AI firms, supporting continuous innovation in water monitoring.	LLM-driven literature reviews and research summaries to support ongoing RandD efforts in climate-resilient water management strategies.

subtle environmental changes, anticipate flood or drought risks, and monitor water usage patterns.[43] Additionally, GIS spatial mapping combined with predictive analytics supports localized decision-making by revealing usage hotspots, potential pollution sources, and the impacts of climate shifts on water resources.[44] As these applications evolve, RS-GIS continues improving water resource management, facilitating a deeper understanding of spatial trends and helping stakeholders make precise, data-driven decisions at local and regional levels.

LLMs are emerging as revolutionary tools for knowledge dissemination and stakeholder engagement in water management. By processing vast datasets, including research articles, policy documents, and real-time reports, the LLMs, such as GPT-based models, BLOOM, Llama, LaMDA, and Bard, provide natural language summaries that translate complex findings into accessible information for diverse audiences. This capability is particularly remarkable for bridging communication gaps, as LLMs can assist water managers, policymakers, and community leaders in understanding technical insights without requiring advanced technical knowledge.[45] Future applications of LLMs include drafting policy recommendations, facilitating training programs, and supporting public awareness initiatives around water conservation and climate resilience. Additionally, integrating LLMs with AI-ML and RS-GIS frameworks creates a cohesive management system, transforming predictive understandings and spatial data into actionable guidance aligned for specific regions or issues. As these models advance, they could also aid in identifying trends in public sentiment or policy needs, empowering data-informed engagement strategies. In the evolving water management landscape, LLMs hold significant potential for boosting collaboration, streamlining decision-making, and fostering more informed, inclusive approaches to resource governance.

Table 2 synthesizes the advanced capabilities of AI-ML, RS-GIS, and LLMs within water resource management, each uniquely suited to tackle climate-induced water challenges through innovative, data-driven, and community-oriented approaches. It envisions the potential of AI-ML models for resource optimization and precise forecasting, RS-GIS's spatial intuitions for monitoring and prioritizing resources, and LLMs' role in stakeholder engagement and policy support. When deployed together, these interconnected technologies promise an adaptable, scalable framework for sustainable water management globally. This approach advocates for a synergistic application of technology, improving resilience against climate variability and fostering cooperative solutions across stakeholders, communities, and nations.

4. Strategic recommendations for climate-smart water governance

Effective climate-smart water governance hinges on clear, actionable policies integrating advanced technologies. To address water resource challenges sustainably, policy frameworks must emphasize the adoption of AI-ML, RS-GIS, and LLMs, each serving as a pillar of data-driven governance. Policymakers should prioritize guidelines for AI-ML deployment in predictive modelling for water demand and scarcity, which can help plan resource allocations based on real-time climate variations. RS-GIS can be utilized to monitor water resources spatially, supporting conservation and efficient resource use policies. Additionally, LLMs can facilitate knowledge dissemination by simplifying complex water data, allowing more inclusive policy formation processes. Implementing these technologies requires scalable strategies to adapt to various regional contexts, especially in developing areas with resource constraints. Policies should also advocate for interoperability between technologies, creating frameworks that foster data sharing and collaboration across departments and regions. Such guidelines would ensure a cohesive and scalable governance model that is adaptable to evolving climate scenarios and responsive to community needs, ultimately fostering resilience through innovative and practical policy solutions.

A fundamental challenge in climate-smart sustainable water management is translating research into practical applications. To bridge this gap, targeted strategies are needed to ensure that technological advancements in AI-ML, RS-GIS, and LLMs reach operational levels within governance structures. Knowledge transfer programs can be pivotal, encouraging collaboration between research institutions and policymakers through workshops, pilot programs, and knowledge-sharing platforms. For example, case studies of regions where AI-driven water scarcity prediction models have been integrated into local governance provide critical visions for scaling such initiatives.[23] Additionally, technology incubators within public-sector water agencies can help accelerate the integration of research outputs, allowing for iterative development that adjusts to feedback from real-world applications. Practical training for water resource managers and government officials can also foster a better understanding of these technologies, ensuring they can effectively implement research findings. Bridging this research-to-practice divide is crucial for creating robust water governance systems that respond to current and future climate challenges, aiding science-driven solutions to impact communities and ecosystems positively.

Sustainable water governance requires the active participation of communities alongside decision-makers, a goal that technology can greatly facilitate. Engaging communities through accessible platforms powered by LLMs allows for interactive Question and Answer (Q&A) sessions, nurturing a two-way exchange of information on water usage, conservation practices, and climate adaptation strategies. AI-ML can support community-driven initiatives by providing localized forecasts of water availability and scarcity, which help residents make informed decisions about water use. Furthermore, RS-GIS provides tools for visually representing water resource data, such as maps displaying real-time conditions, that can empower communities to understand their local water systems. Decision-makers also benefit from this approach; by directly engaging with communities and using transparent technologies, they can build trust and encourage collective accountability in water management practices. Empowering local stakeholders ensures that water policies are not only data-driven but also socially inclusive, responding to the needs of the very populations they serve. This participatory approach promotes resilience, fostering a shared sense of responsibility and collaboration in water governance at all levels. To this end, Table 3 is designed to provide a multifaceted approach to climate-smart water governance, addressing policy formulation, practical application, and community empowerment. Each dimension underscores sustainable water governance's collaborative and adaptive nature, advocating for solutions that engage and benefit all stakeholders involved.

Aligning with Table 3, Fig. 1 proposes a cohesive framework for climate-smart water management, integrating AI-ML, RS-GIS, and LLMs across four main stages. Starting with data collection and pre-processing, it combines spatial data and machine learning to ensure clean, actionable data. Predictive analytics and resource mapping employ AI and RS-GIS to anticipate water demand and environmental risks. AI-driven insights support inclusive, scalable policies and community engagement in strategic planning and multi-sector coordination. Finally, policy implementation and community impact use continuous monitoring and feedback to refine approaches, aligning policy with practical, sustainable water management goals.

In summary, addressing water resource challenges demands collective commitment and technological innovation at this critical juncture in combating climate change. AI-ML, RS-GIS, and LLMs provide advanced, scalable solutions, facilitating a revolutionary approach to climate-smart water resources conservation and management amidst changing climate.

Table 3 Strategic recommendations for climate-smart water governance.

Dimension	Policy recommendations	Research to application	Community empowerment
Technology Integration	Develop policies for seamlessly adopting AI-ML, RS-GIS, and LLMs in water governance systems.	Facilitate technology transfer partnerships between research institutions and local governments.	Introduce community-accessible technology hubs with simplified interfaces.
Data Accessibility and Sharing	Encourage open data policies to allow real-time data sharing among agencies and stakeholders.	Support initiatives to curate and translate complex data for operational use.	Develop community dashboards to display local water data and forecasts.
Scalability and Adaptation	Implement scalable guidelines that are adaptable across varying regions and resource levels.	Pilot region-specific research applications, adjusting for environmental variables.	Provide training for local stakeholders on adapting practices based on data insights.
Capacity Building	Invest in training programs to increase technical proficiency in AI-ML, RS-GIS, and LLM use.	Launch field workshops to bridge theory and practice in water management.	Educate communities on sustainable water practices using AI-powered insights.
Interdisciplinary Collaboration	Form cross-departmental task forces combining engineering, policy, and data science expertise.	Support collaborative projects between government, academia, and private sectors.	Facilitate local collaborations to enable direct community involvement in governance.
Ethical and Privacy Standards	Establish guidelines on data privacy and ethical AI-ML applications in water governance.	Incorporate ethical review boards for AI and RS-GIS projects in public sectors.	Educate the community on privacy standards to build trust in technology use.

(continued)

Table 3 Strategic recommendations for climate-smart water governance. (*cont'd*)

Dimension	Policy recommendations	Research to application	Community empowerment
Monitoring and Accountability	Design accountability frameworks that track policy impact using AI-ML and RS-GIS metrics.	Implement regular audits of applied technologies and publish accessible reports.	Encourage community feedback loops to enhance transparency and trust.
Climate Resilience and Sustainability	Integrate climate resilience as a core policy goal, using tech-driven insights to guide decisions.	Showcase case studies of climate-resilient models adaptable to different areas.	Promote resilience training and knowledge sharing at the community level.
Stakeholder Engagement	Involve diverse stakeholders in policy formulation to enhance relevance and acceptance.	Create platforms for continuous dialogue between researchers and local governments.	Establish community councils to provide input on water governance priorities.
Financial Support Mechanisms	Introduce incentives for adopting climate-smart technologies in water management.	Offer funding grants for research that demonstrates practical applications.	Provide subsidies or grants to communities to adopt sustainable practices.

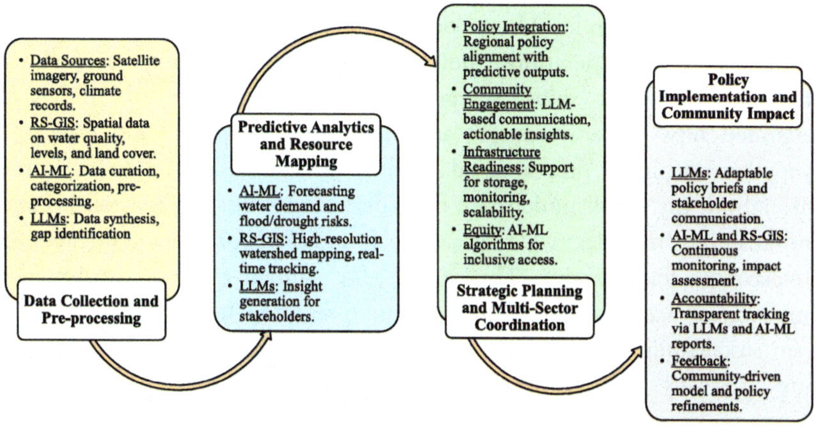

Fig. 1 Climate-smart water management framework utilizing AI-ML, RS-GIS, and LLMs for data-driven, community-centered, and policy-aligned water governance.

These technologies present new pathways for precise predictive analytics, efficient resource mapping, and improved decision-making, encouraging scientists, environmental advocates, educators, and students to contribute their knowledge and creativity toward a more resilient, sustainable future. Sustainable Development Goals (SDGs) call for urgent, meaningful action to protect critical resources, and these tools provide a powerful means to support that mission. Climate enthusiasts, policymakers, and academic professionals alike have a vital role in advancing AI and geospatial applications for environmental stewardship, ensuring that such advancements reach and benefit all communities. From nurturing cross-disciplinary research to promoting community engagement, each effort strengthens a collaborative response to environmental challenges. Embracing these innovations opens doors for impactful change, inspiring a vision where climate resilience and sustainable practices flourish. Together, these collective efforts can shape a climate-smart, equitable future, demonstrating that the journey toward sustainability is possible and within reach for future generations.

5. Concluding remarks and the road ahead

This concluding chapter reflects on the transformative journey from traditional to climate-smart water management, tracing the evolution from conventional practices reliant on empirical methods to advanced, technology-enabled approaches. Historical understandings reveal that while traditional methods grounded in local knowledge and incremental improvements served their purpose for generations, they now face limitations amid rapid climate and socio-economic shifts. Integrating AI-ML, RS-GIS, and LLMs has redefined how water resources are monitored, managed, and forecasted, empowering higher precision, adaptability, and inclusivity. Lessons learned throughout this book underscore the value of interdisciplinary collaboration and adaptive strategies, yielding a solid foundation for future climate-smart governance frameworks. The vision for climate-resilient, technology-empowered water resource management presented here emphasizes long-term sustainability, resilience, and equity. By utilizing AI's predictive capacities, RS-GIS's spatial intuitions, and LLMs' communicative potential, water management systems can become more accurate and more responsive to the needs of diverse stakeholders. This future calls for continuous research, agile policy adjustments, and partnerships that span governmental, private, and community sectors. Such a holistic approach holds the potential to address global water challenges and encourage innovations that support both human well-being and ecosystem health. In sum, this book advocates for a paradigm shift toward water conservation and management amidst mitigating and adapting to changing climate, positioning it as an imperative in today's environmental and social landscape. By embracing these technological advancements and promoting collaborative frameworks, stakeholders can build a resilient water future. The call to action is clear: further exploring, innovating, and committing to sustainable water governance.

Author's contribution

Both authors contributed to the study's conception and design. AS collected the data, reports, documents, and research papers and conducted the analysis, while RM guided the material preparation, data collection, and analysis. AS wrote the first draft of the manuscript, while RM commented on previous versions of the manuscript. Both authors read and approved the final manuscript.

Conflict of interest

The authors declare that the research was conducted without any commercial or financial relationships that could be construed as a potential conflict of interest.

Declaration of generative AI and AI-assisted technologies in the writing process

While preparing this work, the authors used the 'OpenAI - Cloud Computing Platform Version 4′ to improve language. After using this service, the authors reviewed and edited the content as needed and took full responsibility for the publication's content.

References

1. WHO. World Health Organization; 2023. https://www.who.int/news-room/fact-sheets/detail/drinking-water, Accessed 11 November 2024.
2. IPCC. Summary for policymakers. In *Climate change 2023: synthesis report. contribution of working groups I, II and III to the sixth assessment report of the intergovernmental panel on climate change;* Lee H, Romero J, Eds.; IPCC: Geneva, Switzerland, 2023; pp. 1–34. https://doi.org/10.59327/IPCC/AR6-9789291691647.001. Core Writing Team.
3. Srivastava A, Maity R, Desai VR. Assessing global-scale synergy between adaptation, mitigation, and sustainable development for projected climate change. In *Ecological footprints of climate change. Springer climate;* Chatterjee U, Akanwa AO, Kumar S, Singh SK, Dutta Roy A, Eds.; Springer: Cham, 2023, https://doi.org/10.1007/978-3-031-15501-7_2
4. UN Water. United Nations Water; 2023. https://www.unwater.org/water-facts/water-scarcity, Accessed 11 November 2024.
5. Jain S, Srivastava A, Vishwakarma DK, Rajput J, Rane NL, Salem A, et al. Protecting ancient water harvesting technologies in India: strategies for climate adaptation and sustainable development with global lessons. *Front Water* 2024;**6**:1441365. https://doi.org/10.3389/frwa.2024.1441365.
6. Jain S, Srivastava A, Khadke L, Chatterjee U, Elbeltagi A. Global-scale water security and desertification management amidst climate change. *Environ Sci Pollut Res* 2024:1–25. https://doi.org/10.1007/s11356-024-34916-0.
7. Sahu H, Purohit P, Srivastava A, Singh R, Mishra AP, Arunachalam K, et al. Comparative assessment of soil parameters and ecological dynamics in the Western Himalayan wetland and its surrounding periphery. *Environ Qual Manag* 2024;**34**(1):e22283. https://doi.org/10.1002/tqem.22283.
8. Elbeltagi A, Srivastava A, Li P, Jiang J, Jinsong D, Rajput J, et al. Forecasting actual evapotranspiration without climate data based on stacked integration of DNN and meta-heuristic models across China from 1958 to 2021. *J Environ Manag* 2023;**345**:118697. https://doi.org/10.1016/j.jenvman.2023.118697.
9. Kumar S, Srivastava A, Maity R. Modeling climate change impacts on vector-borne disease using machine learning models: case study of Visceral leishmaniasis (Kala-azar) from Indian state of Bihar. *Expert Syst Appl* 2024;**237**:121490. https://doi.org/10.1016/j.eswa.2023.121490.
10. Maity R, Nagesh Kumar D. Basin-scale stream-flow forecasting using the information of large-scale atmospheric circulation phenomena. *Hydrological Processes: An Int J* 2008;**22**(5):643–650. https://doi.org/10.1002/hyp.6630.
11. Maity R, Bhagwat PP, Bhatnagar A. Potential of support vector regression for prediction of monthly streamflow using endogenous property. *Hydrological Process* 2010;**24**(7):917–923. https://doi.org/10.1002/hyp.7535.
12. Maity R, Srivastava A, Sarkar S, Khan MI. Revolutionizing the future of hydrological science: impact of machine learning and deep learning amidst emerging explainable AI and transfer learning. *Appl Comput Geosci* 2024:100206. https://doi.org/10.1016/j.acags.2024.100206. in press.
13. Srivastava A, Maity R. Assessing the potential of AI–ML in urban climate change adaptation and sustainable development. *Sustainability* 2023;**15**(23):16461. https://doi.org/10.3390/su152316461.
14. Sun AY, Scanlon BR. How can big data and machine learning benefit environment and water management: a survey of methods, applications, and future directions. *Environ Res Lett* 2019;**14**(7):073001. https://doi.org/10.1088/1748-9326/ab1b7d.
15. Al-Mukhtar M, Srivastava A, Khadke L, Al-Musawi T, Elbeltagi A. Prediction of irrigation water quality indices using random committee, discretization regression, REPTree, and additive regression. *Water Resour Manag* 2024;**38**(1):343–368. https://doi.org/10.1007/s11269-023-03674-y.

16. Abdelhaleem FS, Basiouny M, Ashour E, Mahmoud A. Application of remote sensing and geographic information systems in irrigation water management under water scarcity conditions in Fayoum, Egypt. *J Environ Manag* 2021;**299**:113683. https://doi.org/10.1016/j.jenvman.2021.113683.
17. Liladhar Rane N, Purushottam Choudhary S, Saha A, Srivastava A, Pande CB, Alshehri F, et al. Delineation of environmentally sustainable urban settlement using GIS-based MIF and AHP techniques. *Geocarto Int* 2024;**39**(1):2335249. https://doi.org/10.1080/10106049.2024.2335249.
18. Pande CB, Srivastava A, Moharir KN, Radwan N, Mohd Sidek L, Alshehri F, et al. Characterizing land use/land cover change dynamics by an enhanced random forest machine learning model: a Google Earth Engine implementation. *Environ Sci Europe* 2024;**36**(1):84. https://doi.org/10.1186/s12302-024-00901-0.
19. Rane NL, Achari A, Choudhary SP, Mallick SK, Pande CB, Srivastava A, et al. A decision framework for potential dam site selection using GIS, MIF and TOPSIS in Ulhas river basin, India. *J Clean Prod* 2023;**423**:138890. https://doi.org/10.1016/j.jclepro.2023.138890.
20. Zagade ND, Umrikar BN. Drought severity modeling of upper Bhima river basin, western India, using GIS–AHP tools for effective mitigation and resource management. *Nat Hazards* 2021;**105**:1165–1188. https://doi.org/10.1007/s11069-020-04350-9.
21. Du X, Tariq A, Islam F, Aziz S, Waseem LA, Ahmad MN, et al. Integrated study of GIS and remote sensing to identify potential sites for rainwater harvesting structures. *Phys Chem Earth, Parts A/B/C* 2024;**134**:103574. https://doi.org/10.1016/j.pce.2024.103574.
22. Zhu JJ, Jiang J, Yang M, Ren ZJ. ChatGPT and environmental research. *Environ Sci & Technol* 2023;**57**(46):17667–17670. https://doi.org/10.1021/acs.est.3c01818.
23. Egbemhenghe AU, Ojeyemi T, Iwuozor KO, Emenike EC, Ogunsanya TI, Anidiobi SU, et al. Revolutionizing water treatment, conservation, and management: harnessing the power of AI-driven ChatGPT solutions. *Environ Chall* 2023;**13**:100782. https://doi.org/10.1016/j.envc.2023.100782.
24. FAO. Food and Agriculture Organization (FAO); 2023. https://www.fao.org/water/en/, Accessed 11 November 2024.
25. Ahmed AN, Othman FB, Afan HA, Ibrahim RK, Fai CM, Hossain MS, et al. Machine learning methods for better water quality prediction. *J Hydrol* 2019;**578**:124084. https://doi.org/10.1016/j.jhydrol.2019.124084.
26. Pandey M, Costache R, Dash P, Geesupalli PD, Siddiqui MA, Pandey PC, et al. Optimized ensemble-based flood hazard mapping in low altitude subtropical riverine terrane. *Discov Geosci* 2024;**2**(1):1–32. https://doi.org/10.1007/s44288-024-00037-x.
27. El-Rawy M, Wahba M, Fathi H, Alshehri F, Abdalla F, El Attar RM. Assessment of groundwater quality in arid regions utilizing principal component analysis, GIS, and machine learning techniques. *Mar Pollut Bull* 2024;**205**:116645. https://doi.org/10.1016/j.marpolbul.2024.116645.
28. Acharki S, Singh SK, do Couto EV, Arjdal Y, Elbeltagi A. Spatio-temporal distribution and prediction of agricultural and meteorological drought in a Mediterranean coastal watershed via GIS and machine learning. *Phys Chem Earth, Parts A/B/C* 2023;**131**:103425. https://doi.org/10.1016/j.pce.2023.103425.
29. Tulla PS, Kumar P, Vishwakarma DK, Kumar R, Kuriqi A, Kushwaha NL, et al. Daily suspended sediment yield estimation using soft-computing algorithms for hilly watersheds in a data-scarce situation: a case study of Bino watershed, Uttarakhand. *Theor Appl Climatol* 2024;**155**(5):4023–4047. https://doi.org/10.1007/s00704-024-04862-5.
30. Srivastava A, Jain S, Maity R, Desai VR. Demystifying artificial intelligence amidst sustainable agricultural water management. *Curr Directions Water Scarcity Res* 2022;**7**:17–35. https://doi.org/10.1016/B978-0-323-91910-4.00002-9.

31. Pandey M. Artificial intelligence algorithms in flood prediction: a general overview. In *Geo-information for disaster monitoring and management;* Pandey PC, Kumar R, Pandey M, Giuliani G, Sharma RK, Srivastava PK, Eds.; Springer: Cham, 2024, https://doi.org/10.1007/978-3-031-51053-3_12
32. Singh RK, Khand K, Kagone S, Schauer M, Senay GB, Wu Z. A novel approach for next generation water-use mapping using Landsat and Sentinel-2 satellite data. *Hydrological Sci J* 2020;**65**(14):2508–2519. https://doi.org/10.1080/02626667.2020.1817461.
33. Safaei M, Longo J. The end of the policy analyst? Testing the capability of artificial intelligence to generate plausible, persuasive, and useful policy analysis. *Digital Government: Res Pract* 2024;**5**(1):1–35. https://doi.org/10.1145/3604570.
34. Elbeltagi A, Srivastava A, Ehsan M, Sharma G, Yu J, Khadke L, et al. Advanced stacked integration method for forecasting long-term drought severity: CNN with machine learning models. *J Hydrology: Regional Stud* 2024;**53**:101759. https://doi.org/10.1016/j.ejrh.2024.101759.
35. Rajput J, Kushwaha NL, Srivastava A, Pande CB, Suna T, Sena DR, et al. Development of machine learning models for estimation of daily evaporation and mean temperature: a case study in New Delhi, India. *Water Pract & Technol* 2024. https://doi.org/10.2166/wpt.2024.144. wpt2024144.
36. Vishwakarma DK, Kumar P, Yadav KK, Ali R, Markuna S, Chauhan S, et al. Evaluation of CatBoost method for predicting weekly Pan evaporation in subtropical and sub-humid regions. *Pure Appl Geophysics* 2024;**181**(2):719–747. https://doi.org/10.1007/s00024-023-03426-4.
37. Brunner MI, Slater L, Tallaksen LM, Clark M. Challenges in modeling and predicting floods and droughts: a review. *Wiley Interdiscip Reviews: Water* 2021;**8**(3):e1520. https://doi.org/10.1002/wat2.1520.
38. Grubinger T, Chasparis GC, Natschläger T. Generalized online transfer learning for climate control in residential buildings. *Energy Build* 2017;**139**:63–71. https://doi.org/10.1016/j.enbuild.2016.12.074.
39. Rolnick D, Donti PL, Kaack LH, Kochanski K, Lacoste A, Sankaran K, et al. Tackling climate change with machine learning. *ACM Comput Surv (CSUR)* 2022;**55**(2):1–96. https://doi.org/10.1145/3485128.
40. Masood A, Hameed MM, Srivastava A, Pham QB, Ahmad K, Razali SFM, et al. Improving PM2. 5 prediction in New Delhi using a hybrid extreme learning machine coupled with snake optimization algorithm. *Sci Rep* 2023;**13**(1):21057. https://doi.org/10.1038/s41598-023-47492-z.
41. Pattyam SP. AI-driven data science for environmental monitoring: techniques for data collection, analysis, and predictive modeling. *Australian J Mach Learn Res & Appl* 2021;**1**(1):132–169. https://sydneyacademics.com/index.php/ajmlra/article/view/135 Accessed 11 November 2024.
42. Amitrano D, Di Martino G, Guida R, Iervolino P, Iodice A, Papa MN, et al. Earth environmental monitoring using multi-temporal synthetic aperture radar: a critical review of selected applications. *Remote Sens* 2021;**13**(4):604. https://doi.org/10.3390/rs13040604.
43. Neog DR, Singha G, Dev S, Prince EH. Artificial intelligence and its application in disaster risk reduction in the agriculture sector. In *Disaster risk reduction and rural resilience. disaster risk reduction;* Mitra S, Shaw R, Eds.; Springer: Singapore, 2024, https://doi.org/10.1007/978-981-97-6671-0_15
44. Janga B, Asamani GP, Sun Z, Cristea N. A review of practical ai for remote sensing in earth sciences. *Remote Sens* 2023;**15**(16):4112. https://doi.org/10.3390/rs15164112.
45. McTear M, Ashurkina M. *Transforming conversational AI;* Apress: Berkeley, CA, 2024, https://doi.org/10.1007/979-8-8688-0110-5

Printed and bound by CPI Group (UK) Ltd, Croydon, CR0 4YY
14/09/2026
02216284-0002